Y0-CAX-674

TV Repair for Beginners

4th Edition

TV Repair for Beginners

4th Edition

George Zwick
Homer L. Davidson

TAB

TAB BOOKS
Blue Ridge Summit, PA

FOURTH EDITION
FIRST PRINTING

TAB Books is a division of McGraw-Hill, Inc.

Library of Congress Cataloging-in-Publication Data

Zwick, George, 1910–
TV repair for beginners / by George Zwick and Homer L. Davidson. -
- 4th ed.
p. cm.
Rev. ed. of: Beginner's guide to TV repair. 2nd ed. c1979.
Includes index.
ISBN 0-8306-2180-6 (pbk.) ISBN 0-8306-2181-4 (hardbound)
1. Television—Repairing—Amateurs' manuals. I. Davidson, Homer L. II. Zwick, George, 1910– Beginner's guide to TV repair.
III. Title.
TK9965.Z89 1991
621.388'87—dc20 91-10803
CIP

Acquisitions Editor: Roland S. Phelps
Production: Katherine G. Brown
Book Design: Jaclyn J. Boone
Cover Photograph: Susan Riley, Harrisonburg, VA. EL1

Contents

12 Symptoms and causes *313*

Introduction

TO SERVE A USEFUL PURPOSE, AN INTRODUCTION OFTEN SERVES AS A SORT OF abstract or summary stating what a book is about as well as what it is not. For the sake of clarity, we shall first state what this book is not. First, it is not a complete course in TV repair. It does not teach you how to do jobs that can be done only by a competent, experienced TV servicer. This includes many and varied tasks, for example, the replacement of the picture tube and subsequent adjustments—a task that is definitely not in the beginner category.

Second, this book does not teach or recommend any repair involving personal hazard. There are such repairs in the complex machine called a TV set, particularly a color TV set, and no one, however handy or mechanically inclined, should attempt these.

Finally, this book does not suggest any repair in which there is a low probability of success. To do so would likely do more harm than good because it is liable to condition the TV owner to accept a mediocre, substandard level of performance, rationalizing that this is "the best he or she can get."

What *does* the book intend to accomplish? It is the purpose of this book to enable anyone interested enough to learn the basics of TV repair to keep his or her receiver performing at its best by following a safe, simple, preventive, and corrective maintenance program that requires no specialized tools or equipment. This maintenance includes such things as adjustments to compensate for aging, deterioration, image distortion, etc. It also tells how to correct such defects as improper illumination (brightness/contrast), picture instability (rolling/tearing), and overloading (smearing, etc.). In addition, this book includes procedures on how to remedy such catastrophic defects as loss of vertical (decrease in picture height to a thin horizontal line), loss of sound, loss of picture, loss of both sound and picture, as well as such frightening symptoms as smoke, crackling and frying sounds, etc.

In the case of color receivers, the uninitiated will find that this book solves the mystery of color without creating new technical mysteries by showing that the color TV receiver is basically a black-and-white set with added capabilities.

With the advent of solid-state technology in TV receivers, whether hybrid (part tube, part transistor) or all solid-state, the do-it-yourself TV owner must of necessity acquire a modicum of familiarity with the infrastructure of this type of equipment. This involves not only the gradual disappearance of vacuum tubes and their replacement by semiconductors—transistors, diodes, varactors, etc.—but also the almost-revolutionary change in construction through miniaturization. Bulky discrete parts with their point-to-point wiring have been almost totally replaced by miniaturized subassemblies, printed-circuit wiring and integrated circuits (ICs). As a logical consequence, the handling of such components has also undergone a radical change. The old familiar screwdriver, soldering iron, and long-nose pliers have been superseded by semiminiature, almost jeweler-type tools. Their manipulation requires new techniques and considerable circumspection. This does not mean the end of do-it-yourself. In fact, in many instances it has actually become easier. Yet there still are areas best left to the professional repairman, which are identified in the book. But we also guide the would-be home do-it-yourselfer through the new required procedures in handling transistors, ICs, and printed-circuit (PC) boards. Chapters 10 and 12, Solid-State Circuitry and Simple Solid-State Repairs and Adjustments, can help the do-it-yourselfer, novice, and TV owner to understand how the latest TV chassis functions.

To establish correlation between functional circuits and physical components—a vital step in identifying actual TV parts and subassemblies by the TV owner—we have shown parts of actual set schematics of modern receivers, sectionalized on the diagram in correspondence with physical circuit boards. This follows the very common practice of TV manufacturers in their service manuals and data; thus, when you look at such a partial diagram, you also see the physical PC board or subassembly of that diagram. Following the same logic, we have also shown some actual sample procedures for removal of circuit boards from a complete TV set.

As for the *don'ts* for the TV owner, we identify those tasks that, for reasons of safety (as in the case of picture tube replacement) or of technical complexity including such functions as IF and tuner alignment—rarely needed, but extremely complicated, should not be attempted so that you will not go out on a limb, so to speak. While the more courageous TV owner might attempt, following manufacturers' instructions and admonitions, to replace the picture tube, he or she would still risk the predicament of a possible ''stuck'' tube that refuses to slide out of the yoke assembly. In case of tuner or IF adjustments, although not involving the danger of an implosion as in case of the picture tube, we caution the do-it-yourselfer against almost 100-percent certain failure, unless you sport an elaborate and expensive array of sophisticated test equipment. In such cases, and wherever feasible, I direct the TV owner to an alternative procedure, as in some cases of tuner replacement.

The technical level of this book is necessarily different for different users. While the more experienced reader might find some of the detailed explanations somewhat naive, these same explanations may be nothing short of astounding revelations to the uninitiated; however, the overall level of the subject matter is such that the average reader will be able to progress from page to page without difficulty, so that what might at first glance seem too technical and overwhelming will in due course become fairly simple and relatively easy to accomplish.

Introduction to the fourth edition

BESIDES KNOWING HOW THE VARIOUS CIRCUITS IN THE TV CHASSIS WORK AND why they fail, there are many people who would like to make a few simple TV repairs. There are many easy TV repairs you can do throughout this book, like changing the fuse, setting the circuit breaker, cleaning the tuner, connecting the antenna cable, and troubleshooting many circuits with a multimeter.

Before tearing into any electronic chassis, you should know the hazards and safety factors. Never remove parts or components with the power cord plugged into the ac receptacle. Likewise, never solder any connection with the power on. Only have the power on when taking voltage measurements or waveforms. Remember, the 120-volt ac power line is dangerous when crossed or mistreated. Respect all voltages while servicing the TV chassis.

Be careful around the picture tube when the TV chassis power is on. Keep hands and body away from the deflection yoke and high-voltage anode connection. The HV can draw an arc up to a *couple of inches*. Do not use a multimeter to try and measure the high voltage at the CRT. Use only the correct high-voltage test instrument in taking HV measurements. Remember, the picture tube can hold a high-voltage charge for several days after the TV is shut off. Do not be afraid of the TV chassis, just respect it!

There are many new TV circuits that have appeared on the scene in the last ten years. Examples are the x-ray protection circuits, scan-derived voltages, saw-filter networks, high-voltage shutdown, chopper power supplies, sand-castle generators, switched-mode power supplies, stereo sound, on-screen display, surface-mounted components, and program review (also called ''picture in a picture''). All of these new circuits are described in this fourth edition.

Each chapter has been expanded and revised to cover the new circuits in the solid-state chassis. Besides adding new circuits and block diagrams, there are over 300 illustrations, with over 125 photos to illustrate the various components found in the latest TV chassis. Learning how the TV set operates and fails can be quite rewarding. Besides, electronic servicing is a lot of fun.

1
The TV system

BEFORE EMBARKING ON A DETAILED, NONTECHNICAL DESCRIPTION OF VARIOUS portions of a modern TV system, it is desirable to present a brief synopsis of the overall process, from end to end. This can provide a continuity of visualization so that as you read about a particular step in the system, you will have in the back of your mind, so to speak, the complete picture of the overall purpose and end result to be achieved.

Basic sound and picture systems

The TV system, whether monochrome (black-and-white) or color, consists of two distinct transmitter-receiver systems: the sound (audio) system and the picture (video) system. The interconnections and any commonalities between the two systems is entirely incidental. In other words, it is not essential that there be any common connections or functions between the two systems. It is done merely for economy and convenience. Let it be thoroughly understood that a physically separate sound system alongside a corresponding totally independent picture system would be just as feasible from a strictly technical viewpoint.

The sound system is essentially the same as the FM portion in AM/FM receivers. The FM sound part of the TV signal picked up by your antenna has the same general characteristics of an FM signal—namely, high immunity to noise (static, electrical interference, etc.), relatively short-range reception capability and (potentially) a higher sound quality. Figure 1-1 shows, in block diagram form, the major components of the TV sound system.

Figure 1-2 shows, in similar form, the building blocks of the video (picture) system. This is an AM signal (similar to the AM music stations) and is subject to the same general conditions accompanying AM station reception—fading, freak long-distance reception, interference (streaks, etc., across the screen) and so on.

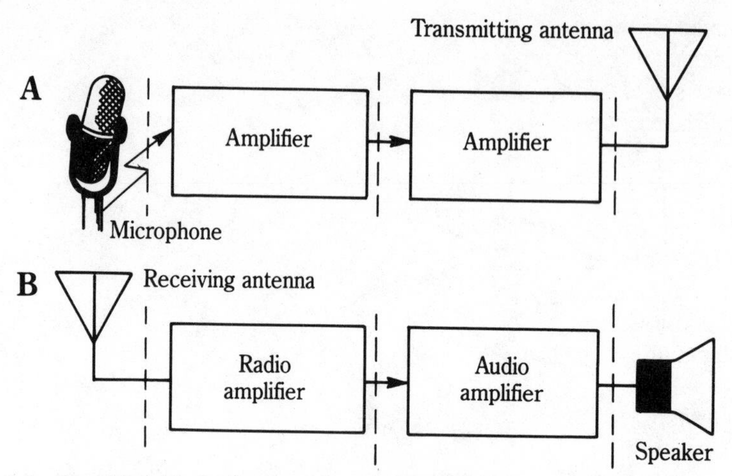

1-1 Simplified block diagram of a sound transmitter (A) and a sound receiver (B). Notice that the two systems are virtually mirror images of each other.

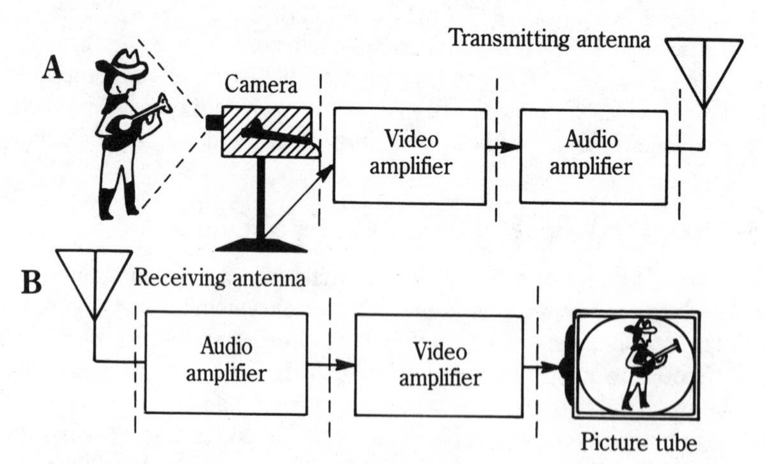

1-2 Simplified block diagram of a picture transmitter (A) and a picture receiver (B).

Terminology

A few words about the terminology just used might not be out of order here. AM stands for *amplitude modulation* and signifies a system of radio transmission and reception in which the magnitude of the signal varies directly with the loudness of the voice or music or with the brightness of the image in case of a TV picture. Since electrical noise can add to or subtract from such a signal, it is only logical that such noise is reproduced by the receiver as physical noise (static, clicks, etc.) or as streaks, dashes, etc., on the TV screen. In other words, AM is not immune to noise.

FM stands for *frequency modulation*. This system ignores any variations in size of the radio signal as might be caused by the addition of noise to the signal and transmits and receives sound information by means of a variation in the *spacings* (modulation) between adjacent waves. Technically, both picture and sound could be AM or FM, but there are other reasons for the present use of FM sound and AM picture.

Transmitting and receiving systems

In Figs. 1-1A and 1-2A, the vertical dashed lines define the three distinct major functions of each transmitting system. On the extreme left of Fig. 1-1A is the microphone which changes sound waves to electric currents. In Fig. 1-2A the camera performs the corresponding function of changing light into electric energy.

The second sections of Figs. 1-1A and 1-2A serve the major functions of amplification (enlargement) of the faint signals. The third sections in both illustrations convert the amplified signals into a form suitable for transmission via the antennas. They also further amplify them before feeding them to the antennas. The right-hand sections in both illustrations actually radiate the signals into space to be received by many receiving antennas (Fig. 1-3).

Receiving systems (Figs. 1-1B and 1-2B) are virtually the mirror opposite of the transmitting systems just described. On the extreme left are the receiving antennas, located so as to ''get in the way'' of the radiated energy from the transmitters. (That is

1-3 A yagi type antenna picking up channels 2 through 13. The small end of the antenna points toward the station being received.

why receiving antenna location is so important.) The second section of each receiver amplifies or builds up the minute signals to the levels necessary for further processing. The third section in each illustration converts the signals back to a form suitable for display as picture or sound, as well as amplifying them. The right-hand sections reproduce sound waves from the electrical energy in the sound signal and form pictures from the electrical energy carried in the picture (video) signal (Fig. 1-4).

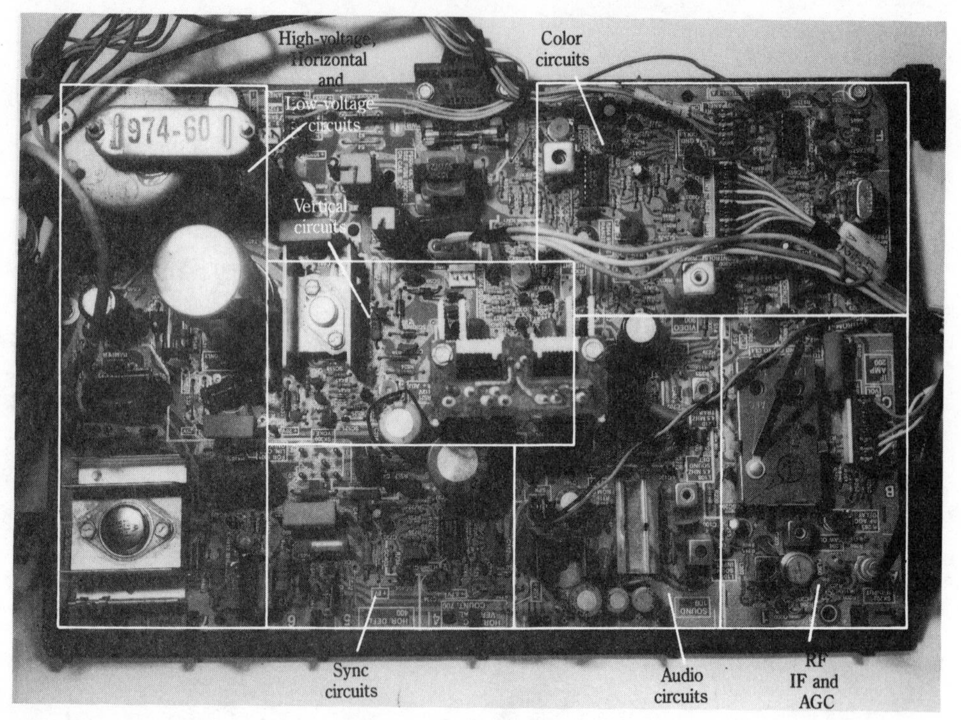

1-4 The top view of a solid-state chassis with all components except the tuner mounted on one large PC board.

Original sound to reproduced sound

All sound (voice, music, or noise) is produced by a physical force that sets the air into vibration. The amount of force applied to the air determines the loudness of the sound, while the different pitch (from bass to a whistle) results from the rate (speed) of vibration of the air.

Microphone

The microphone is a device that converts mechanical energy (air pressure or vibration) into electrical energy. The amount of the electrical energy produced depends on the physical force applied (sound loudness). The pitch of the original sound determines the frequency (numerical rate of vibration) of the electrical energy produced. For sound, the frequency might be as low as 20 vibrations per second (a deep, organ note) and as high as 20,000 vibrations a second (a very high-pitched whistle). Incidentally, ordinary household ac has a frequency of 60 vibrations (cycles or *hertz*) per second.

Amplifier

The amplifier is an electrical device, using either tubes or transistors, that has but one function—to faithfully enlarge the feeble electrical currents emanating from the microphone. No one tube or transistor can provide all the amplification that is required; so there usually are a number of amplifiers, following each other, each amplifying the output of the preceding amplifier or stage until the required maximum is obtained (Fig. 1-5).

1-5 The audio section parts in the solid-state chassis are usually clustered together. Notice the discriminator coil and audio output chip.

Audio and radio frequencies

To attain a useful understanding of TV transmission and reception, it is essential to have a clear concept of frequency. As was mentioned earlier, electrical energy exists as a wave-like phenomenon, just as sound energy consists of air waves, not unlike the waves or ripples caused by a disturbance in water. The term *frequency* is used to denote the number of times a wave recurs in a certain period of time, usually 1 second. Thus, in all future reference to frequency in this book, the meaning will be the same: the number of waves per second, or preferably, cycles per second. As stated before, audio frequencies

(AF, sound frequencies audible to the ear) are assumed to extend from about 20 cycles per second (deep organ note) to about 20,000 cycles per second (for a very shrill whistle). Above 20,000 or 30,000 cycles, they are called radio frequencies (RF for short), extending through AM and FM radio, TV and radar transmitting and receiving frequencies to the near-optical (light) portion of the spectrum. Here are the first infrared (invisible heat waves), followed by the visible light spectrum (red through violet), the ultraviolet (again invisible) and on to x-rays, etc. Video (picture) frequencies are related to television. They stand for those frequencies that carry picture information and extend from about 20 cycles per second to about 4,000,000 cycles per second.

Until several years ago, frequencies in the entire spectrum from audio through invisible light were referred to as cycles per second, thousands of cycles per second (kilocycle), millions of cycles per second (megacycle), etc. Now by international standards, the term *hertz*, abbreviated Hz, is used instead of *cycles per second*. Heinrich Hertz, a German physicist, was the first to demonstrate the production and reception of electromagnetic or radio waves. Thousands of cycles is referred to as kilohertz (kHz), etc.

A final note on waves and frequencies: the higher the frequency, the shorter the length of the wave. For example, the sound wave of a telephone bell (with a frequency of, say, 1000 hertz) is approximately 186 miles long. By contrast, the radio wave transmitted by a UHF television station is little more than 1 foot long. This explains why the elements or metal rods of a UHF TV antenna are so much shorter than the corresponding VHF types. While any frequency can be transmitted from place to place over wires, only those classed as radio frequencies (RF) can be sent through space (wireless transmission).

RF and AF amplifiers

The middle sections of our simple diagrams in Figs. 1-1 and 1-2 consist mainly of amplifiers, accessory devices, and controls. Those immediately following the microphone are audio amplifiers, as stated previously; however, in order to transmit this audio economi-

1-6 The IF and video section in the RCA color chassis is under a shielded area. Remove the metal cover to get at the various components.

cally and efficiently over great distances a temporary change is made. The audio wave is superimposed (that is, it modulates) on a radio wave piggyback style. This combined RF and AF energy is further amplified and finally applied to the antenna (Fig. 1-6).

Antennas

A transmitting antenna has but one function: to radiate the energy applied to it in the desired direction (or in all directions) in the most efficient manner possible. While the obvious purposes of the receiving and transmitting antennas seem to be quite different from each other, they actually behave very much alike. Incidentally, despite many claims to the contrary, there is no substitute or shortcut worthy of its name for a high-grade, elaborate antenna, whether for transmitting or receiving.

At the receiving end, the antenna does not radiate energy—it collects. Located in the path of the radiated energy, the receiving antenna collects or acquires a very small sampling of the original energy or signal as radio waves cross it. By means of a transmission line (sometimes called lead-in) the intercepted signal is carried to the input of the receiver (Fig. 1-7).

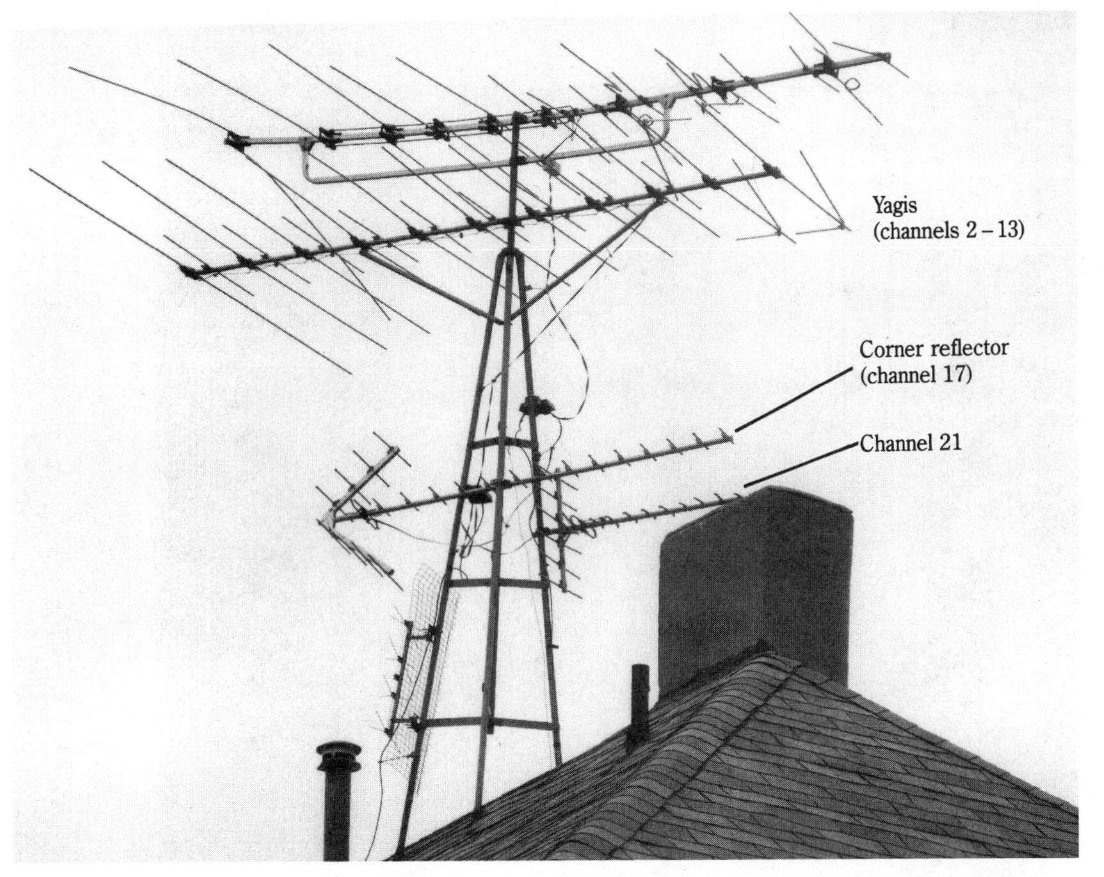

1-7 The corner yagi antenna for channel 17 is coupled to the booster through a matching UHF-VHF block.

It is unfortunate but true that while a properly designed receiving antenna will favor the stations (frequencies) for which it was designed and discriminate against others, it is fairly helpless in rejecting undesirable electrical energy, such as noise, static, etc. The transmission line, or lead-in, is equally susceptible to noise or static pickup, although something can be done here. Shielding, or protecting the lead-in from surrounding electrical noise, is feasible. The antenna itself cannot be shielded (or it won't receive any signals); the only recourse lies in the selection of a suitable location where noise is at a minimum.

Receiver

The amplifier sequence in the receiver is virtually a mirror reflection of that in the transmitter. Amplifiers immediately following the antenna amplify the combined (RF plus AF or modulated RF) signal to the required level. Next, and in the same general portion of the receiver, a reversal of the modulation process takes place by a device called a demodulator or detector. This reversal consists of stripping the RF, which has served its purpose, from the signal, leaving only the audio-frequency signal. Further amplification now takes place until the AF signal is sufficiently strong to operate a speaker or headphones (Fig. 1-8).

1-8 The antenna lead-in wire from the VHF antenna and separate UHF antenna connect to each separate tuner in the TV set.

Speaker

In function, a speaker or loudspeaker is the exact opposite of the microphone. Electrical energy in the form of audio waves is applied to the speaker, where it is converted to the mechanical motion of a surface called the speaker *cone*. The cone vibrates faster or slower, depending on the frequency (pitch) of the corresponding original sound waves, and sets the air in similar motion. The loudness of the sound depends on the distance the cone moves, which in turn depends on the strength or size of the electrical wave applied to it (Fig. 1-9).

1-9 Ten- and 12-inch speakers are often used in large tube console sets while smaller speakers are used in today's TVs. After replacement, check the speaker polarity with a battery if more than one speaker.

Original picture image to radio signal

In the following paragraphs, I shall trace the sequence of steps necessary to convert a visual scene into an electromagnetic energy wave suitable for transmission in space. After the conversion from light to electrical energy has been accomplished in the camera, the remaining functions are very similar to the sound-to-RF-to-sound process just described. The functions to be described apply equally to monochrome and color transmission reception and should give you a clear nontechnical concept of what is going on so that you will be able to deduce what went wrong and where in the event your TV set performs unsatisfactorily or fails to perform altogether.

Physical image

All images—persons, scenes, or whatever—are seen by the human eye because of the light they emit or reflect. It is quite obvious that all such images are color images. In addition the various colors reaching the eye are of different light intensity; thus, we commonly say dark brown, or dazzling white, or bright yellow, etc., suggesting that we see, in addition to different colors, different degrees of light intensity. The best example of this is a so-called black-and-white picture made by a camera, where all colors become different shades of gray. A person's face is brighter than his or her shoes, but not as bright as a white shirt, etc. In a manner of speaking, our descriptions of different colors are given in terms of brightness or dimness, with the notion of color being a separate characteristic of the object or scene we are viewing.

Photographic recording

To the human eye, a scene or object presents a continuous range of color and illumination. Actually, however, this seeming continuous image is a composite of many discrete little areas, each of which may differ from its neighbor either in color, brightness, or both. In a black-and-white photograph, the image is actually made up of tiny specks of the image spaced from each other. The concept of grain (fine-grain photograph, for instance) refers to just this particle structure. While the color photograph is claimed to be grainless, this is but a relative term in contrast with the chemical grain structure of the black-and-white picture.

Halftone

A practical example of a discrete particles composition of an image is the familiar newspaper or magazine photograph. Viewed with the naked eye, such a photograph looks smooth and continuous. Under a magnifying glass, however, the picture is revealed as consisting of individual dots with clear spaces between dots. Shading in the picture depends on the weight of the black dots in comparison to the adjacent white spaces. While this composition is due to the technique of mass reproduction (newspaper printing, etc.), the fact remains that images of very high quality can be and are reproduced by the discrete dot structure.

Conversion of image dots into electric signals

The first step in producing a television signal is the conversion of light (optical energy) into electricity. It begins with a photographic-type camera, lenses, focusing, viewfinders and all; however, here is where the similarity ends. Instead of projecting the image onto a ground glass (as in studio cameras) or on a photographic film (any camera) it is projected onto a special electrochemical surface called a photomosaic. Incidentally, the projected image might be quite small, often not much larger than a postage stamp.

Next, an electron beam or stream scans the mosaic image, one spot at a time, as if reading the bits of the image. Each time the electron stream impinges on one of the spots of the mosaic image, an electric current is generated that has characteristics peculiar to that particle. That is, the electric pulse or current from that image particle contains information necessary to reproduce this image particle on the home TV screen.

There is also included with each such electric pulse the position or location information, both in the vertical and the horizontal directions. This positioning information is transmitted and received by the home TV antenna, faithfully amplified and reproduced and finally used to guide the electron beam in the TV picture tube for proper positioning of the image particle.

A simple example of an electron gun used to produce a scanning beam is found in the neck portion of a TV picture tube. It is composed of a glowing heater—a sort of electron generator—and a number of metallic structures (hollow cylinders, baffles, etc.) that shape and guide the electrons toward the screen in the form of an extremely thin, sharp beam. In the home receiver, the electron beam "writes" on the picture tube screen in step with the "reading" done by the beam in the TV camera. The location and brightness of each written bit is automatically known from information transmitted with each bit so that the completed written image is an exact duplicate of the original. The scanning sequence is arranged in accordance with an established code (National Television Standards Committee, NTSC). In simplified form, it is as illustrated in Fig. 1-10 and can be described as follows. (This scanning sequence can be considered, for the sake of illustration, analogous to the writing of a letter, double-spaced, then filling in the spaces between the lines so that the complete letter is read line after line, from top to bottom):

Beginning at the upper-left corner, the electron beam scans line 1 from left to right. It then skips back to the left side and scans the second line underneath the first. This is line 2. This sequence continues until approximately 235 lines have been produced. In Fig. 1-3, all these lines are shown solid for a reason that will become clear presently. The beam continues to scan an additional $27^1/_2$ lines, for a total of $262^1/_2$, with but one

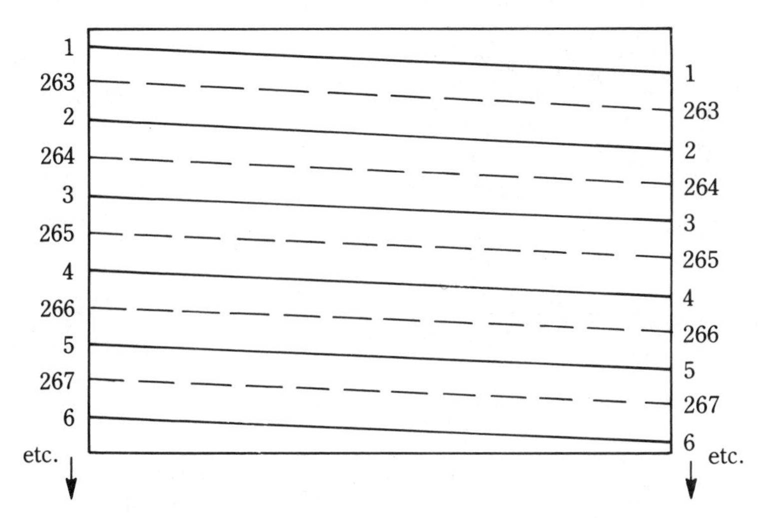

1-10 How a TV picture is scanned by electronic beams. The solid lines represent one field, and the dashed lines show the alternate scan field. Line spacing and slant are exaggerated for clarity. Dashes are used to distinguish between the two fields. The typical sequence is 1, 2, 3, and so on up to 262. Then it returns to the top of the screen and continues with 263 to 525.

difference—these additional lines are not visible on the home screen—their location is just below the visible bottom of the picture. We later explain the purpose of these "invisible" lines. The beam now skips back to the upper-left corner and starts scanning again, but this time between the lines just finished, producing lines 263, 264, etc., until about line 498, then continues scanning invisible lines again until 525 are produced. The complete process now repeats, starting with line 1.

The picture tube screen has now been scanned twice, each scan lasting 1/60 of a second, the period of which is called a *field*. The two fields, lasting a total of 1/30 of a second, are called a *frame*. The process of alternate line/field scanning is called *interlacing*. In case of a still scene, more than one frame of the same scene would be scanned; in action scenes, there will always be at least one frame (two fields) scanned, this being required for a correct picture of the scene.

High-definition TV (HDTV)

High-definition TV is a new system that improves picture resolution, with twice the horizontal and vertical resolution. Besides picture definition, it has improved audio, somewhat like the compact disc. The new system does not interfere with existing transmission of the present-day television channels set by National Television Standards Committee (NTSC) of 30 years ago. Your present TV will not be obsolete.

The FCC provides standards that require transmission systems to operate within specific widths and frequencies used in present-day VHF and UHF TV bands. Within these parameters, there are gaps where inactive channels exist. HDTV will use these inactive channels and at the same time broadcast over existing broadcast channels. To enjoy the benefits of HDTV, you must purchase a new TV with HDTV features.

The present-day standard TV receiver has 262.5 scanning lines in the odd and even field scanning lines. The interpolated-definition TV (IDTV) receivers introduced by foreign manufacturers has 525 odd and even field scanning lines. The interpolated memory digital circuits provide additional scanning lines between the broadcast signal, showing the entire image upon the TV screen (Fig. 1-11).

Several American and Japanese firms plan to manufacture HDTV receivers by 1995. Zenith Electronics and American Telephone & Telegraph Co. are working on a $24 mil-

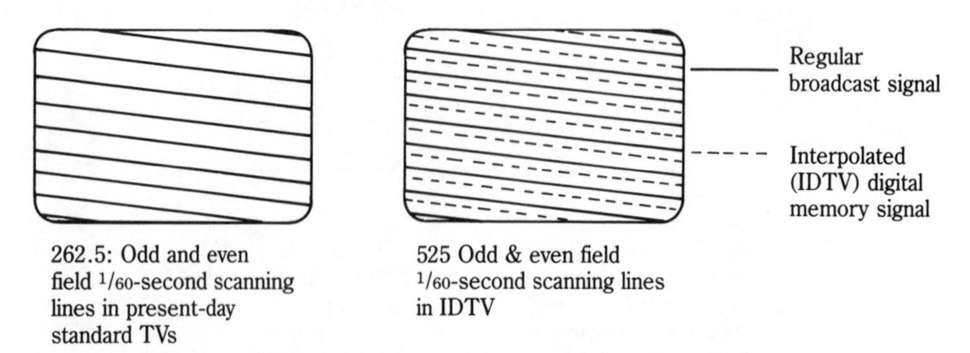

1-11 The 262.5 odd and even field with 525 odd and even fields with an interpolated (IDTV) digital memory signal between.

lion research and development system of the HDTV system. On December of 1989, the Duran-Leonard fight was televised in high-definition television.

Blanking and synchronizing

In describing the scanning sequence it was stated that at the end of each line, the beam skips back to the beginning of the next line, and at the end of a field, it skips to the upper-left corner of the beginning of the next field. The very important activity during the blanked interval, *blanking*, simply means making the lines invisible because they are not part of the image and would only add meaningless lines, bars and streaks to the picture. This activity consists of transmission of *synchronizing* pulses, which are guidelines for starting and ending of each line. These pulses occur at the end of each line, while the timing of each field is sent during the blank period between fields. Any deficiency or degradation of these sync pulses would be cause for picture instability, such as horizontal tearing or vertical rolling. There are other signals sent during the blanking period, such as color correction (VIR) in the most recent expensive TV sets (referred to later in the book).

In connection with this discussion of blanking, it should be mentioned that in addition to the brightness and contrast controls on the home TV set, which are adjustable at will, there are some definite normal levels of brightness and contrast, preset and identified by the transmitted signal, incident to the blanking interval information. Thus, it is possible for the home TV viewer, by misadjusting the brightness control while at the same time misadjusting the vertical hold control so that the picture begins to roll slightly, to actually see the vertical blanking and synchronizing pulses on the screen, located between adjacent picture frames. Of course, such excessive brightness setting is detrimental to normal picture light distribution, causing light areas to be burned out with loss of detail. It similarly affects the opposite end of the illumination range, preventing black areas from being actually black. This applies equally to color as to black-and-white sets; thus, indirectly, the appearance of any such blanking lines on the screen is a sign of excessive brightness in some TV sets, assuming an otherwise normal set.

Interlacing and pairing

It was mentioned previously that the alternations of lines from the first and second fields of each frame is called *interlacing*. Later, the related phenomenon of *pairing*, an operational defect that causes picture degradation is explained. It is therefore desirable to explain this relation a bit further at this time. I shall do so with the aid of an analogy to the familiar photograph.

If an ordinary photograph were to be cut into a large number of very thin strips, the original image could still be seen, no worse than before, provided the individual strips were kept from moving, sliding, or shifting. If, however, some or all of the strips were moved so that they partially or completely overlap the picture would naturally suffer, depending on the degree of overlap. With regard to the TV picture, it is quite correct to visualize it as being made up of such very narrow strips (called lines) arranged in two steps. In step one, every other strip is temporarily omitted. This produces a picture with a number of gaps. In step two, these gaps are filled with the remaining set of strips, each exactly between the two adjacent strips. The picture is now complete.

If, however, the second set of strips were placed so that they partially or even totally overlap the first set, the picture will not only be incomplete because some of the strips are covered over, but also because there is a mixing or blurring of whatever is still visible. This phenomenon of overlapping and blurring is called *pairing* because there no longer are any single, individual lines, but pairs of such lines. As explained later, such pairing is caused by a failure of the horizontal sync, either due to a defect in the TV set or because of transmission difficulties, making each line double-width and resulting in half of the normal number on the screen.

Automatic brightness control

In recent years, manufacturers have added a number of so-called automatic features to their receivers, usually for the sake of simplification of operation by the user. Sometimes, however, such a feature is a protective measure against improper adjustment and possible consequent damage to the set. Such is the case with ABC (automatic brightness control) or ABL (automatic brightness limiter). From a technical viewpoint, this feature protects some internal parts, especially the picture tube, from premature failure from excessive cathode current; it also keeps the image on the screen from becoming extremely overbright due to careless setting of the front-panel brightness control. As a consequence, the above-mentioned trick to display the blanking and sync bar might not work without a bit of trying and careful misadjustment. It is assumed that after such a trick, the set owner will return the controls to normal by first reducing them to below normal, then gradually bringing them up to normal.

Conversion of electrical impulses into radio waves

Electrical impulses representing individual image elements and timing (synchronizing) pulses are combined with or superimposed upon a radio wave (carrier), amplified to the power level required for transmission over the intended distance (a TV station is licensed for a certain maximum power level to meet the needs of the geographic area it is to serve) and applied to the antenna.

Incidentally, frequencies of TV stations in the United States are much higher than those used for ordinary AM radio broadcasts for technical reasons (number of stations to be accommodated, shorter distances covered, etc.). While ordinary AM stations are centered around 1 MHz (megahertz), TV frequencies range from just over 50 MHz to almost 900 MHz. This affects not only the location of TV stations in the radio spectrum but also poses different requirements on design, complexity, etc.

TV sound

There remains one additional signal required for a complete TV transmission: the sound (or noise) accompanying the picture. As stated before, TV sound is transmitted as FM for certain technical reasons as well as for some advantages (noise immunity, for instance). TV sound could be transmitted over a completely separate system from microphone to loudspeaker. This, however, would be unwieldy, more expensive, and technically less satisfactory. For these and other reasons, TV sound is sent as part of the overall station transmission with many functions, circuits, and actual components shared by the picture and the sound portions of the receiver.

Figure 1-12 shows, in simplified form, the sequence of functions and their combination into a single final signal for transmission. Notice that the sound part of the system resembles the ordinary radio (AM) mentioned previously; however, instead of being applied to the antenna system directly, the sound signal (an FM system) is combined with the TV picture signal. Then both are amplified as an integrated signal and applied to the transmitting antenna. For all practical purposes, this is now a single radio signal containing all the information necessary to reproduce a TV picture, both sight and sound.

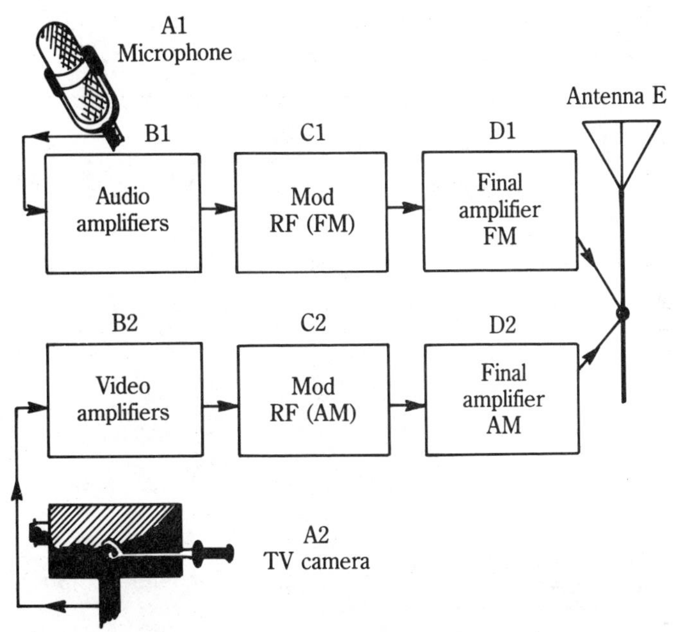

1-12 Simplified block diagram of a TV transmitter showing how the separate sound and picture paths combine before reaching the transmitting antenna.

Audio processing

In the typical mono present-day audio system, the video signal passes through a 4.5 MHz bandpass filter that separates the video from the sound. One large IC processor contains the limiter, detector, attenuator, and audio out. Inside the processor, the output of the limiter passes on to the sound detector. Here the tuned LC network recovers the audio signal.

The attenuator circuit (controlled by the volume control) and audio signal is passed on to the audio AF circuit. In most receivers, the audio is coupled to a preamp audio transistor or IC component. Some TV audio circuits consist of push-pull audio amp transistors (Fig. 1-13).

Stereo processing

Within the typical stereo audio system, the stereo demodulator contains mono and stereo audio signals. A baseband audio signal is developed and applied to the stereo mod-

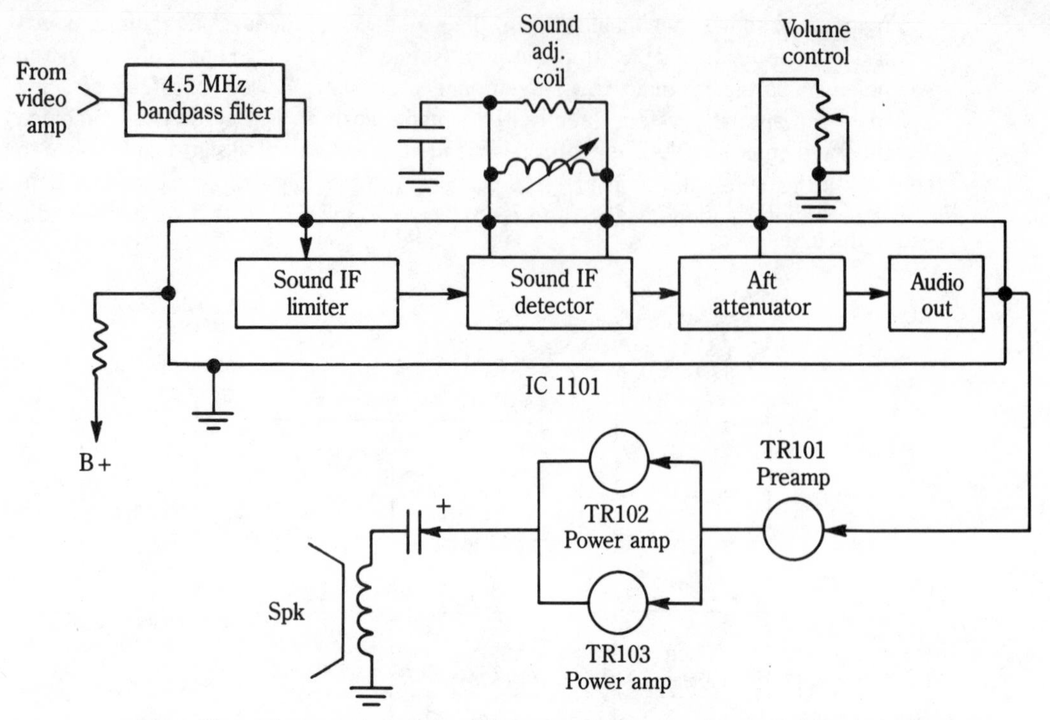

1-13 The audio mono processing circuits found in present-day color TV receivers.

ule. The output signal of IC1 is amplified and passed on to the matrix and differential amp. The left and right signals are applied to the logic monitor IC processor from the matrixing transistors (Fig. 1-14).

The monitor logic IC contains input from the matrix circuits and also has auxiliary audio stereo input jacks. IC3 switches the mono and stereo audio signals. Now, the stereo audio output signal is fed to the volume control attenuation section. The controlled volume is applied to the audio power amplifier output IC. IC5 provides stereo signal to both stereo speakers. Often, IC processor components are used throughout most stereo processing audio circuits.

Complete TV signal

To summarize the characteristics of a typical TV broadcast signal as received by the home TV set, the TV signal is a composite of two independent signals or stations—an FM sound station and an AM picture station. The FM sound station contains just the sound that accompanies the TV picture. The AM picture signal consists of the following:

- Picture brightness information, bit by bit.
- Picture background illumination, the overall scene.
- Picture bit location information in the form of cueing signals for starting and stopping each line, field, and frame.

- Picture repetition rate timing signals, which ensure exactly the same line, field, and frame frequency at both the transmitter and receiver.

The third and fourth item above are, in effect, two components of control information serving the same general purpose; that is, to ensure that the reproduction of the picture on the home TV set is precisely in step with the production of the original picture in the pickup camera. During the description of a TV receiver in chapter 2, reference is made to the various components of the timing signals, particularly in connection with the subject of vertical hold and horizontal hold.

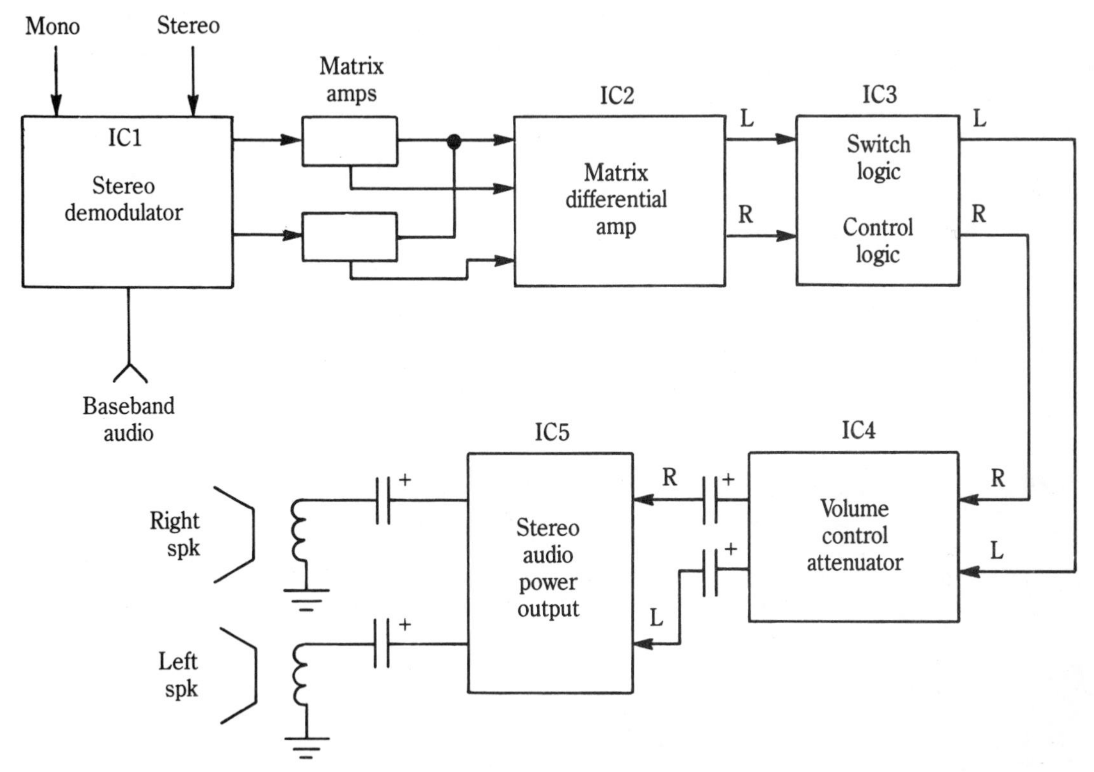

1-14 The stereo audio processing circuits found in today's stereo TV sets.

2
How a TV receiver works

SOME LOOK AT A WIRING DIAGRAM OR A SCHEMATIC OF A TV AND SEE A MASS OF meaningless lines and symbols, but to an experienced TV servicer, every line and every symbol is significant. They help the servicer follow the path of each signal through each section of the receiver in order to determine why the set won't work as it should. A schematic also tells the technician what should happen to the signals as they pass through each section. To help you get some idea what's going on in those circuits, consider each part of the TV receiving system (Fig. 2-1).

Antenna system

A home TV antenna differs from that used by a TV station for two major reasons. One is the fact that the receiving antenna must be receptive to a large number of stations, each of a different frequency. In other words, the receiving antenna is a broadband device; therefore, some compromise is necessary in comparison to an antenna designed for one particular station only. The latter is the case with the transmitting antenna of any one station. The second difference stems from the fact that the transmitting antenna is intended to radiate in all directions (it is omnidirectional) while the home antenna, even if it has to be aimed in different directions to receive stations from different geographic locations, is in fact a highly directional system. Its multidirectional receiving capability is provided by rotating it to the desired position. It may be said that all the elements (rods) of the antenna point in the same direction for maximum signal pickup from that direction (Fig. 2-2).

The transmission line or lead-in wire might be a flat-ribbon type or a shielded coaxial cable. Ideally, a flat 300-ohm lead-in wire matches the impedance of the antenna and TV receiver terminals (Fig. 2-3). The 75-ohm shielded coaxial cable must have a 75- to 300-ohm matching transformer at each end of the transmission line for a correct impedance match.

2-1 The schematic is a must item when servicing TV sets, especially in horizontal output, chroma/luminance, and video circuits.

2-2 Point the narrow part of the antenna towards the TV station. The different antennas found on apartment houses, yagi, corner, and UHF types.

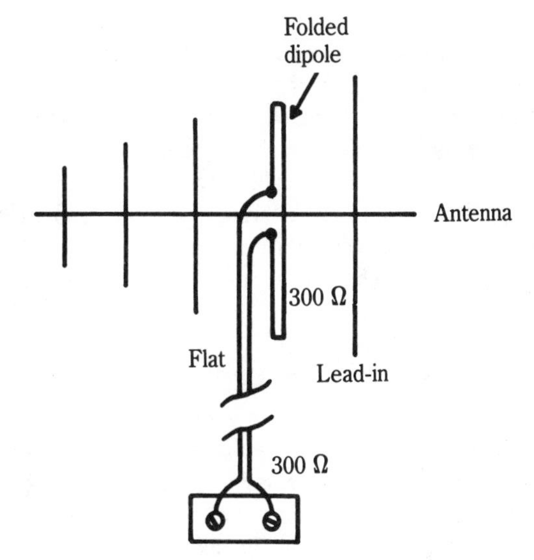

2-3 The first, most effective transmission line or lead-in wire was the 300-ohm flat ribbon cable. This lead-in wire is the ideal match between antenna, lead-in, and TV receiver.

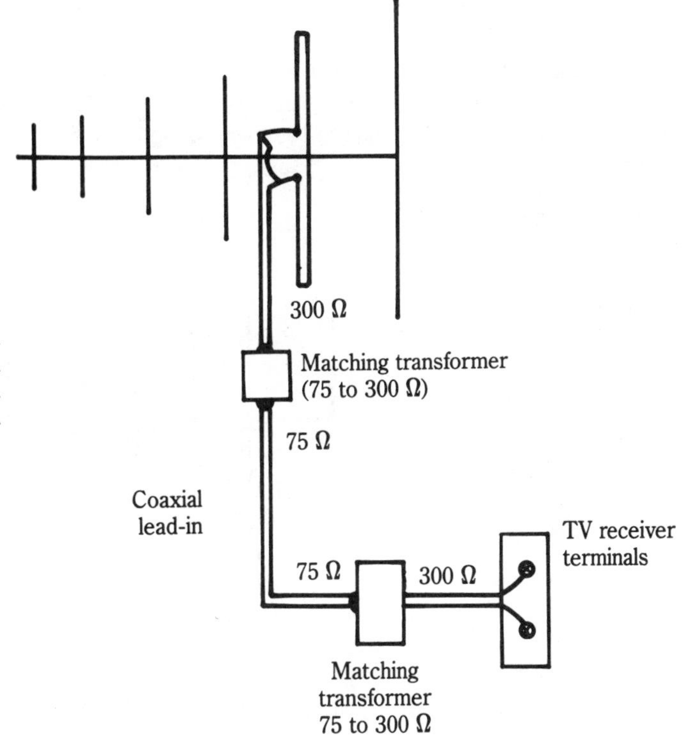

2-4 The 75-ohm shielded lead-in cable prevents unwanted signals from being picked up with the lead-in wire. This 75-ohm cable must have a matching transformer at the antenna and TV set.

Shielded cable is an excellent choice when poor TV reception is caused by standing waves (stationary wave distribution of current or voltage) or in areas where excessive noise is generated. Using shielded cable keeps noise and spurious signals from FM, police, and CB broadcasts from entering the lead-in wire (Fig. 2-4). Long runs of shielded cable lead-in can cause excessive attenuation of the broadcast signal. A booster system (RF amplifier) could solve the problem in weak fringe areas or where two or more TV receivers are sharing the same transmission line. Low-quality TV reception could also result from poor installation or improper location of the transmission line. Inferior transmission line problems and their solutions are considered in the section on troubleshooting.

Common type antenna

The yagi antenna is one of the most common directional antennas seen on towers and rooftops today. A yagi antenna is constructed from rods of various lengths that are spaced in a straight line along a central boom. (Fig. 2-5). The rods are the antenna elements. Elements that do not directly receive energy from the transmission line are called *parasitic elements*. The element connected for the efficient transfer of energy between it and the transmission line is called a *driven element*. In a yagi antenna system, a folded dipole is the driven element connected to lead-in wire. The parasitic elements that are shorter than the driven element are called *directors*. They contribute to the antenna's directional characteristic by improving signal reception from the direction they are pointing. Directors are located on the front end of the boom. Remember that if you ever install this type of antenna, point the end of the boom with the shortest element in the direction of the desired signal source. The longest element located at the opposite end or rear of the antenna is called a *reflector*. It aids in the antenna's directional characteristic by canceling or reducing signal reception from the antenna's rear. It is not uncommon to lose at least half of the desired signal strength if this parasitic element is missing or damaged.

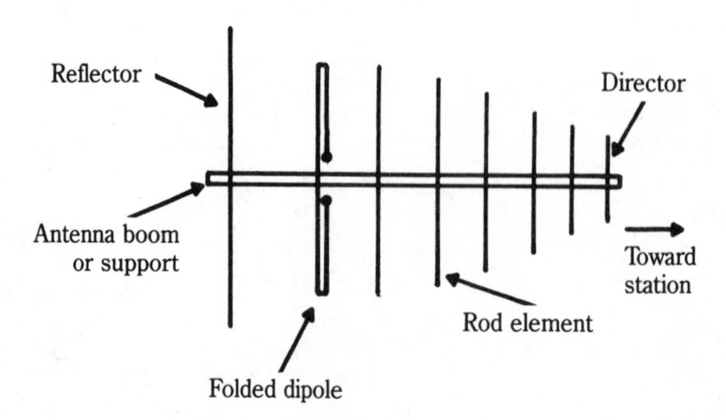

2-5 The most common directional TV antenna used today is the yagi type. Point the short element towards the station, with the longest reflector rod at the rear of the antenna.

Transmission line

The lead-in or transmission line's sole purpose is to transmit the energy picked up by the antenna to the TV set input with a minimum of loss and, perhaps just as important, with no pickup along the way. Since any such pickup is likely to be almost all noise and hardly any signal, the importance of a proper high-quality transmission line cannot be overemphasized.

TV receiver

Figure 2-6 is a block diagram of a TV with each of the major functions of the receiver represented by a box. Let's look at each of these boxes primarily from the service viewpoint (Fig. 2-7).

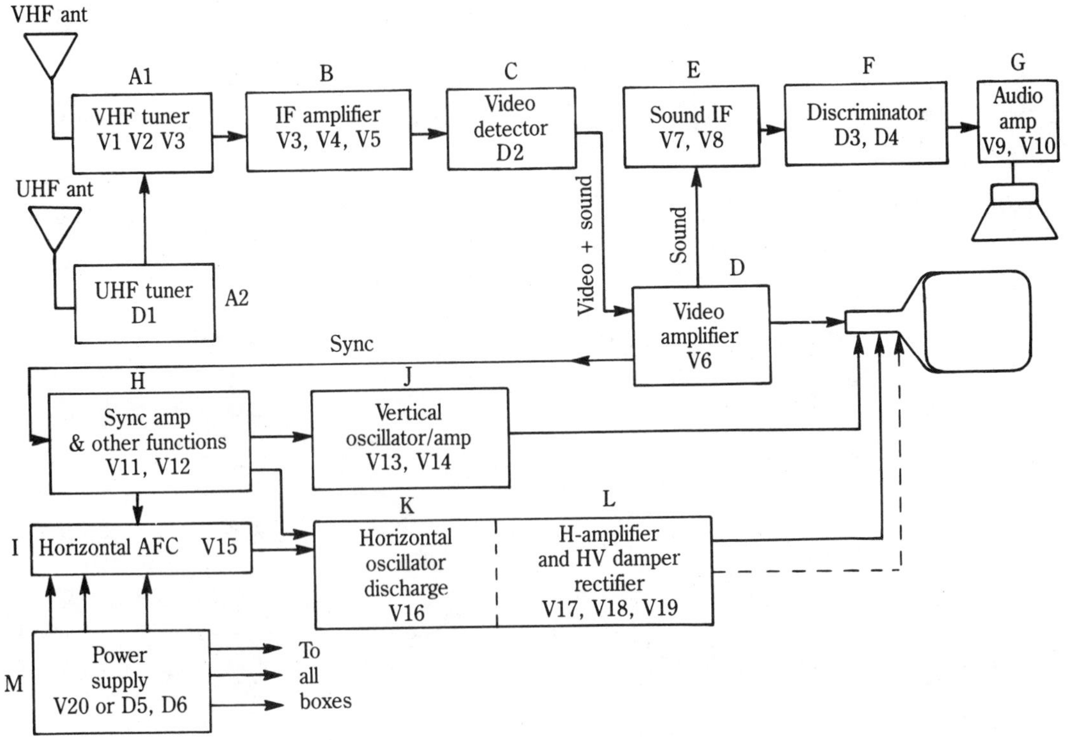

2-6 Block diagram of a typical TV receiver. The lines and arrows indicate signal paths.

Front ends

Boxes A1 and A2 represent devices called *tuners* or *front ends*. Their purpose is to select the desired station and amplify the relatively feeble signal from the antenna. For the models of the past few years, there are two tuners in each TV set, one VHF (very high frequency), covering channels 2 through 13, and UHF (ultrahigh frequency), covering channels 14 through 83. Both tuners are complex, rather sophisticated assemblies,

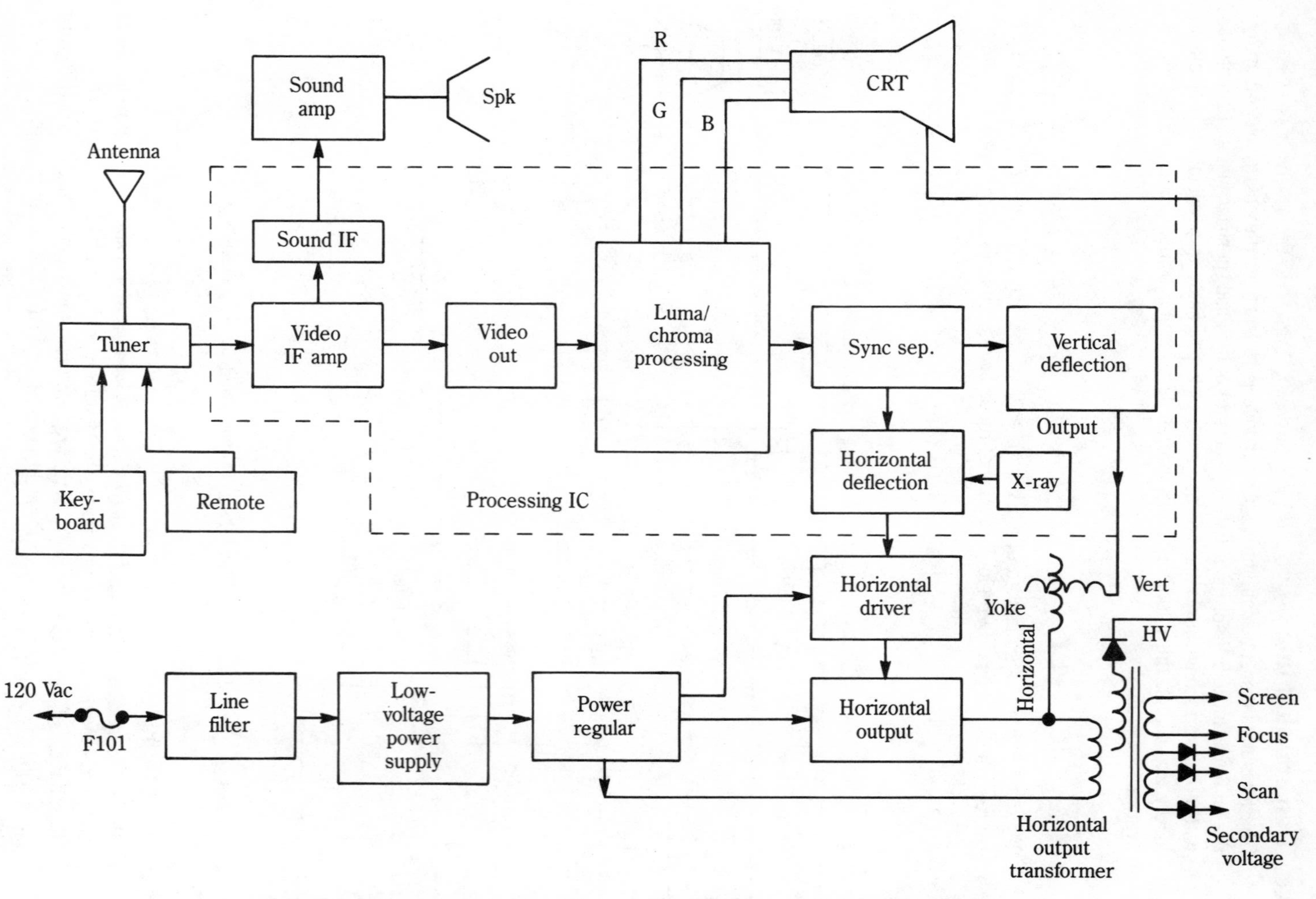

2-7 Block diagram of a keyboard for a remote-control present-day solid-state TV set.

requiring test equipment and technical know-how well beyond the capability of not only the beginner but even a good segment of the TV repair industry. In fact, many high-grade TV repair shops have TV tuners repaired by special service stations fully equipped to do the work. The intent here is to caution the beginner against any rash action in attempting to correct a malfunction in this portion of the receiver. There are, however, certain tasks that are within the ability of the beginner that can be performed satisfactorily, and these are described in chapters covering troubleshooting.

Construction and location of the tuner are of particular interest, more so than any other section of the set (except perhaps the high-voltage cage, which is described later). Both the station selector and the fine-tuning adjustment are, for technical reasons, physically built into the tuners. This requires that the tuner be located at an accessible position on the cabinet. For this reason, as well as for some strictly technical factors, the tuners are physically separate units that are relatively easy to disconnect from the remainder of the TV set (Fig. 2-8).

2-8 The VHF and UHF tuner located in the left top corner of a black-and-white TV receiver.

Interconnections between the tuners and the main TV chassis and the antenna are as follows:

- The signal input from the antenna terminals to the tuners is by way of twinlead or sometimes concentric (coaxial) cable.
- The signal output, almost universally in the form of a shielded coaxial cable (a center wire surrounded by insulation with a braided copper outer sleeve) terminated in a single contact plug.

- A three- or four-wire cable connecting the various voltages in the main chassis. These wires might end in individual lugs fastened by screws, or in a common, multipin plug that fits a mating socket on the main chassis.

Finally, there must be some provision for rigidly fastening the complete tuner assembly to the cabinet. Figure 2-9 is one example of such a tuner; others differ in minor detail, but essentially they are of the same general construction and interconnection arrangement.

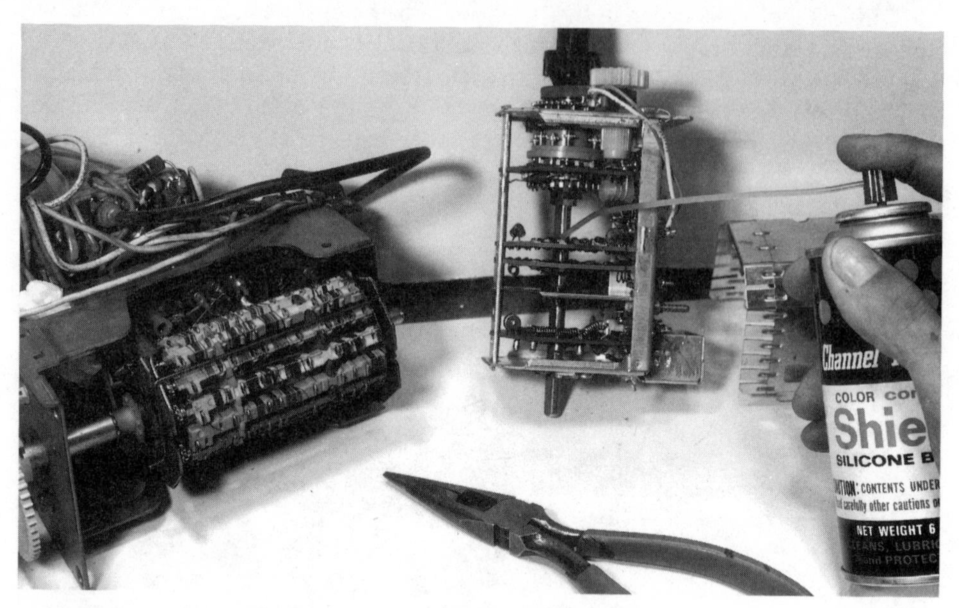

2-9 Typical VHF tube-type TV tuner, left, and an early solid-state tuner, right.

Tube-type tuners

The VHF tuner, if it is of the tube type, usually has two tubes of the miniature (all glass, no separate base) kind, and they are clearly marked for location, type, and function; thus, a tuner might have the following stamps near the tube sockets: 6AH5, V1; 6EA8, V2. Sometimes when more than one tube type is suitable, dual markings appear such as 3HA5 or 3HM5 (3HA5/3HM5) and 6EA8. In other cases, the designations V1 and V2 do not appear on the tuner but will be shown on the tube location chart somewhere inside the TV cabinet. The correlation between tube markings on the tuner and the tube location chart inside the TV cabinet is used extensively to identify tubes by function, malfunctions, symptoms, and suggested corrective action.

In those tuners where transistors have replaced tubes, there is the high probability that the transistors are soldered into place instead of plugged into a socket, in which case the advice given above regarding tuners in general applies very much here. To repeat, tuner work, other than tube replacement, is not for the amateur. Even if a beginner could locate the suspected transistor, any attempt to desolder it could be disastrous. Removal of the entire tuner, where such a procedure is indicated, is far easier and safer.

Of the two tubes in a tube-type tuner, at least one is a dual-purpose type, so there are three distinct functions performed here. The first tube, almost always located nearer the incoming antenna lead-in, is an RF amplifier that increases the signal level of the station being received. A bad tube in this position seldom results in no picture and sound whatsoever. Instead, it will account for a very feeble picture with heavy snow and weak sound very much resembling a weak or distant station. The remainder of the TV set is doing its best with the very small input signal that manages to get through the defective amplifier. Removing a bad RF amplifier tube from its socket might produce very little added degradation of the picture. In those sets where removing one tube causes all other tubes to stop glowing, another procedure is suggested under the discussion of transformerless TV receivers.

The second tube of the tuner is almost always (except in those rare cases where the tuner has three tubes) a dual-function tube. It will invariably be identified on the location chart as V2 and sometimes with an added optional description such a MIX-OSC (for *mixer-oscillator*). You need not know the whole story about mixer-oscillators to know their functions. If this tube fails, there is nothing on the screen (except the white *raster*), no faint picture, no snow, just a total blank, and no sound. On a nontechnical basis, you might consider a defective V1 (RF amplifier) as a gate that is barely cracked open, while a defective mixer-oscillator is shut tight (100 percent closed) so that nothing can get through.

In the case of the rare possibility of encountering an older vintage TV set with a three-tube tuner, these tubes probably will be identified on the chart as V1-RF, V2-MIX, V3-OSC. It should be apparent from this that the combined functions of V2 in the modern tuner are separated in the old three-tube unit. The comparison is quite simple. The first tube is identical, both in function and malfunction, on both tuners. The tube marked V2-MIX in the old version, when defective, might still permit some noise (snow) to get through, but seldom a picture, even a faint one, or sound. A failure in V3-OSC might produce similar symptoms with one exception: no picture or sound at all. In troubleshooting, it is not automatically necessary to replace both mixer and oscillator tubes if only one tests unsatisfactorily, but it is most practical to suspect both tubes when the symptoms are as described above.

The semiconductor tuner

In recent years, the change from tube to solid-state TV receivers was made in two steps. The first was to the hybrid (tubes and semiconductors) receiver, and the second to all-semiconductor set. Strictly speaking, there are no 100 percent solid-state receivers because at least the picture tube, and often one or two others, are still of the glass type. But these sets are commonly described as all solid state. While a few hybrid sets are still being manufactured, the tuners in virtually all TV sets are of the solid-state variety. The reasons are many and varied, but as far as the nontechnical consumer is concerned, this feature is all the better. These tuners are more stable, more rugged and longer lasting. Figure 2-10 is a view of a modern solid-state tuner. Greater compactness and freedom from protruding tubes is obvious even if the durability and greater freedom from repairs might not be visually evident.

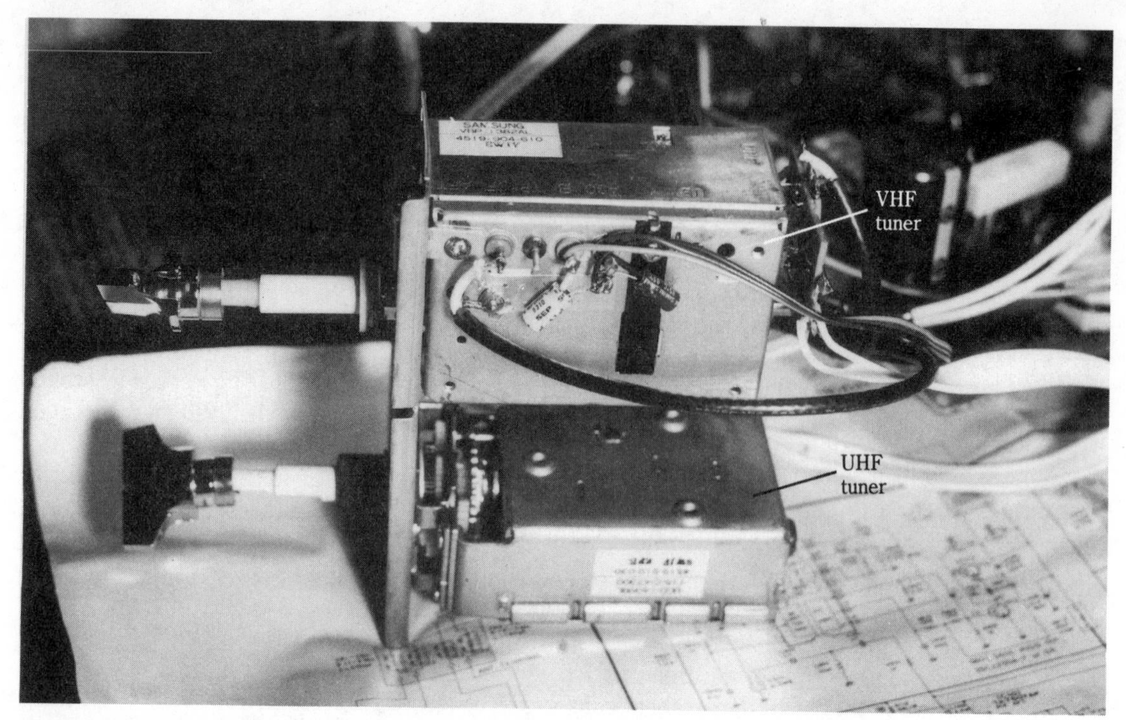

2-10 Modern solid-state tuners. The top tuner is a manual VHF tuner and the bottom one is a manual UHF tuner.

Varactor tuner

A radical departure from conventional tuners, whether tube type or all solid state, is the so-called *varactor* type, or *voltage-controlled* type. It has many advantages over the conventional type—in stability, freedom from noise, minimal degradation with use, not to mention greater simplicity in construction and maintenance. For an understanding of this, however, it is necessary to have a little background information of tuned circuits in general, and TV tuners in particular.

The term *tuning*, whether applied to automotive adjustment or musical instrument adjustments, refers to setting or adjusting a device or a group of devices to a predetermined condition. In electronics, tuning is used exclusively to indicate adjustment of one or two components in a radio receiver (station tuning) or TV set (channel selection). In repair procedures, a function corresponding to tuning is called *alignment*. In each of these examples, tuning is accomplished by a coil (a loop or spool of wire) and a capacitor. In the case of TV tuners, each channel has its own little coil (actually a number of such coils) cut to size and connected to the circuit by means of a multiple-contact switch actuated by turning the channel selector. Since the exact configuration of such a coil or coils is essential for repeatable accurate tuning (getting the correct station reception every time) these coils are made as sturdy and as rigid as possible. Nevertheless, long use, shock, vibration, and switching cause a gradual deterioration of the whole tuning system, eventually requiring replacement of the tuner, or rebuilding—a tedious and expensive job at best. But the weakest component is the switch, operated by the channel-selector knob. The switch contacts, being part of the functioning coil-capacitor assembly, cause the worst degradation with use due to contact corrosion, accumulation of dirt and oily film, etc. To the user, this evidences itself in the form of a *noisy tuner*—erratic pictures, frequent need for jiggling the knob to restore the picture, and ultimately complete loss of stability (the station will not stay on channel). The varactor tuner goes a long way toward elimination of most of these problems (Fig. 2-11).

2-11 The varactor tuner in a Sharp 19D80 model color TV receiver.

In simple terms, the varactor is a semiconductor, a diode in fact, which not only acts as a capacitor (hence a tuning device), but one whose capacitance can be changed over a considerable range through the application of a voltage. The device is sometimes referred to as Varicap (trademark), meaning a *vari*able *cap*acitor. The advantages in TV tuners are obvious. Although station switching by means of the channel selector is still required, switching can now be done in less critical points, making for much greater stability. Simplification of tuner design and construction is a further advantage of this new technique in semiconductor application to modern TV receivers. While the advantage might not be obvious to the average user, especially in a new or fairly new receiver, it will become apparent later, as evidenced by the absence of chronic misbehavior of the tuner, as is the case with the coil-switching type.

Tuner module The varactor solid-state tuner can operate in a manual, push-button keyboard, or remote control mode. The tuner module is easily exchanged by simply unplugging the wires and IF cable to the TV chassis, then remove two or more mounting screws to free the tuner. Varactor tuners of the same type may be found in other configurations with the push-button or remote control TV receiver (Fig. 2-12).

2-12 The tuner modules in the General Electric chassis receive up to 112 channels.

Push-button keyboard In the keyboard TV operation, you simply push the correct numbers to select the desired TV station. To select channel 13, push buttons 1 and 3. Likewise, the UHF station channel 27, may be selected by pushing buttons 2 and 7 in that order. All VHF and UHF channels are directly accessed regardless of frequency band or numerical designation (Fig. 2-13).

Besides having a tuner module, the push-button operation may have a separate synthesis or memory module. Usually these modules are plugged into each other and receive their operating power from the TV chassis. The antenna cable and IF receiver cable plug into the tuner module. These modules are easily replaced by removing the plugs and metal mounting screws. Manufacturers may call the various modules by different names, but the operation is the same.

Remote control tuner modules Since cable TV has become so popular, TV manufacturers have designed the TV antenna terminals to accept the cable connection. The cable system may be connected directly to the tuner module through a 75-ohm cable connector, or a standard 300-ohm antenna connection may be used. Today, more remote control TV receivers are sold than any other chassis, since all of the stations may be easily selected at random.

2-13 The push-button tuning board of a Goldstar CMZ4122 model portable TV.

2-14 The remote-control TV receiver has a tuner, memory, and control module that can be removed by unplugging the connecting cable and wires. This solid-state control can select the various TV channels instantly and silently. These three modules are in the RCA CTC108 chassis.

In older remote control TV receivers, noisy and cumbersome motors were used to rotate the tuner and TV controls. With the new tuner, memory and control modules, the TV receiver can be tuned to any TV channel electronically within seconds and with silent operation (Fig. 2-14). The remote tuning is accomplished with a small hand-held transmitter. A receiver is in the control module located within the TV cabinet.

Remote transmitter The present day simple hand-held transmitter can select any channel, turn the receiver off and on, adjust volume up and down, and when the telephone rings, the mute button can be used to silence the TV (Fig. 2-15). A more elaborate remote transmitter might have the capabilities to allow quick scanning of the entire TV frequency band within seconds of individual channel selection by pushing the correct channel designation. Besides controlling the TV, a more elaborate remote transmitter may control a video disc player and VCR.

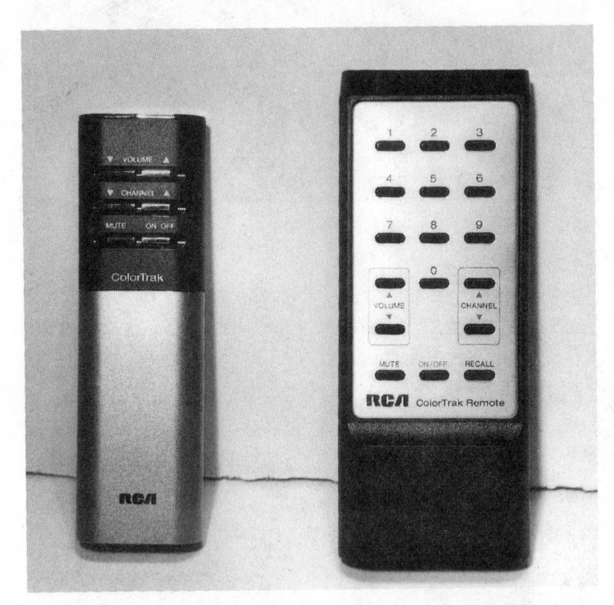

2-15 A hand-held remote transmitter is capable of selecting any available channel, adjusting the volume up and down, turning the TV on and off, and muting the sound. These small transmitters are battery operated.

One type of remote transmitter radiates a unique RF signal for each operation mode, and when more than one TV of the same type is operating in the home, the remote control transmitter could operate other TV sets. This problem could even occur in an apartment complex between neighbors. Today, the infra-red transmitter will only operate the TV receiver if the remote transmitter is pointed at the TV. These infra-red transmitters are very directional and radiate a short range infra-red signal.

The portable hand-held remote transmitter may operate from two small penlight cells or a regular 9-volt transistor battery. The batteries will normally operate satisfactorily for six months to one year; when replacement is required heavy-duty leakproof types are recommended. Weak batteries may cause the remote hand-held transmitter not to operate at all, or weaken the radiated infra-red signal, making it impossible to operate the TV when the transmitter is held more than a foot or two, away.

Check the remote transmitter batteries any time the TV set fails to respond to the infra-red signal; try cleaning the battery terminals by rubbing each terminal with a cloth or on the carpet. If the remote control will only operate after tapping or rapping the case, check for loose batteries or a broken connection. The hand-held remote transmitter is often dropped, dislodging components and cracking the PC board. Remote control transmitters may be sent into the tuner repair depot that services TV tuner assemblies for repair.

Frequency synthesis tuning system

The most expensive and latest RCA chassis have a channel-lock frequency synthesis tuning system. The tuner may provide up to 127 channels. Also, the tuner covers all of the UHF band. The tuning system can be controlled manually or remotely with a hand-held remote transmitter. A cable/normal switch lets the viewer decide between regular TV broadcast or cable reception.

The frequency synthesis board may consist of a synthesizer IC which contains a PLL step generator, band switch decoder, AFT, digital sync, presence detector, serial decoder, aux switch and video blanking. The microcomputer section provides correct IR decoding, keyboard decoder, sync communicator clock, and data signal for the synthesizer IC. It also provides a display driver to the channel information buffer. The synthesis tuner may appear in a module component.

Manual preset tuner

The manual preset tuner is operated manually with stations already tuned into a varactor tuner. The varactor tuner tunes in the TV broadcast stations by applying different voltages to a varactor diode within the tuner. The voltage applied must be varied by a variable reostat control or with preset resistance. In this tuner a small variable resistance control is provided, one for each station, including 6 on the UHF bands (Fig. 2-16).

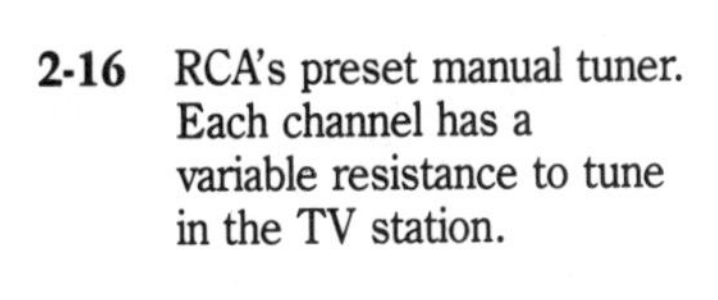

2-16 RCA's preset manual tuner. Each channel has a variable resistance to tune in the TV station.

Since there are only 12 channels, a small variable resistor is included at the front of the tuner for adjustments. Slip off the plastic plate to get at these controls or remove tuner. These adjustments can be made from the front of the TV set. The numbers A through F can be removed and the UHF number placed in each section. Because most areas have only a few UHF stations, the rest of the letters will remain. The six large variable resistors at the bottom represent the UHF stations. If one of the stations slides off or cannot be tuned-in manually, remove the plate and tune the correct variable control with channel number.

IF amplifiers

The RF amplifier is the first tube of the tuner. It and the amplifiers to be described here have similar functions—to amplify a small TV signal to a larger one. The difference between the two types of amplifiers is in the degree and level of amplification. IF means *intermediate frequency* and refers to the amplifiers after the tuner (block B, Fig. 2-6). When IF tubes malfunction, the severity of the snow resulting is much less, but they still behave generally as gates which are usually open and more or less shut during a malfunction.

There are, however, some noteworthy differences from the beginner's viewpoint. First, because the signal levels are so much greater here, the chances of a total absence of the picture and sound are somewhat less; a dim or thin picture will get through more easily. Second, because of the use made of IF amplifiers (some common to both sound and picture, others are strictly for sound) a failure in one IF amplifier may cause a loss of sound without affecting the picture. Third, the physical configuration is quite different. While the RF amplifier is almost universally located on a separate, integral unit (the tuner), the IF amplifiers very generally are located in some sort of string fashion on the main chassis of the TV set.

In Fig. 2-6, the paths traveled by the picture and sound from the antenna to the picture tube and speaker are shown by lines and arrows. It can be seen that for VHF (channels 2 through 13) the signal path is from the antenna, via the VHF tuner to the IF amplifiers. For UHF (channels 14 through 83) the path is from the antenna through the UHF tuner, through the VHF tuner, then to the IF stages. This difference is of importance in troubleshooting, as we shall see later, because the RF amplifier on the VHF tuner is no longer the first stage in the path of the signal from the antenna.

As was mentioned previously, a TV receiver is really a two-in-one device, combining a picture (AM) receiver with a sound (FM) receiver. As such many components are common to both. Block B in Fig. 2-6 is a multiamplifier system for the combined video and audio signals. Although these signals are of comparatively greater strength than those entering the front end (A1 or A2) a failure in any stage of this amplifying system can be just as catastrophic for either picture or sound.

There are usually three separate amplifiers or stages in sequence, as links in a chain. A defect or total failure in any one stage breaks the chain and prevents the signal from continuing on its way. There is usually some feedthrough of signal even when one stage goes dead, as in a case of tube or transistor burnout or other failure, so that some small portion of the normal signal is transferred to the next stage. One symptom of such a failure might be a very weak picture on any station; although some stations may come in better than others largely because of the original difference in signal strength between stations.

Symptoms of snow mentioned in regard to the front end are much less apparent here and may even be completely absent. Another result of such a failure might be picture instability, such as tearing of the picture in a generally horizontal (actually diagonal) direction and possibly a rolling picture (vertical direction). This is due to the fact that the amplifiers in block B also must pass the synchronizing pulses. A failure in these amplifiers will invariably degrade the quality of these pulses to a level below that required for picture holding. In the section on troubleshooting, I shall present concrete procedures designed to help locate trouble in the IF portion of the set.

At this point it is imperative to clearly indicate what the beginner may not and should not attempt as far as the IF transformer adjustments. You may see a half-dozen either rectangular metal cans or uncased spool-like coils or transformers, some of them with a hole or slot that seems to invite a screwdriver blade for turning. These are in fact adjustments, but they cannot be properly done by eye or ear. Required is some very sophisticated equipment and very specific knowledge and expertise, different for each and every adjustment in all TV sets. Worse yet, unlike the adjustment of some circuits recommended later in this book that can be either repeated or restored to their original position, the adjustment of any of the IF coils is, from the very beginning, a point-of-no return case. Not only is it virtually impossible to know what the adjustment is accomplishing, but it is equally impossible to go back to the starting point.

To further illustrate the futility of such adjustments, the professional service uses complex equipment that gives a visual presentation of the operation of all the circuits at the same time. Only on such a visual display can an expert servicer observe the effect of each adjustment on the overall picture. Looking at the front of the TV set while twiddling one of the adjustments is almost certain to destroy the normal quality of the picture without showing any immediate change as the twiddling is made. Finally, it is a fortunate fact that these adjustments, except for some very minor effects, are virtually permanent and do not require any correction, except by the professional servicer after replacing a transformer in case of a burnout or other catastrophic failure. To sum up, the best advice is leave the IF adjustments alone. They are probably as they should be.

Transistorized SAW filter and PIX IF

The IF signal is applied to the PIX IF and surface acoustic wave (SAW) filter component coming from the tuner system. Usually, the IF signal is amplified and then applied to the SAW filter network (Fig. 2-17). The output of the SAW filter is applied to the IC processor.

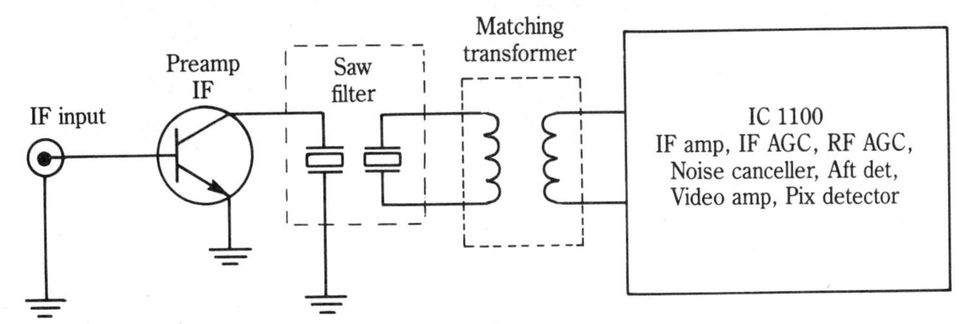

2-17 Block diagram of a SAW filter network between IF input and IF amplifier IC.

The IC processor might have two or three stages of IF amplification, video detector, AFT, and AGC circuits. The surface acoustic wave filter features a very compact unit. This device is accurate in establishing correct IF response. The length of the surface wave differs somewhat according to the material element of the piezoelectric element. The interdigital transducer is so arranged in length and width to provide desired bandwidth for the PIX IF circuits. This small device very seldom causes any problems and takes the place of several transistor or tube IF stages.

Video detector

In the video detector stage (block C, Fig. 2-6), although relatively simple and consisting of but one diode (or one tube) in most cases, a number of functions occur. A semiconductor diode (D2) converts the video IF signal to a form suitable for application to the picture tube. Also, the combined picture and sound signals coming from the common IF amplifiers—V3, V4, and V5—are separated. A failure (rather rare) of this diode stops both picture and sound signals. In some receivers the sound is "picked off" before it reaches the video detector; so a failure of the diode in such sets would not affect the sound. We shall discuss this again later under the sections covering troubleshooting (Fig. 2-18).

2-18 The IF and video sections of a tube chassis.

Video amplifier

The video amplifier tube (seldom more than one) in block D V6, Fig. 2-6 is a video (picture signal) amplifier, building up the picture signal to the level required for application to the picture tube. In addition this tube also amplifies the sound IF signal to block E in many receivers and the various picture control (sync) pulses to block H. Should this tube fail all three signal components would be affected, as in the case of the defective diode in block C. In case of tube deterioration due to old age, etc., the effects may not be catastrophic; that is, the picture and sound may still be there but on degraded levels. In such a case the picture control (sync) pulses may be inadequate to keep the image from rolling and/or tearing. In normal operation the picture signals goes from V6 to the picture tube, as shown in Fig. 2-6.

Solid-state video section

The transistorized video section may be included in one or two transistors, single IC, or included in one large IC with IF amp, sound detector, sound amp, video detector, video amp, noise canceler, AFT detection and AGC detection circuits (Fig. 2-19). In some recent color TV chassis, only two large ICs are in the whole chassis, except the sweep output stages.

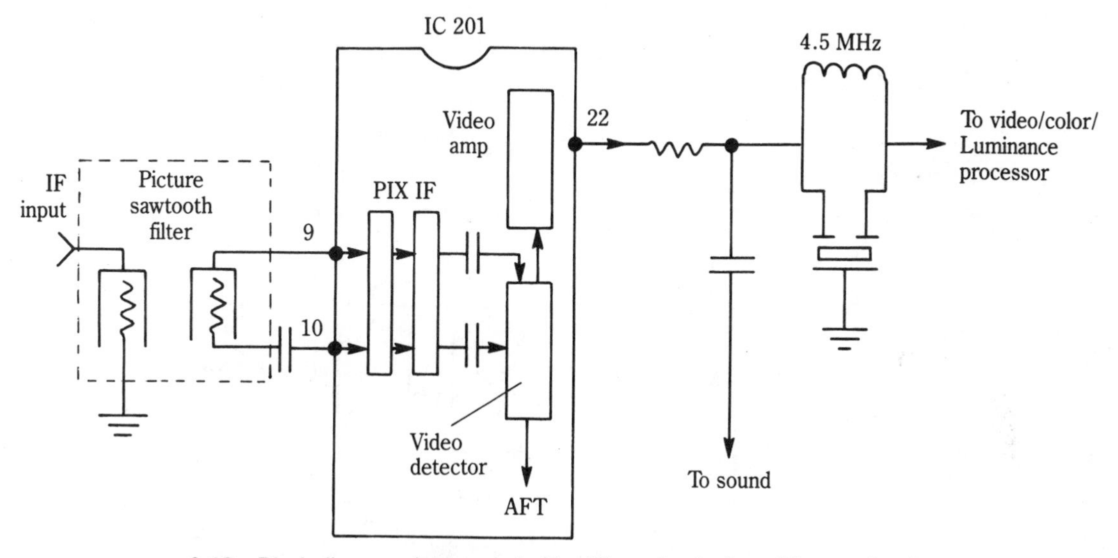

2-19 Block diagram of the typical video IF section in the solid-state chassis.

The IF signal from the picture IF filter (SAW) may have one preamp trasnsistor, or in recent color receivers the SAW filter couples directly into a large IC. Internally, the IF signal goes through an IF amp or buffer stage, video detector, and then to the video amp. The output of the video amp is fed to a coil and 4.5 MHz trap ceramic filter where the sound carrier component (4.5 MHz) is removed and applied to the video and color processor IC. The sound signal is capacity coupled before the ceramic trap filter and fed to the sound circuits.

Sound IF system and detector

The sound IF amplifier system (block E, Fig. 2-6), including V7 and V8, is functionally very similar to the IF amplifiers in block B, except that only sound signals (FM) exist in this unit. Sometimes the amplifier consists of but one stage (one tube), depending on the design of the particular manufacturer. Another similarity to the IF amplifier in block B is the fact that the sound amplifier is also followed by a detector—an FM sound detector, often called a *discriminator*. The discriminator (block F, Fig. 2-6) may be either a tube or a pair of semiconductor diodes, depending on the particular make of a TV set, and a transformer or coil. Failure of the discriminator tube (or diode) will, of course, cause a complete loss of sound.

Solid-state audio-detector-amp circuits

All sound circuits, except the audio output, may be located in one large IF video IC. In some earlier audio circuits, a separate IC may process the IF sound and include a driver sound amp (Fig. 2-20). FM sound detection is tuned with a small coil that can be touched up with the adjustment tool. When the sound becomes erratic or muffled, try adjusting the sound detection coil for clarity. Sometimes, a change in temperature will cause the sound coil to drift off frequency.

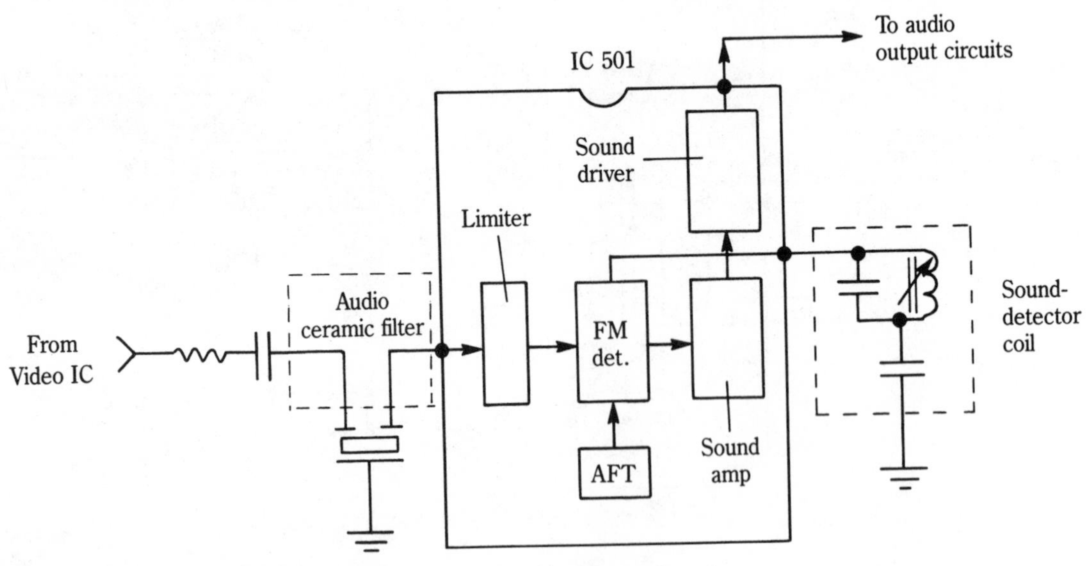

2-20 Block diagram of a solid-state IF sound section.

Audio amplifier and speaker

Beginning at the discriminator output and all the way to the speaker, the signal is known as *audio* for it consists only of audio frequencies. Where the discriminator consists of a pair of semiconductors, the audio amplifier in block G (Fig. 2-6) might consist of two tubes. In some sets a multifunction tube is used as a discriminator audio amplifier combination, followed by only a single tube called the final (output) audio amplifier in block G. In either case the final audio amplifier connects (via an output transformer) to the speaker.

Failure of either or both of the audio amplifiers will cause a total loss of sound, although sometimes a defective first audio tube (V9 in our case) may allow a faint sound to go through. In such a case, the volume control that regulates the amount of signal going into this tube (regardless where the control is physically located) may have some minor effect on the already feeble sound.

Solid-state audio output circuits

The transistorized audio output circuits may be included with a preamp or driver stage within the large video IC or all included within the audio output IC. Some audio output amps have several audio transistors (Fig. 2-21). The audio signal from the driver inside the IF sound IC is fed to another AF transistor that drives two transistors in push-pull operation.

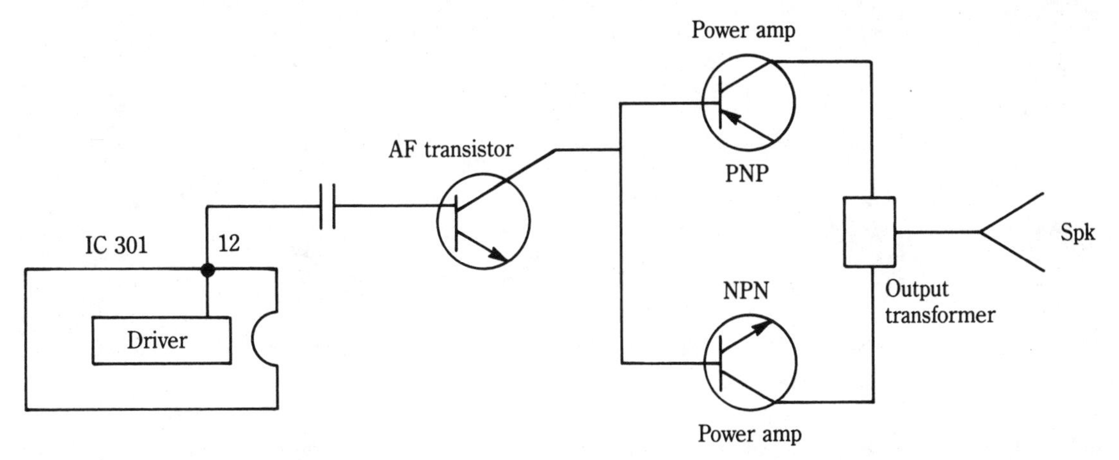

2-21 Block diagram of a typical solid-state transistorized audio-output circuits.

Usually, when audio output transistors are used, a matching output transformer is coupled to the PM speaker. When the IC chip is found as audio preamp, driver and output circuits, the speaker is coupled with a fairly large electrolytic capacitor (Fig. 2-22). Check this coupling capacitor when intermittent or dead sound is the symptom. Stereo sound may be found in the high end console TVs.

The audio output transistors in the signal ended push-pull circuit contain one npn and the other pnp. If the voltage at the collector of Q300 is decreased to below the specified limit, Q301 is turned off while Q303 is turned on. Thus, the current is limited at the emitter of Q303 and is connected through the transformer primary winding. Then, if the voltage at the collector of Q300 is increased to a limit, Q301 is turned on while Q303 is off producing current through the primary winding of output transformer.

Sync circuits

The sync circuits (block H, Fig. 2-6) contain two or three tubes, some of them being sections of dual types. The purpose of the sync section is to provide the proper timing

2-22 The solid-state sound and discriminator coil section of an RCA solid-state receiver.

and placement of the image bits on the screen. As stated earlier this synchronizing information is transmitted by the TV station as part of the picture signal. There are two main types of pulses here: those responsible for the proper starting time of each line on the left side of the screen (horizontal sync pulses) and those controlling the exact start of the picture at the upper-left corner of the tube (vertical sync pulses).

Thus, the first function of the sync circuits is that of a sync amplifier; that is, to increase the sync signal to its required level. There is also a sync separator stage that separates the vertical sync pulses from the horizontal pulses. And there is, additionally, a third function called sync clipping, which, in simplified terms, means the extraction of the sync information from the combined picture-sync signal. For practical reasons, such as economy, most of the sync tubes are dual tubes. An additional function, not necessarily pertaining to sync, is performed by such a "spare" half-tube. Frequently the spare is utilized as a vertical oscillator, to be discussed later.

The sync circuits and tubes have no effect on sound at all, and only indirectly on the picture. Actually the picture information is completely independent of sync performance; however, since the sync signal controls the sequence and location of each picture element, any malfunction in the sync section is certain to produce chaos in the picture, either an up or down rolling of the picture or some tearing in the horizontal direction. It is, therefore, a safe general procedure, whenever the picture will not stand still, to suspect one or more of the sync tubes.

Solid-state sync and AGC circuits

Within the tube chassis, a separate tube was used for sync and AGC circuits, while in the transistorized chassis, separate transistors were used for each circuit. Today, the AGC and sync circuits are all found inside the PIX IF and video IC. Internally, the RF and IF AGC circuits are taken from the IF amp and buffer circuits. The RF AGC is fed from terminal B to the varactor tuner. The AGC delay control is fed from pin 12 of the RF AGC circuit. This RF AGC delay control adjusts the input voltage to the tuner circuits (Fig. 2-23).

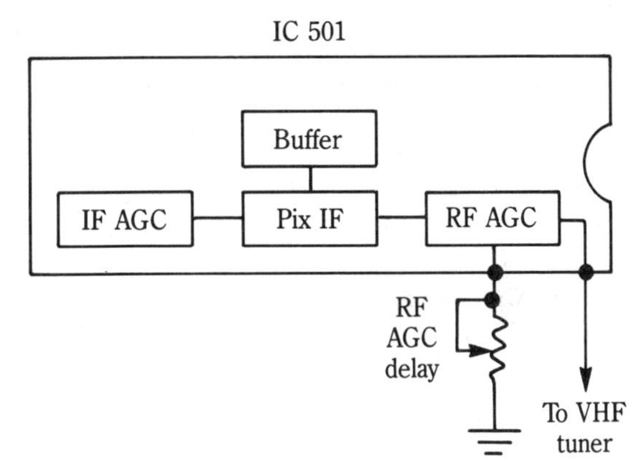

2-23 Block diagram of typical AGC circuits within the IC.

Vertical circuits

We know that the picture on the TV screen is "painted" by an electron beam moving in zig-zag fashion, one line at a time, from left to right. It is apparent from this action that as the beam sweeps from left to right it is also pulled downward at a constant rate. This downward movement of the beam is accomplished by a system known as the *vertical sweep*, consisting of a vertical oscillator and a vertical amplifier (tubes V13 and V14 in block J, Fig. 2-6). The two tubes in the vertical system follow each other in sequence so that either one, if defective, will cause a failure in the downward pull of the beam on the screen. The first tube, V13, is called the vertical oscillator.

In nontechnical terms, the vertical oscillator could be called a *timer* because it determines how long it takes the beam to traverse the screen from top to bottom. Since the time must be precisely the same as the corresponding time at the transmitter, the vertical sync pulse is used to cue the timer. In other words, the vertical sync pulse from the transmitter sets the precise rate for the vertical oscillator in the receiver.

Figure 2-24 shows what happens when the vertical sweep circuit fails. With a single bright line across the center of the screen the electron beam is painting all the lines on top of one another. There are other picture defects attributable to trouble in the vertical sweep system. A crowded (compressed or squashed) picture at the top and a stretched picture at the bottom are often due to a defect (nonlinearity) in the vertical amplifier and its associated components.

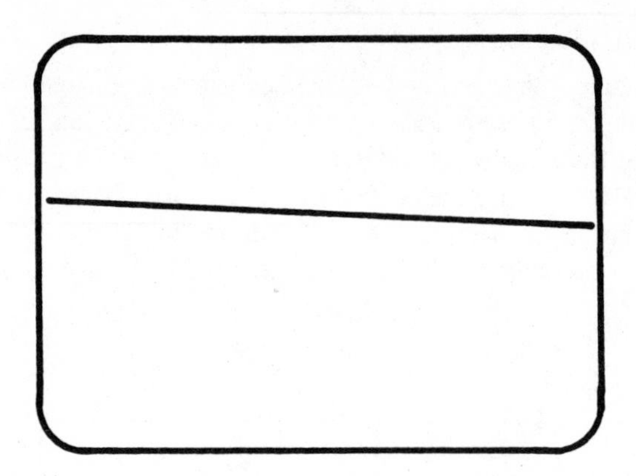

2-24 A simple horizontal line across the screen is an indication that the vertical deflection system has failed.

There are three adjustments associated with the vertical sweep system which may require resetting in case of malfunction. These are vertical hold, vertical size, and vertical linearity. The first of these in most sets is accessible from the front of the cabinet. The last two almost always are located in the rear because they seldom require adjustment. Each is described under troubleshooting, where cause-and-effect relationships are outlined.

As the name implies, the vertical hold is used to stop this picture from moving up or down; that is, to prevent rolling vertically. The vertical size is an adjustment to make the picture cover the full height of the visible screen. The vertical linearity adjustment serves to adjust the picture for minimum distortion, as when the lower half of a circular object looks flattened while the upper half is egg-shaped. The vertical linearity control is adjusted to make a test pattern circle perfectly round.

Solid-state vertical circuits

Separate transistors may be used in the early transistorized chassis within the vertical circuits. A sawtooth signal is fed from the luminance/color IC and applied to the base terminal of the vertical drive transistor. This causes the vertical output transformer to conduct. Both vertical output transistors are conducted by the sawtooth signal produced at the vertical drive transistor. The output is capacitor coupled through the electrolytic capacitor to the vertical deflection yoke winding (Fig. 2-25A).

You may find the vertical sync, ramp generator and vertical driver in one large luminance/color IC. The ramp generator signal is taken from the second count down circuit. The vertical driver signal is taken from pin 23 and applied to the input vertical output IC. The vertical output pulse from IC401 is connected directly to the vertical yoke winding (Fig. 2-25B). In some TV chassis, the vertical output may be included within the deflection IC component, while in others two vertical output transistors are tied to the vertical driver circuit within the deflection IC.

Horizontal circuits

While the sync circuits are responsible for the precise timing of the picture reproduction process, timing of horizontal synchronization is more severe and more rigorous than the

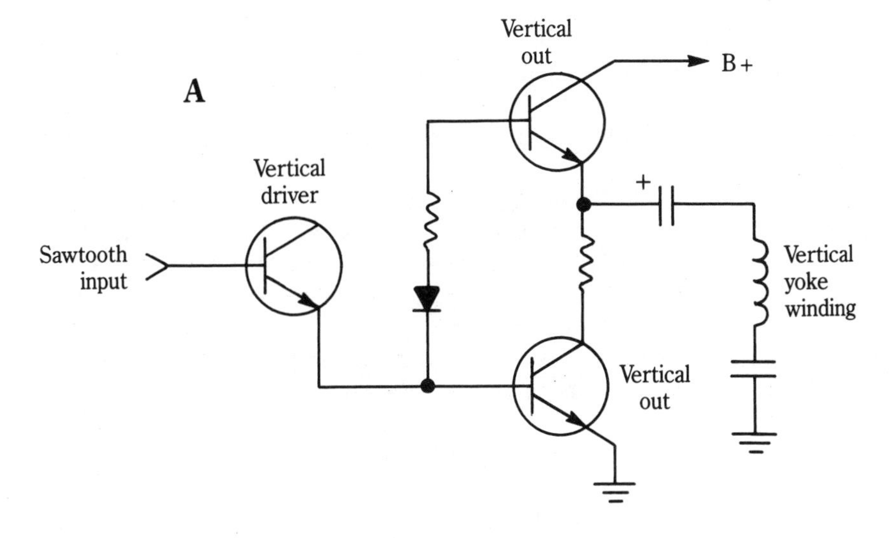

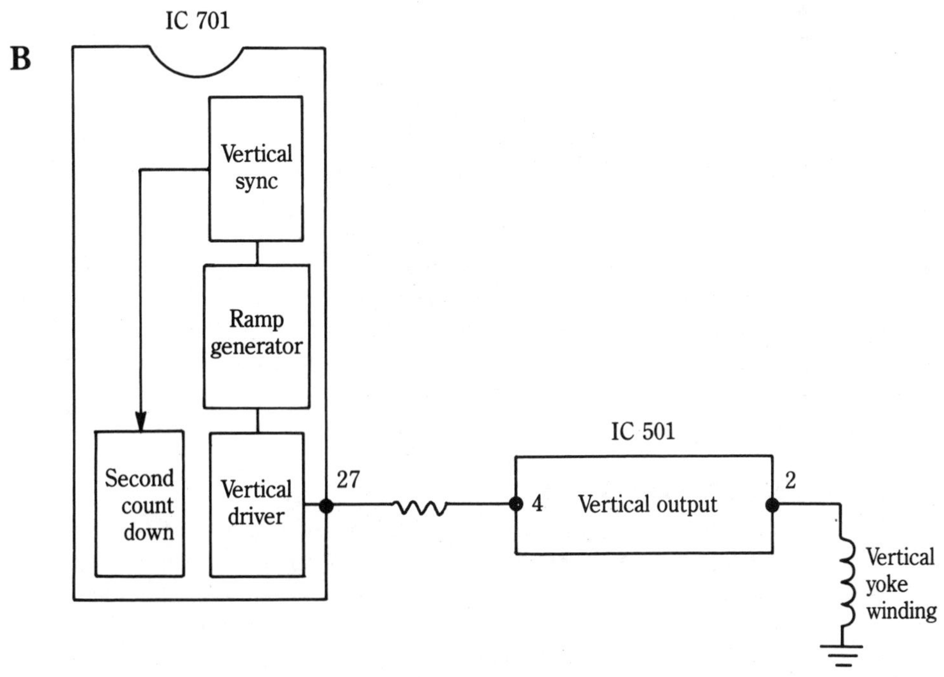

2-25 (A) shows a block diagram of a typical vertical output circuit found in the transistorized TV set. (B) shows a block diagram of an IC countdown vertical deflection circuit and vertical IC output.

vertical. It is for this purpose that an additional safeguard is provided in the form of horizontal AFC, or automatic frequency control (block I, Fig. 2-6). Briefly, this is a self-correcting system using both the incoming sync pulse and a feedback system in the receiver to maintain horizontal synchronization at all times.

Horizontal deflection system

The primary purpose of the horizontal deflection system (block K, Fig. 2-6) is for the left-right movement of the beam. For this reason we first describe this function only, although, as we shall see later, this deflection system has a most important secondary function.

Because of the added functions of this portion of the set, as well as because of more critical accuracy demands, the horizontal deflection system contains more tubes than the relatively simple vertical circuit. The first tube, V15, acts as the horizontal AFC (automatic frequency control). It receives the horizontal sync pulse from the sync separator and produces an automatic, precise timing pulse for the horizontal oscillator. Horizontal oscillator and discharge tube V16 performs a dual function: the first is similar to that of V13 in block J; the other being required because of the more rigorous demands of this portion of the TV set.

Following the oscillator is the horizontal amplifier, V17, a heavy-duty, high-power tube which performs, in addition to the obvious function of a horizontal amplifier, that most important second function mentioned previously. The output of the horizontal amplifier goes to the picture tube for the basic function of deflecting the electron beam from left to right. As such, its task is quite comparable to the vertical deflection amplifier just described; however, because of the very important added functions this unit performs, the whole horizontal deflection portion of the TV set is much more complicated and elaborate.

At least two of the tubes—horizontal damper V18 and high-voltage rectifier V19—perform functions entirely different from deflection; however, since these two functions are the result of the basic sweep circuit operation and because of the interdependence of these diverse functions, they are grouped together. Special precautions here include a local fuse (does not affect the rest of the TV set) in some makes for the horizontal circuits only, and a protective enclosure or cage because the very high voltage generated here is a potential safety hazard. Even with the cage, and even with the TV on-off switch in the off position, a severe electric shock and consequent secondary injury may result from carelessness in this area.

Solid-state horizontal circuits

The solid-state horizontal circuits consist of a horizontal oscillator, driver, and output stage. In countdown deflection, IC processor circuits the horizontal oscillator feeds into the first countdown circuit and feeds a horizontal predriver stage (Fig. 2-26). The horizontal sawtooth waveform is fed from the predriver (pin 17) to the driver transistor (Q402). The horizontal pulse is transformer coupled to the horizontal output transistor (Q402).

The horizontal output transistor generates horizontal scan and high voltage to the CRT. Actually, the horizontal output transistor acts as a switch during horizontal scanning. The output of the horizontal output transistor applies a high peak pulse to the damper diode and horizontal deflection coil. The flyback or horizontal output transformer is connected to the output of horizontal transistor (Fig. 2-27).

Most problems within the solid-state TV chassis are caused by the horizontal output transistor. The transistor will either short, or appear leaky or open. The flyback trans-

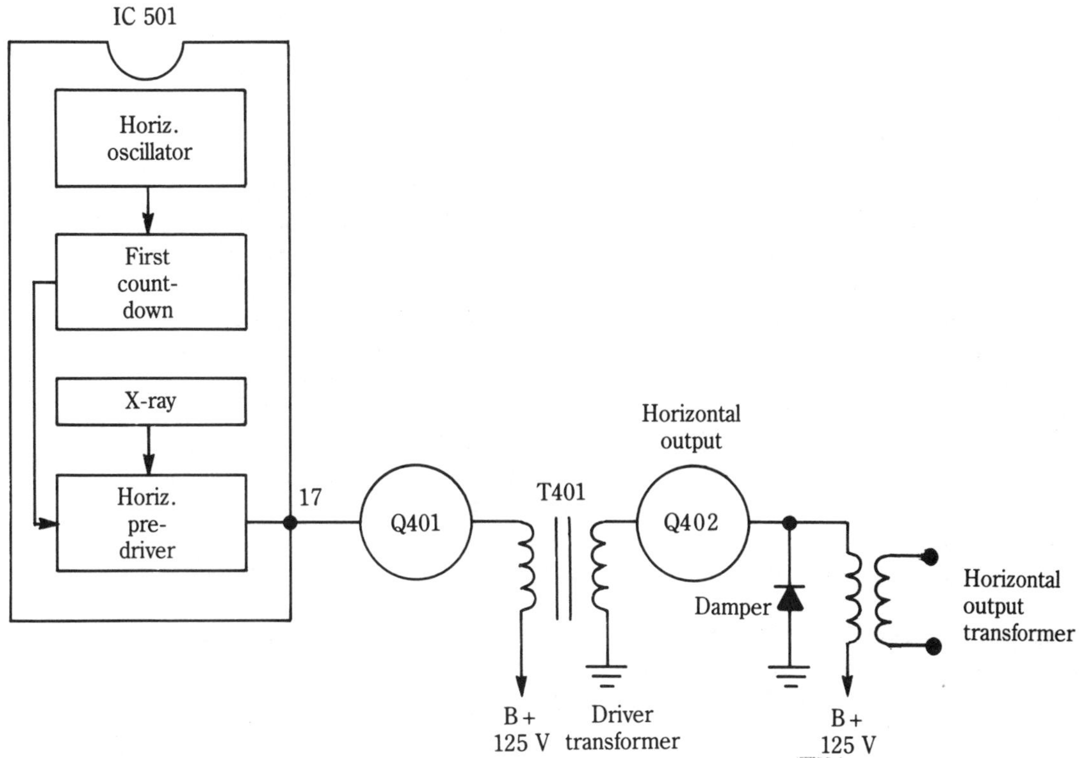

2-26 Block diagram of a horizontal oscillator, driver and output circuits in the solid-state TV.

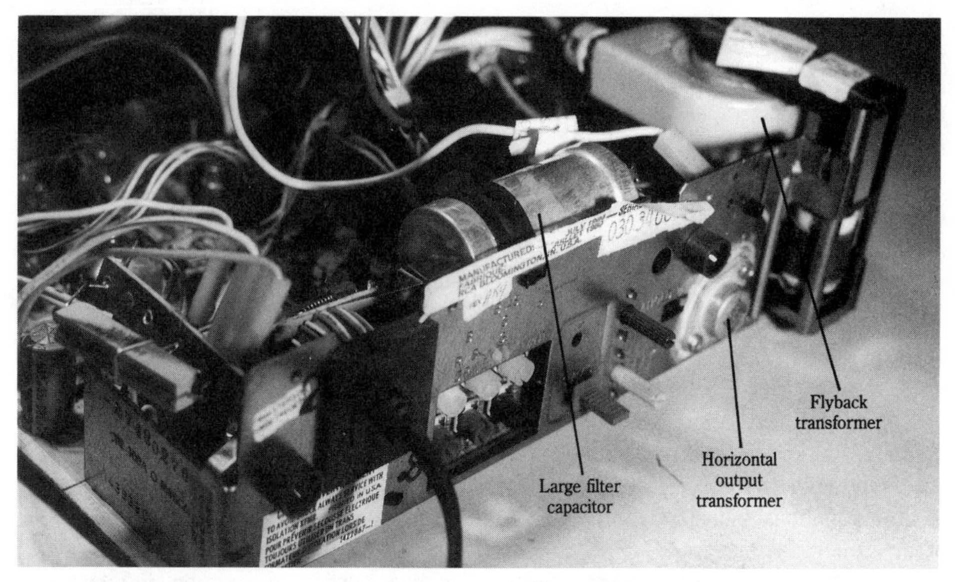

2-27 The location of horizontal output transistor and flyback transformer in an RCA 13-inch chassis.

former and drive pulse should be checked when the output transistor appears leaky or shorted.

High-voltage circuits

The high-voltage (HV) (block L, Fig. 2-6) of any TV set is physically distinct and separated from the open part of the chassis primarily for reasons of safety. It is here that the extremely high potential (25,000 volts in many color TV sets) for the picture tube is developed. While it is true that the energy behind this extra high voltage is not enough to be lethal, the danger from this source is nonetheless very grave. Even secondary effects, such as the reaction to the shock, a resultant fall, etc., can be serious. Caution labels in the vicinity of the high-voltage cage have their purpose, but a clear understanding of what is involved is even more valuable.

In some cases, the cage contains only three tubes; the fourth (oscillator-discharge tube) is left outside the cage for practical considerations. Using the four-tube configuration shown in Fig. 2-28, the functions are as follows: V16B is a sort of intermediate tube between the precisely timed horizontal oscillator tube and the horizontal amplifier, V17. Nontechnically speaking, the horizontal discharge tube serves as a transfer stage for the horizontal sweep voltage from oscillator to amplifier. Horizontal amplifier V17 performs primarily the function of amplifier, which brings the level of the horizontal sweep up to that required by the picture tube. Tubes V18, the damper, and V19, the high-voltage rectifier, are part of what was earlier called the important second function of the horizontal amplifier.

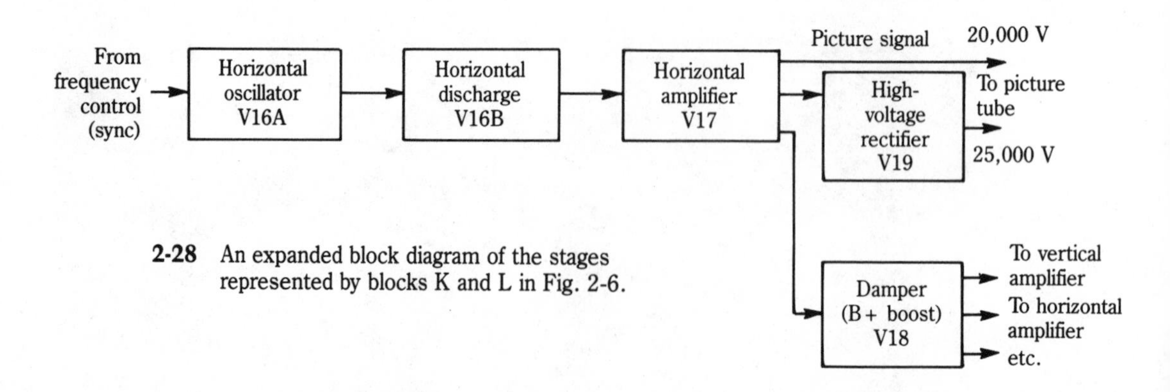

2-28 An expanded block diagram of the stages represented by blocks K and L in Fig. 2-6.

At this point, there is a need to elaborate somewhat on the mechanism of ''picture painting,'' in order to make some troubleshooting procedures and corrective adjustments more understandable. The picture tube of any set, monochrome or color, consists of three functionally distinct components: the electron gun, the positioning of deflection structure and the screen.

The gun is physically located in the rear or neck portion of the tube, in which the glowing heater can be seen when the set is on. It performs the function of generating, shaping and focusing the electron beam to a pencil-point sharpness at the point where it hits the screen. The image brightness adjustment is also connected to this portion of the tube. It may be worthwhile to repeat here that this is an invisible beam of electrons, not light. The light you see on the screen is the result of the electron beam striking a phosphor coating inside the picture tube.

Positioning of the beam, including its zig-zag movement across and down the screen, is accomplished by the outputs of the horizontal and vertical amplifiers connected, respectively, to the vertical deflection yoke (coils placed around the picture tube neck). It is, therefore, obvious that if the beam stays on a single horizontal line instead of moving gradually down to create a complete picture, the vertical deflection system is at fault. By logical deduction one might conclude that a failure in the horizontal deflection system would produce a single up-and-down line on the screen. Logical as this may seem, this is not the case, because of that second important function of the horizontal amplifier referred to earlier. We shall now see why this is so.

The third portion of the picture tube, the screen, is an electron-to-light converter. The coating on the inside of the tube has this capability, but it will produce light only if the electron beam strikes the coating at sufficiently high speed. This high speed is imparted to the beam by a high voltage (actually many thousands of volts) generated by the horizontal amplifier under normal operation; therefore, whenever the horizontal amplifier is not performing, no such high voltage is generated and no light whatsoever appears on the screen.

We can now return to the description of the tubes in block L and their functions. As stated previously, the horizontal amplifier generates, almost as a by-product, a high voltage ranging up to about 25,000 volts. Since this happens to be an ac voltage—dc voltage is required for the acceleration of the electron beam—a rectifier tube (V19, Fig. 2-28) is used to convert the ac to dc. This may be a 1 × 2, 1B3, 1A × 4, etc. Unlike most glass tubes in a TV set, this one glows very dimly, and cannot easily be seen; however, failure of this tube alone, even if the horizontal amplifier operates normally, will result in no light whatsoever on the picture tube screen.

The damper tube (V18) serves another useful by-product function. In fact, this tube is sometimes marked on the cabinet chart as B+ boost. (B+ is a supply voltage used by all stages in the receiver.) Since all tubes in the TV set require a voltage (up to 350 volts or so) for operation, advantage is taken of an available surplus voltage in the horizontal amplifier to boost the nominal 150- and 200-volt supply to about 350 volts where this is needed. The damper or boost tube is also a rectifier, changing ac to the required dc. What is important to the beginner is the fact that if this tube fails, some vital voltages are interrupted and the picture again disappears completely. In fact, the horizontal amplifier which generates this boost voltage is dependent on it for its operation, a sort of pulling-yourself-up-by-your-own-bootstrap scheme that works very well.

Another key device in block L in some sets is a fuse, not for the TV set as a whole, but only for the horizontal deflection system. Failure of this fuse, sometimes without apparent cause, will disable the horizontal amplifier and indirectly the B+ boost and the high-voltage power supply. The symptoms will then be the same as a failure of any other vital link in block L, that is, the loss of all light on the TV screen. Failure of this fuse is rather infrequent, and because it is a part of the rather dangerous high-voltage system, it is not accessible from the outside rear of the cabinet, as is the main 115-volt ac fuse. In addition it is often soldered in place. Fortunately, satisfactory replacement can be made in almost all TV makes without the use of a soldering tool.

There are a few operating adjustments associated with the horizontal deflection system. These adjustments are reasonably accessible, quite safe, and with use of the manufacturer's instructions where available can be easily made. The adjustments are made at the factory and may require minor touching up after long periods (2 years or more) of operation; or they may need readjustments after replacement of a defective part. The adjustments and their functions follow.

Transistorized HV circuits

In the early transistorized TV chassis, the flyback transformer was quite small with a tripler unit to build up the high voltage. Today, the horizontal output transformer provides picture scanning through the horizontal yoke, high-voltage and several derived-voltage sources. High voltage is developed with integrated diodes molded right inside the transformer (HVT).

The flyback transformer is connected to the horizontal output transistor and B+ power source. The secondary winding produces very high voltage with the enclosed diodes between windings. Besides furnishing HV, this winding is tapped to produce screen and focus voltage for the picture tube (Fig. 2-29).

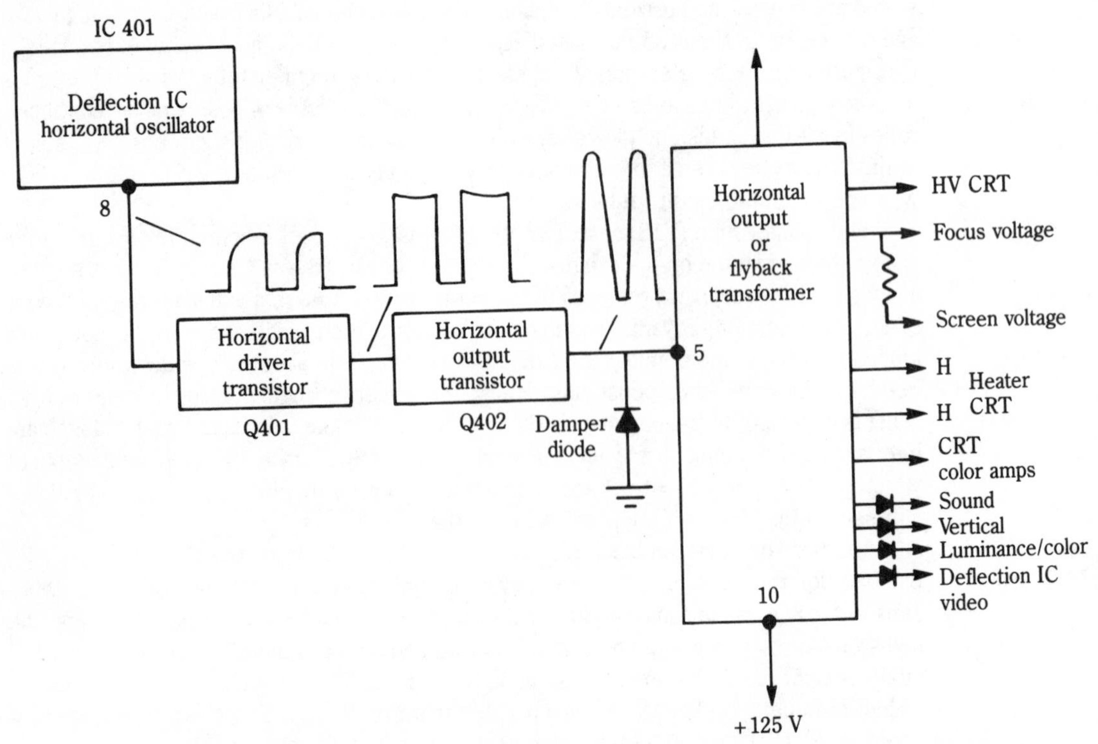

2-29 Block diagram with waveforms in a solid-state TV chassis of the horizontal and high-voltage circuits.

Besides the HV winding, additional secondary windings may provide a high voltage (+250 V) for the picture tube and color amps. Sometimes another winding provides rectified voltage sources to the sound, vertical, luminance/color processor, and IF/video stages. Each secondary winding includes a small silicon diode with filter capacitor for these secondary derived voltage sources. You may find the heater voltage for the picture tube loosely wound around the ceramic frame of the flyback transformer.

Horizontal width

This is a slotted-shaft adjustment for obtaining the correct picture width on the screen. It is a multiturn adjustment and should not be necessary except on rare occasions, as when replacing the horizontal amplifier tube (V17) due to old age. It should not be necessary to turn this adjustment more than four or five full turns either way. If this does not produce the desired result some other defect is probably the cause. Incidentally, turning this adjustment too far in either direction is likely to completely disengage the moving core (slug)—it will actually fall out. If this happens it may be beyond the beginner's ability to recover and replace (Fig. 2-30).

2-30 The white vertical line represents no horizontal sweep in the portable TV chassis.

Horizontal hold

This adjustment is physically very much like the width adjustment. Its purpose is to preset the circuit controlling the movement of the beam across the screen (left to right) to its approximately correct position, so that the automatic frequency control (AFC) can take over and make the timing precisely correct. We shall see later that in some cases of picture tearing a slight adjustment of this circuit will restore the picture to stability. But the horizontal hold control cannot correct any sweep faults. Further, when an adjustment is made it must be done very gradually, sometimes only a fraction of a full turn, for results.

Horizontal drive

In some older sets, an adjustment marker H drive, for horizontal drive, may exist. Hopefully most of these sets have gone to that big trash dump in the sky. This adjustment has the effect of stretching one side of the picture and should only be made when poor horizontal sync exists. Don't try this adjustment unless you really feel that it will help. Most TV servicers only attempt this with proper test equipment; however, there is an alternate method of adjustment that required only a screwdriver and good eyesight.

Tune in a very weak station, one with lots of snow. Turning the adjustment in one direction will produce a white vertical (up and down) line in the center on the screen. Now turn the adjustment until the line just disappears, and you've got it.

Horizontal lock-in

This adjustment, like the previous one, is found only on some sets. Its behavior is somewhat similar to the hold control described earlier. Turning it from its normal position toward either extreme will produce a diagonal tearing of the picture, in one direction or the other, depending on the direction of rotation of the adjustment. Figure 2-31 is an example of such tearing. The lock-in control's correct setting is the point where channel changing or adjustments will not produce any tearing due to the automatic control by the horizontal sync pulse.

2-31 A display on your TV screen similar to this drawing is a case of horizontal tearing.

Centering controls

These functions used to be rear-of-chassis adjustments on TV sets a number of years ago. They are no longer in use today, as they are replaced by a mechanical centering device which is part of the picture tube deflection yoke; however, in those rare cases where one of these adjustments is found on the back of the chassis, the procedure is simple: With a test pattern on the screen (or an equivalent circular emblem of some sort), the control is adjusted for best centering of the image on the screen.

Power supply

The power supply (block M, Fig. 2-6) is a rather simple yet very important part of the TV set. It supplies power to all the tubes or transistors in the set. A failure here is certain to disable the complete TV set, except, in case of the tubes, it may not prevent them from lighting. The tube, V20, is called a rectifier, and its function is to convert ac to dc, as required by the various tubes and transistors. Sometimes two tubes, side by side, serve the function, in which case failure of one may not completely disable the set; however, since the remaining tube is probably overloaded, it is most prudent to check both and replace both even if one seems to function. Some basic precautions are to be observed when handling rectifiers and tubes, and these do's and don'ts are detailed in the section covering troubleshooting.

In connection with the power supply, it is appropriate to explain the two main types of tube hookups currently used by manufacturers. These are the parallel system and the series system.

Parallel tube wiring

With this arrangement all tubes are wired independently of each other, similar to the wiring of a string of modern Christmas tree lights in which burnout of one or more does not affect the others. You can tell at a glance which bulbs are burned out. In a TV set using this wiring method it is simpler to locate a defective tube. Should one burn out, it alone will not light (and feel cold to the touch) and is easily identifiable.

An easy way to determine whether the set has parallel tube wiring is to examine the tube chart inside the cabinet. In a parallel tube set all the tube numbers (with two exceptions given below) will *begin* with the same number, usually a 6. For example, the tubes may be numbered 6BC6, 6CG8, 6BM8, etc. The reason why 6 is a common first number is that they are all 6-volt tubes; in other words all tube filaments operate on 6 volts.

The two exceptions are the picture tube where the first number indicates the tube diameter (21GP4B indicates a 21-inch tube, a 17LP4B stands for a 17-inch tube, etc.), and the high-voltage rectifier, which in the vast majority of black-and-white receivers is a 1-volt tube.

Series tube wiring

This type of tube filament wiring is a characteristic of the so-called transformerless TV sets. As the name implies this TV receiver has no power transformer. There are three ways to identify a series filament string. First, and this never fails, if one tube burns out; all tubes go out, like the old-fashioned Christmas tree light string. Even when the set operates normally, pulling out one tube disables all the others. A second identification, also infallible, is the tube numbering system where the first number or numbers of the tubes vary (4BZ8, 10JT8, 38HE7, for example). The third clue is an inscription or imprint on the rear cover of the TV cabinet (or just a printed label) warning that one side of the 115-volt electric line is connected directly to the metal chassis. The requirement stems from a shock hazard that may exist with such construction, a condition exclusive with transformerless receivers.

For troubleshooting a TV receiver with a tube failure a very simple and fairly fool-proof test is substitution—replacement of a suspected tube by its exact duplicate known to be good. This can be done most easily in the transformer type of parallel filament set, as will be detailed later. In the transformerless set there is first the problem of locating the suspected tube since all are out when one goes. Substitution is still possible but it is a hit-or-miss procedure in which the defective one may be the last one you try. In a parallel filament (transformer) set you actually can see and feel the tube which burned out; therefore, a different procedure has to be followed in a series string TV set.

Out-of-circuit testing

Each tube in the series string must be removed from its socket and have its filament tested for continuity. If the tube checks good, returning that tube to its socket before the next tube is removed and tested will prevent a mixup of tube types. This test will require the use of a continuity tester or ohmmeter (vom or DMM). Operator familiarity with the instrument chosen is assumed to the extent that test results indicating filament failure are recognized.

In-circuit testing

This method requires that the ac potential be applied to the set chassis. This may be accomplished with a cheater cord (ac line cord that defeats the safety interlock device). Be careful because you will be measuring the ac line voltage (120 Vac). First, set the ac voltmeter to the 200-volt ac scale. Next, touch the test probes of the voltmeter to the socket's solder tabs corresponding with the tube's filament pins. Move the test probes from socket to socket until the defective tube is located (Fig. 2-32). The defective tube will have a voltage drop across its filament equal to the line voltage.

Solid-state power supply

There are many different power supplies from half-wave, full-wave, regulated, pulse-width-modulated (PWM) and chopper circuits in the solid-state chassis. In a typical half-wave or bridge rectifier, low-voltage power supply operates directly from the power line, while the power transformer provided ac voltage in the tube receiver. The ac line is fused with another fuse after the half-wave rectifier. A large filter capacitor helps filter out the ac from the dc source.

The rectified dc voltage (145 V) is fed to a high-voltage regulator IC. This regulator is built somewhat like the horizontal output transistor or power sound output IC. Sometimes these regulator ICs are shunted with a high-wattage resistor to the output regulated terminal (115 V). This dc source is fed directly to the horizontal output transistor, sound output and deflection IC circuits (Fig. 2-33).

The pulse-width chopper power supply may consist of a full-wave bridge rectifier, regulator drive transformer, regulator control IC, x-ray protect, overcurrent shutdown

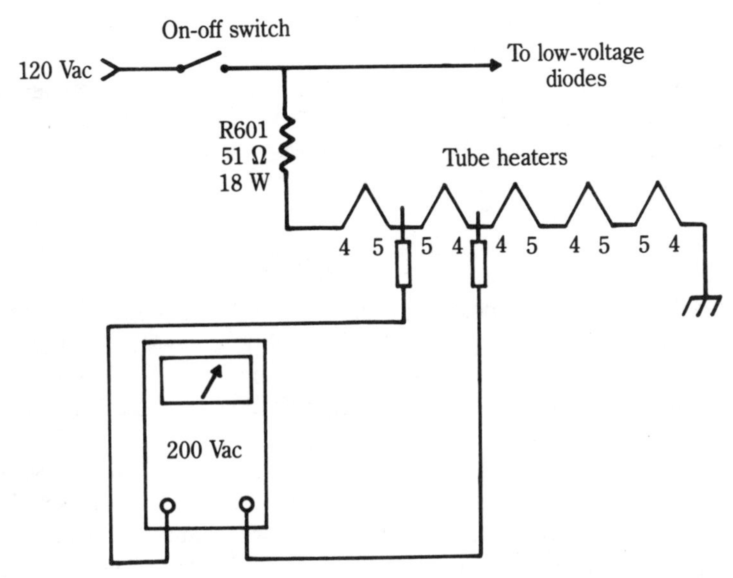

2-32 To locate an open heater, measure the ac voltage between the heater pins of each tube. When you get a reading that indicates the entire power line voltage, you have located the defective tube.

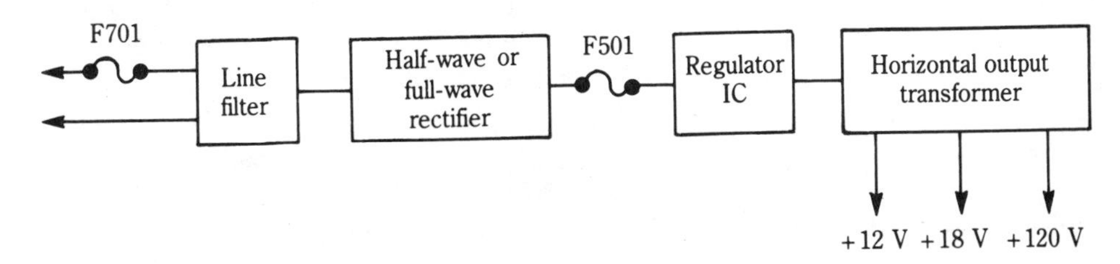

2-33 Block diagram of a typical low-voltage power supply.

transformer, chopper output transistor, and chopper output transformer. The output voltage from transformer provides a 125-, +24-, −24-, and a 16-volt source. The 125-volt source powers the horizontal output circuits.

Hybrid TV set

As more and more semiconductors (transistors and diodes) are finding their way into TV sets a consideration of a third arrangement—part tubes, part transistors—is of importance to the extent that such sets lend themselves to the efforts of a beginner.

There are no 100-percent solid-state TV sets. At least the picture tube and perhaps one or more others are still of the conventional vacuum type. Transistors do not require filaments and, with the exception of a few high-power types, operate cold to the touch. The high-power types are easily identified by their size or the addition of a heatsink (metallic fins or plate used as a heat exchanger) that becomes very hot during operation.

Power transistors are often mounted directly on the metal chassis which then doubles as a heatsink.

When hybrid chassis were designed transistors were not capable of operating at the high voltage potential required by some circuits. Therefore, transistors are found in the low voltage circuits (tuner, sync, AGC, IF, color and video driver stages) of the hybrid chassis. Transistor circuits require less power than the circuits of their tube counterparts.

The tubes in the hybrid chassis have filament voltages rated at 25, 38, 40, and 42 volts ac. If the filament voltages of the tubes in the chassis are added together, their sum will often equal the line voltage of 120 volts ac. In hybrid sets, these tubes are found in the horizontal and vertical sweep circuits. Some hybrid chassis also use tubes in the video output and color demodulator circuits. Since a majority of TV problems occur in the sweep circuits, many troubles can be resolved by the easy replacement of a vacuum tube.

Heat and light play an important role in diagnosing faulty tube circuits. Since transistors do not use filaments or heaters, another method of locating suspected stages is required. Transistor stages are often fabricated into small subassemblies containing a number of associated parts mounted on a printed-circuit board (PC board). Examples of how to diagnose this type of assembly is covered in chapters 10 and 11.

Solid-state TV receiver

Although the block diagram in Fig. 2-6 is representative of practically any TV set it is desirable to familiarize oneself with its solid-state equivalent. This is not only because this type is the standard for sets currently being manufactured, but also because of some significant differences due largely to the nature of semiconductors. A brief explanation of the above statement is in order.

An examination of any block diagram of a tube-type receiver—or the tube layout invariably found inside the TV cabinet—will disclose that many tubes are dual- or even triple-section devices, each section being a stage in itself, all sharing a common glass envelope. Such types are easily recognized by their designations, namely, V1A, V1B, V2A, V2B, V2C, etc. In each case, the number designates the whole physical assembly, while the letters A, B, and C identify the separate functional tubes in the same glass or metal assembly. Quite often the individual tubes are different from each other and perform different functions so that although located in one place on the chassis, they connect to different parts of the circuit.

With the advent of semiconductors, however, the picture has changed, primarily because of the nature of transistors; they do not lend themselves to multifunction construction with ease. Thus, while in tube tuners one of the tubes (the mixer-oscillator) is almost invariably a dual-function assembly, the semiconductor counterpart has separate transistors for the mixer and the oscillator. Furthermore, where in tube tuners two tubes suffice, three or more transistors are usually found performing that function in the solid-state receiver. In addition the solid-state receiver often has added stages, consisting of transistors and associated parts, not because of added functions, but rather because of a peculiar need for performing the same functions with semiconductors. For these reasons, Fig. 2-34 is presented for comparison to Fig. 2-6. We shall briefly com-

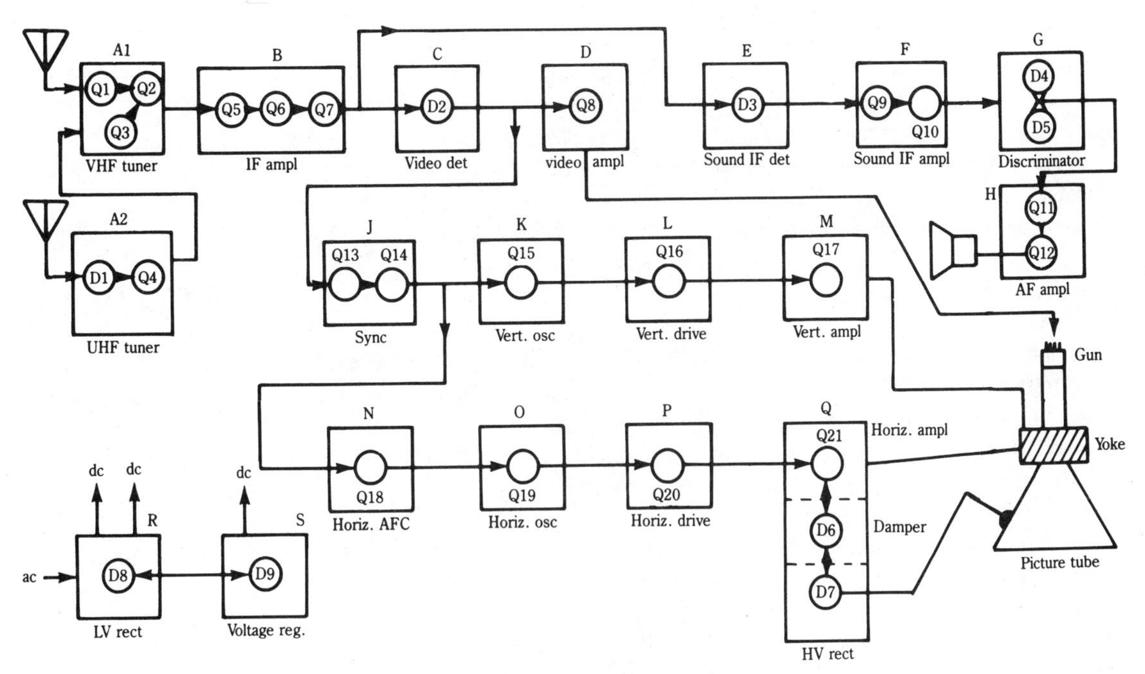

2-34 Block diagram of a typical solid-state receiver.

pare the two block diagrams for the express purpose of illustrating the nature of semiconductors and their performance from the TV owner's point of view.

There are many multifunction assemblies in the solid-state field, as for example, IC (integrated circuits); in fact, these are exclusive with semiconductors. But these are complete assemblies, such as audio amplifiers, multiple IF amplifiers, etc. They are always part of the same circuit, electrically as well as physically. Such was not the case with vacuum-tube construction.

We shall also give an illustration of the application of ICs in suitable portions of the set. This will provide an example of the transition from individual transistor stages to preassembled multitransistor/multidiode entities, which are always replaced as a unit, not only by the most knowledgeable do-it-yourselfer, but even by the most professional repair establishment. These entities are factory assembled and encapsulated.

Tube-type and solid-state receiver comparison

In the following comparison, the particular function descriptions will not be repeated; they are the same for both block diagrams. We shall, however, indicate functional differences.

Blocks A1 are the same, transistors Q1, Q2, and Q3 corresponding to tubes V1, V2 and V3; however, while in the tube version V2 and V3 are generally a single dual-purpose tube, transistors Q2 and Q3 provide separate functions of mixer and oscillator, respectively.

Blocks A2, and UHF tuner, both use a diode mixer, but the solid-state version adds a transistor UHF oscillator, Q4. Block B, the IF amplifier in Fig. 2-34 has three transistors, Q5, Q6, and Q7, corresponding to tubes V3, V4, and V5; however, although in many inexpensive tube-type sets there may be only two tubes instead of the three shown in Fig. 2-6, the transistor version is more likely to have three stages, because of the relatively lower gain or amplification of the transistors.

The video detectors in blocks C are identical, as diodes have been generally used for this function in tube sets. The video amplifiers in blocks D fully correspond. In both cases the output, which is the video information, goes to the picture tube. Sync information in the transistor version comes from the video detector and not the output of the video amplifier.

There is a slightly different arrangement in case of the sound portion of the signal. In case of the tube set the sound component is taken from the video amplifier, block D, and goes to the sound IF, block E. In the transistor version the sound is taken from the final video IF amplifier and is fed to a sound IF detector, D3, block E. This strips, or picks off, the sound IF component from the total (composite) signal. It is not a detector in the usual sense, since its output is not yet a sound signal suitable for hearing. The sound IF, block E in the tube version, corresponds to block F in the transistor version, V7 and V8 corresponding to transistors Q9 and Q10, respectively. In some transistor receivers, there may be three stages of sound IF amplification, where the particular design requires it for adequate gain. Block F in Fig. 2-6 corresponds exactly with block G in Fig. 2-34; both use a pair of diodes as a discriminator, the real sound detector. Block G of Fig. 2-6 and block H of Fig. 2-34 also correspond, each showing two audio amplifier stages, although in some solid-state receivers there may be as many as four transistors performing the same function as the two tubes of block G, Fig. 2-6. In such a case, the first two transistors might be designated as drivers or preamplifiers while the last two might be a push-pull output amplifier.

Block J in the transistor set corresponds to block H in the tube set; all sync functions are accomplished by the two transistors Q13 and 14 corresponding to tubes V11 and V12 in Fig. 2-6. Blocks K, L and M in Fig. 2-34 are the vertical oscillator, vertical driver and vertical output amplifier, corresponding in function to V13 and V14 in the tube version.

In the horizontal deflection circuitry, transistor Q18 in block N of Fig. 2-34 performs the same function as V15 in block I, horizontal AFC, in Fig. 2-6. In some transistor TV sets there are usually two diodes associated with Q18. Tube V16 in block U of Fig. 2-6 corresponds in the transistor set to block O (horizontal oscillator Q19) and block P (horizontal driver Q20). Block L of Fig. 2-6 corresponds to block Q of Fig. 2-34. Horizontal amplifier V17 is replaced by Q21, damper V18 by D6 and high-voltage rectifier V19 by semiconductor diode D7. In both cases these three are shown in one group since they actually work that way. Block M of Fig. 2-6 shown to contain rectifier V20, or two diodes, corresponds to block R in Fig. 2-34, containing diode D8. Different designs may call here for two and even four diodes, all for the same purpose—rectification of the ac. Finally, block S in Fig. 2-34 has no counterpart in the tube type set of Fig. 2-6, although many tube sets have the same function. This is a low-voltage regulator stage, usually consisting of a diode (D9) or a transistor followed by a diode.

Integrated circuit applications

The integrated circuit (IC) is replacing the discrete component circuit in many fields of electronics including television. Briefly, an integrated circuit is an assembly of various parts, including transistors and diodes, so that only the necessary connections are externally accessible; all other interconnections are internal and permanent. Physical size of the IC package depends on the complexity of the circuit. The smaller devices have fewer circuit functions and need just a few external connections (pins) called SSI (small-scale integration). More complex circuits capable of performing a multitude of tasks require a greater number of external connections (pins) called MSI (medium-scale integration) and LSI (large-scale integration) ICs.

In today's TV receiver, it is not uncommon to find just one IC replacing several individual circuits. One IC could replace all of the sound circuit or just the pre-amp and driver stage. Individual ICs could replace the AGC, sync, and video IF circuits, or you could find all of these functions replaced by one IC. The horizontal and vertical drive circuits may require separate ICs in one receiver and have both functions performed by one IC in another receiver. Depending on the make and model, ICs are replacing discrete circuits in many of today's video and color circuits and in the tuner, memory, and tuner control circuits as well (Fig. 2-35).

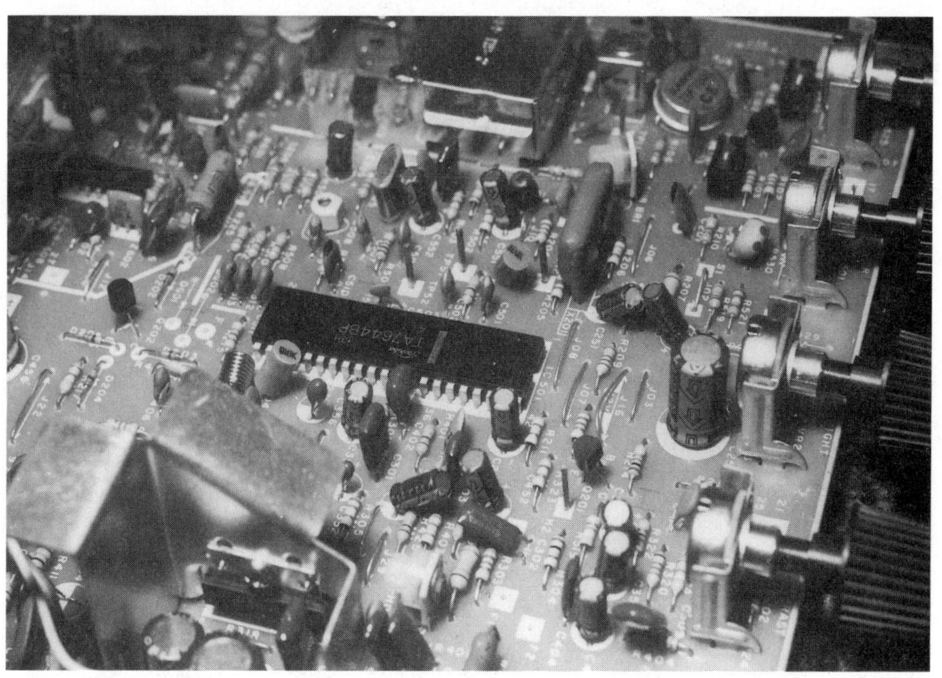

2-35 One large IC (center) contains most of the circuits in one of the latest Japanese TV receivers.

Replacement of a single IC could solve many different receiver troubles, but even the failure of one section of the IC would require the IC to be replaced. Although this may at first seem uneconomical, requiring replacement of a complex assembly just because one item has failed, actually this is not the case. Preassembly and encapsulation save more in the long run, making it more economical (and often better performing).

In chapter 5, we recommended that the TV owner purchase the complete package of data for his or her particular model receiver. In such a package, there always are schematic diagrams and other partial diagrams necessary for servicing the receiver.

Printed-circuit wiring

Today, the main chassis containing components are found upon PC wiring or etched wiring board. In fact, the first TV set to incorporate PC wiring in their TV chassis was Admiral Corporation. The printed circuit is laid upon a very thin copper sheet cemented to a fiberboard. After the circuit is laid out, the copper-clad board is etched with etching solution.

The liquid etching solution is either ferric chloride liquid or ammonium persulphate powder etching powder. The board is etched after passing through this solution in the automatic machine. The board components are placed into the correct holes and passed through a bath of solder. All connections on the board are soldered and processed with one machine.

The PC board may be plugged into sockets or connecting wires attached for other larger components (Fig. 2-36). Some of these boards are double-sided with regulator components on top- and surface-mounted parts underneath. If the PC board is too thin, it may warm and pop some rivets or PC wiring connections. Most of the trouble found in surface-mounted boards are poor connections between parts and PC wiring.

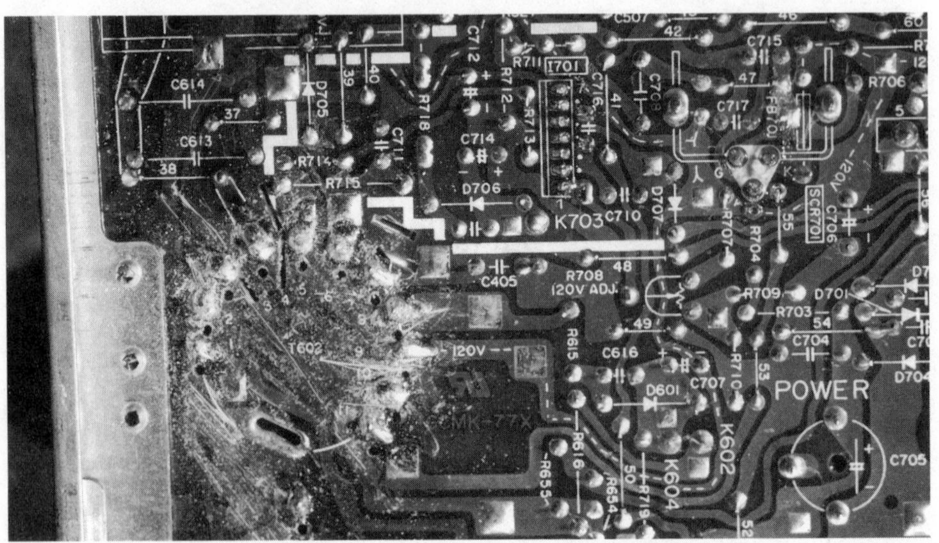

2-36 The bottom PC wiring of a solid-state TV chassis. Notice the flyback transformer has been replaced (lower left).

The PC board may crack or split when the TV is dropped in shipment or when handled improperly. Look closely around heavy components upon board for cracked areas. A poorly soldered connection may be located at the bottom of component wire or ground eyelets. When replacing large components upon the board, be careful not to damage PC wiring. Too much heat applied to the PC wiring may cause it to lift up from the resistant clad board. Clean off all rosin residue from the connections with circuit coolant, contact cleaner, or rosin flux remover.

Surface-mounted components

Surface-mounted components have been used in compact disc players and camcorders for several years and have just entered use with the TV chassis. The IC processors and control ICs come in flat-pack IC packages with terminal leads coming out of the sides of the component. These leads are soldered directly to the PC wiring.

The IC processor may have up to 82 terminal leads in a camcorder while the transistor surface-mounted component has only three terminals. Of course, the resistor and capacitor have only two flat end terminals. The digital transistor has only three leads with base and bias resistance enclosed inside. You may find a dual resistor network with two separate resistors in one chip with solid bare soldered connections at the ends. These leadless chips must be heated at the end with the soldering iron and removed with a pair of tweezers.

With a double-sided PC board, the larger components for the TV circuits are located on top of the chassis and surface-mounted components underneath. The fixed resistors and capacitors are quite small in size. Here in Fig. 2-37 is a card of fixed resistors with five of one value. When a double-sided board is used with surface-mounted components underneath, the overall TV chassis is quite small.

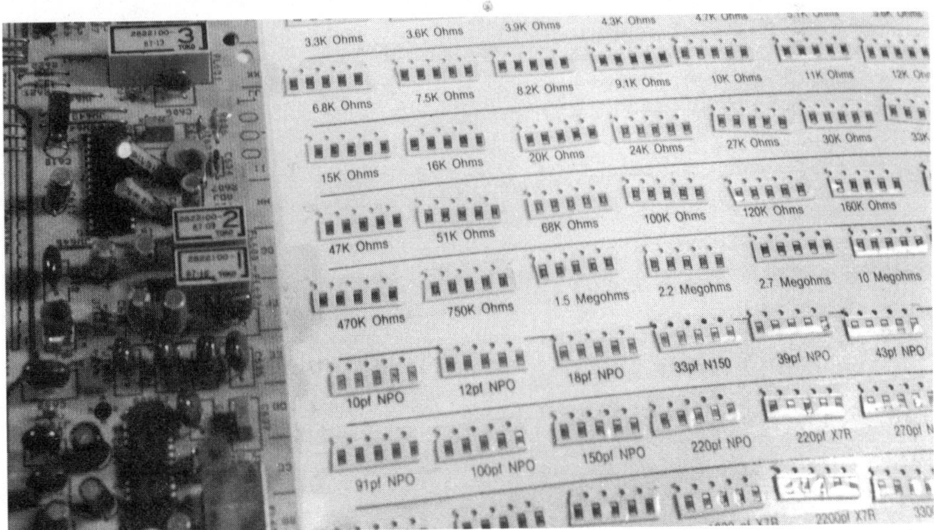

2-37 Comparing the small surface-mounted components with the standard components on the top side of one of the latest TV chassis.

3
Color TV and how it works

UP TO THIS POINT, COLOR TV HAS NOT BEEN INCLUDED IN OUR DISCUSSION, BUT everything that has been said pertains equally to color. In fact, when I discuss the circuits and functions peculiar to color receivers only, you will see that these circuits and functions are additional to the basic TV set circuits. It is but necessary to adjust one control (color) to its minimum (equivalent to off) position in any color TV set to arrive at a black-and-white representation insofar as results are concerned.

Generally speaking all the TV circuits up to and including the video detector-amplifier are common to all signals—sound, video, and color.

The visual spectrum

Figure 3-1 is a graphic depiction of the color components of visible (white) light after passing through an optical device called a prism. Although discrete color blocks are indicated, all colors gradually blend into each other. Figure 3-2 is a much simplified version of the same phenomenon, more familiar to the student of elementary painting. The three primary colors produce intermediate colors as well as white. In the present color TV system, this basic three-color system is utilized to produce the color TV picture as we know it.

Color transmitter

Figure 3-3 is a functional or block diagram illustrating in much simplified form the components of a color picture transmission system. A comparison with the monochrome TV transmitter in Fig. 2-6 shows the basic difference between the two. Let's consider the components of a color system.

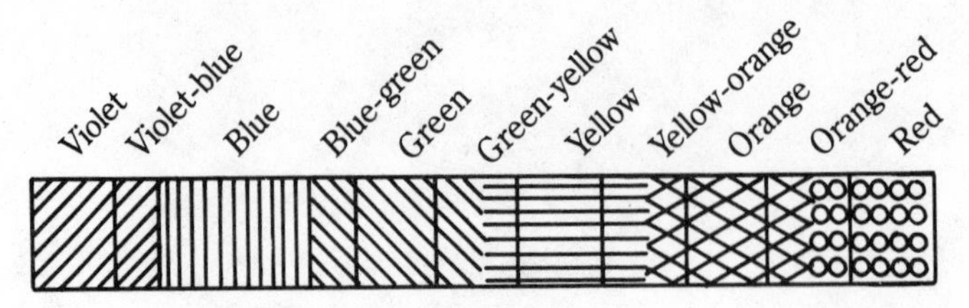

3-1 Pure white light is composed of all the colors depicted here.

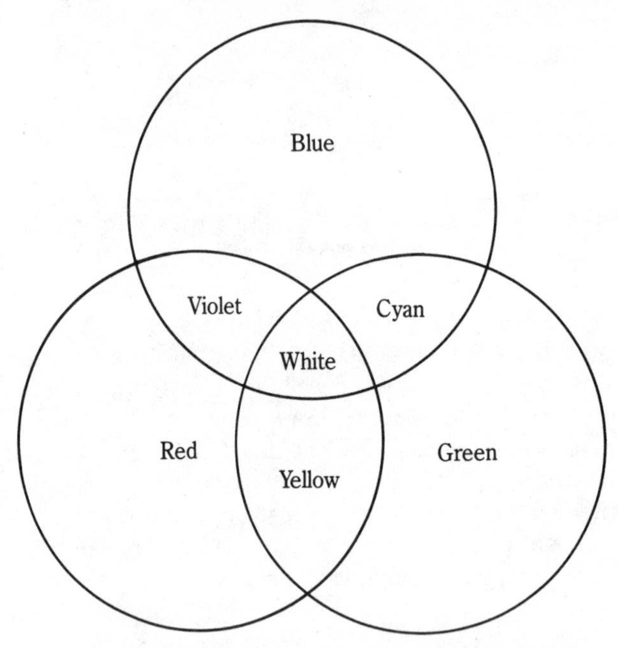

3-2 The color wheel principle shows how the three primary colors blend into these intermediate colors and white.

Camera tubes

Three camera tubes (block A Fig. 3-3), one for each of the primary colors together with their individual amplifiers, combine to produce the three components of color information required to produce a color picture on the home TV set. These three components are:

- *Picture brightness information.* This establishes the overall color picture brightness in the TV set, not just the brightness of any particular color but the background brightness level (proportionately) correct for all colors.
- *Picture color information.* This is the heart of the color signal and contains all the information required by a TV set to reproduce the entire range of colors, corresponding to the original scene.

- *Color sync information*. This corresponds (and is in addition) to the earlier described sync functions in black-and-white television. It is sometimes referred to as the 3.58 MHz color burst.

Modulators

The modulator or combiner portion of the color TV transmitter corresponds to the similar functions in the monochrome transmitter. Here the various electrical components of the color picture are combined (block B, Fig. 3-3), without losing their identities, into an overall color signal, which in turn is combined with the sound signal.

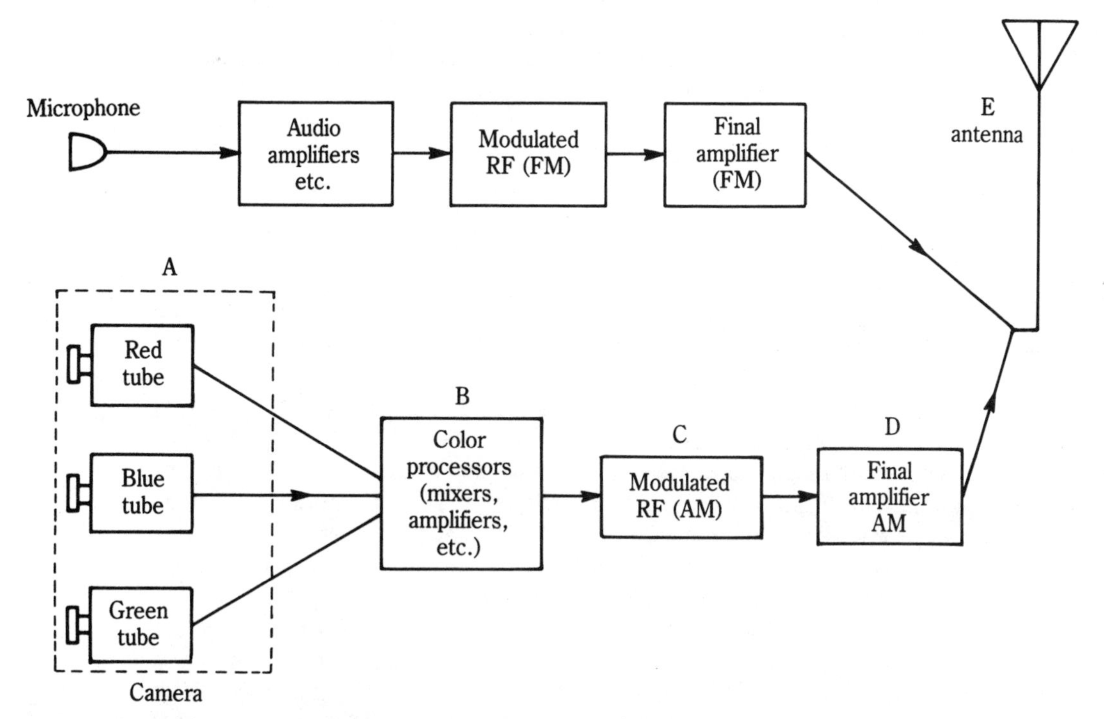

3-3 Simplified block diagram of a color TV transmission system.

Final amplifiers and antenna

The final amplifier and antenna system are essentially the same as the black-and-white transmitter in Fig. 2-6, except for the addition of the color information. Neither the final amplifier nor the transmitting antenna "know" the difference; that is, they amplify and radiate (respectively) a total television signal made up of the required components. This same is true with a receiver. We shall now proceed to examine a color TV receiver on a function-by-function comparison with its black-and-white counterparts.

Color receiver

The color TV receiver is a combination of a basic monochrome receiver plus an add-on section which concerns itself only with the color functions of the set. In fact, it will soon become apparent that this add-on section picks up where the black-and-white portion of the set leaves off. Consequently, and as was briefly stated earlier, the standard procedures for analyzing and troubleshooting a color TV set is to first consider it as a black-and-white set. This is easily done by reducing the color signal (turning the chroma control) to zero, thus leaving a black-and-white picture as would be received by a monochrome receiver tuned in to a color program. It is for this reason also that the block diagram to follow (Fig. 3-4) is divided into two sections by a dashed line. The black-and-white portion is readily recognizable as the familiar monochrome TV set shown in Fig. 2-6; therefore, the black-and-white section receives only a minimum of emphasis in this discussion.

It will be advantageous, however, to view the complete TV receiver as a unit, in order to clearly visualize the continuity between and the transition from the combined monochrome and color functions to the strictly color portions and functions. The overall functional diagram is represented with this purpose in mind.

Figure 3-4 gives a simple, nontechnical means of comparison between a monochrome TV set and the corresponding color set. It shows a complete color TV set, with no reference to black-and-white operation; however, it is obvious that the complete black-and-white TV set of Fig. 2-6 appears with hardly any change inside the framework of the color set diagrams.

In addition to the basic facts of structure this graphic comparison also shows, in greatly simplified form of course, the two main constituents of the color picture. Function blocks O and P are responsible for the three component colors plus the overall color picture brightness information, while function blocks M and N provide the information for the proper color registration and mixing of the three primary colors into the complete range of color shading which makes up the final color image on the home TV screen.

Before discussing individual tube, stage or functional blocks of the complete color TV set let us briefly analyze the complete sequence of functions as indicated in Fig. 3-4 in order to get a concise overall story of what happens. The TV signal, containing picture information and sound information, is intercepted by the antenna. It is a composite (all-in-one) signal. Block A (the tuner) selects any one desired station from the number of stations available in a particular location. The IF (intermediate frequency) amplifiers (block B) amplify the selected station to the signal level required for further processing.

So far the composite signal picked up by the antenna and selected by the tuner has been made larger as required, but it still is riding piggyback on a carrier. Block C, the video detector, performs but one simple main function—it separates the intelligence (picture, sound and sync information) from the carrier, which served its purpose as a vehicle for transmission of the signals. The output of the video detector is the composite signal described previously.

Although the carrier appears to be discarded at this point it is not. As far as the intelligence goes it has done its job. Nevertheless it is utilized to perform a vital function of adjusting the TV receiver's amplification to the required level; that is, the weaker the incoming signal, the more it is amplified, and vice versa. This is known as automatic gain

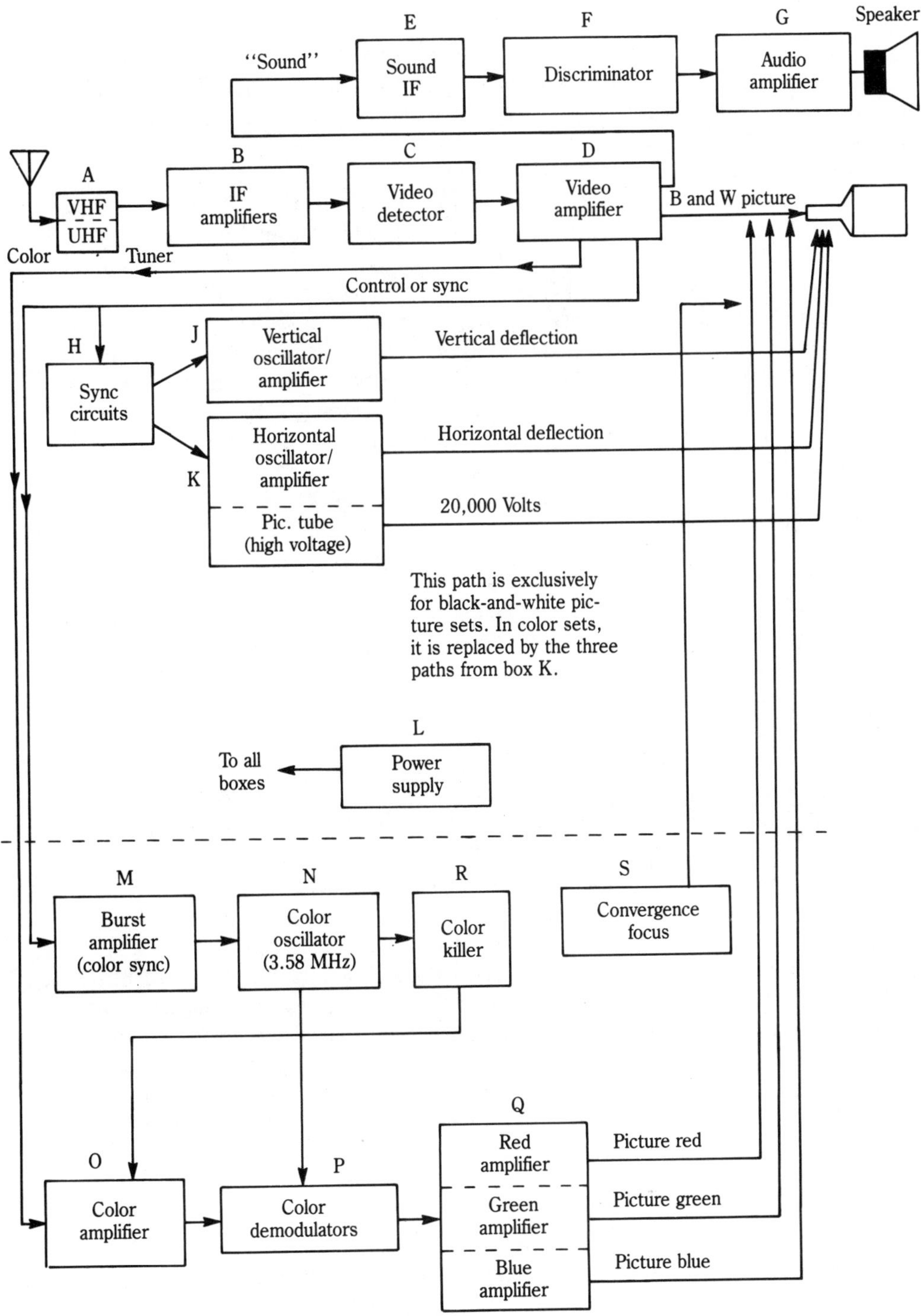

3-4 Block diagram of a color receiver. Those functions above the dashed lines parallel those in a black and white receiver. Below the dashed line are color-only functions.

control (AGC for short). Incidentally, the AGC adjustment is a back-of-the-set adjustment that sometimes required resetting on both color and black-and-white receivers.

Block D in Fig. 3-4 is the video amplifier. True to its name, it amplifies the video signal together with its other components (sound and sync signals) received from the video detector just described. At the output of the video amplifier the three signal components separate, each going its own way.

Picture information (commonly called video) goes to the picture tube for conversion into an optical image. Sound information, a sound-modulated carrier (sound energy piggybacked on an RF carrier), goes to block E—the sound IF amplifier. The signal now is the equivalent of an FM radio signal in an ordinary FM receiver. As such it is a sound signal superimposed on a carrier. This in no way contradicts our earlier statement that a carrier has been discarded by the video detector, although the technical explanation is of no consequence to our purpose. After amplification the signal goes to an FM detector (block F) or discriminator where sound energy only is extracted. Block G is an audio amplifier, which brings the sound energy to the level required by a speaker connected to the output of block G.

At this point, you may be wondering why the same amplifiers appear over and over again, interrupted by other functions in the chain. Wouldn't it be better to have just one amplifier in one place and do all the required amplification at once? It would seem simpler; however, a TV (and radio, too) receiver is designed to receive a faint signal from the antenna. In fact, such a signal may be a million or more times weaker than the voltage of the ordinary flashlight cell (nominally 1.5 volts). Necessary incidental losses within the TV set make this situation even more critical. In other words the incoming TV signal must be amplified millions of times before it becomes the final product (picture or sound).

The most practical and most (technically) efficient way to achieve this tremendous magnification is to do it in small stages and at particular locations in the set. Thus, some initial amplification (RF amplifier) is done immediately after the signal arrives from the antenna. Additional amplification is provided after initial processing (conversion to IF in the tuner), this time in the IF amplifiers. A third and usually final portion of amplification is accomplished, after another conversion (detection), in the video amplifier for the picture signal and in the audio amplifiers for the sound signal. There also are other amplifiers in the TV set, for such nonsignal functions as sync and AGC (automatic gain control).

Meanwhile, back at the video amplifier the third component of the composite signal from the video amplifier is the synchronizing-timing signal (called sync for short). Block H contains sync amplifiers, separators and shapers, with an overall function of making the timing pulses suitable for controlling the accuracy of the sweep voltages via the horizontal and vertical oscillators. Block J is the combined vertical/amplifier which produces the voltage for vertical deflection of the picture tube beam. Block K is the corresponding group for horizontal deflection, as detailed earlier for monochrome. The dotted line in block K separates the strictly sweep function from the incidental (although most important) functions of B+ boost and high-voltage generation. Finally, block L represents the power supply for the entire TV set, including some adjustments and controls for the picture tube.

So far, we have described a complete black-and-white receiver, although hidden within its various stages were color signals—information convertible to a color picture after proper processing. The following paragraphs describe such conversion and processing.

From the video amplifier (block D, Fig. 3-4), the two components of the color signal go to the color-only portions of the receiver. Color picture information goes to block O, a color amplifier. After the signal is amplified to the required level, it goes to block P, the color detector or demodulator, then to the individual color amplifiers in block Q. Here there are three separate amplifiers, one for each of the primary colors, whose outputs are applied, respectively, to each element of the three-color gun in the picture tube.

The second component of the color signal is the *burst* information. It is responsible for the proper color registration in the picture. It is fed to block M, the burst amplifier. From here it is used to synchronize the color oscillator for precise timing or synchronizing of this all-important component of the color process. While block M of the color section is analogous to block H (sync circuits) in the black-and-white portion of the set, the color oscillator is peculiar to the color process. No corresponding function exists in the black-and-white receiver.

The color oscillator signal (3.58 MHz) is used in the color demodulator (block P) to retrieve the three primary color signals. If each color is to be reproduced accurately the color oscillator must operate in step with a similar oscillator at the transmitter. A failure in the color oscillator causes a complete loss of the color in a picture. The color oscillator output is used as a carrier (just like a picture or sound carrier) to detect or demodulate the color signals. The color signal is called a subcarrier because it is within the black-and-white picture carrier.

There are three additional functions in the strictly color portion of Fig. 3-4, namely, the color killer and the focusing and convergence systems (blocks R and S). Of necessity, we shall leave the explanation of these for later.

Solid-state TV chassis

In the early solid-state TV chassis, only the tuner and sound contained transistors. Some of these chassis were known as *hybrid TVs*. Then, the horizontal output section added power transistors. Afterward, the integrated circuits were introduced and placed in the TV chassis (Fig. 3-5). Again, the IF and AGC and sound circuits contained IC components. Next, the integrated circuits were found in color and luminance circuits. Today, you may find only two large IC components with very few outside transistors (Fig. 3-6).

The tuning section in recent color TV chassis may have a few transistors within the tuning, band selector, and AFT sections (Fig. 3-7). Some chassis may have the whole front end with IC components. The preamp transistor may be followed with a SAW filter and tied into a large IC that contains IF amp, sound IF amp, sound det and amp, video detector, video amp, noise-canceller, AFT detector, and AGC detector.

In the output section of the solid-state chassis, three color transistors feed the CRT, with horizontal driver and horizontal output transistors (Fig. 3-8). The vertical output and horizontal deflection circuits may be a separate transistor or included in a large IC (Fig. 3-9).

One large IC processor may contain the video amp, pedestal clamp, chroma amp, color demodulator, ACC, sync separator, x-ray protector, horizontal oscillator, horizontal

3-5 In the early Kmart portable TV chassis, many integrated circuits were used in addition to the many transistors of the solid-state chassis.

3-6 In this portable TV chassis, there are only two ICs and very few transistors.

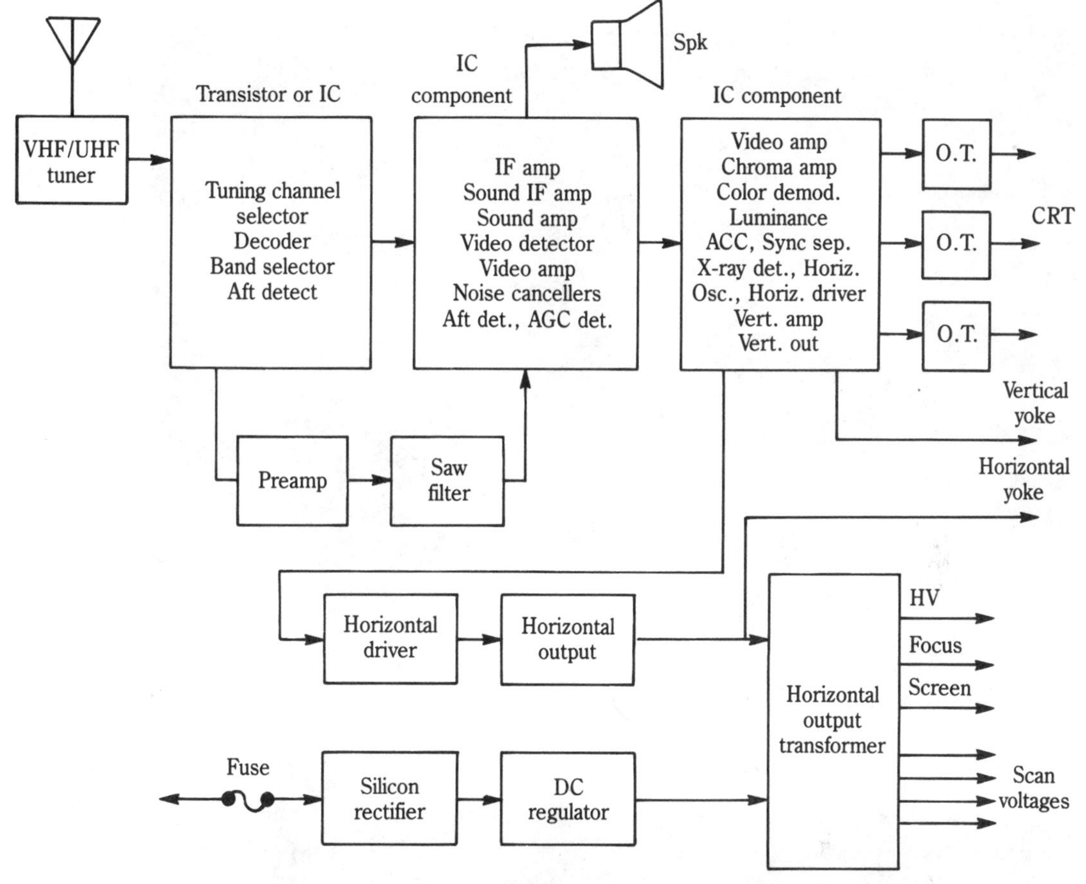

3-7 Block diagram of a typical solid-state color chassis.

3-8 The three color amps are mounted on the CRT PC board in this J. C. Penney portable TV.

3-9 Here one large IC includes the chroma, luminance, color demodulator, ACC, sync, x-ray protection, horizontal and vertical deflection, and vertical amp circuits. The IC must be replaced if any one of the circuits are defective.

drive, vertical oscillator, and vertical amp. You may find the vertical drive and output in one single IC component. Today, most solid-state chassis have more IC components than transistors in the RCA CTC131 chassis (Fig. 3-10).

3-10 A fairly recent RCA CTC131 solid-state chassis in a 26-inch console TV.

Digital system control

In some TV chassis, the digital system control circuits contain a microcomputer system that controls data in and out, clock timing, and system control reset features (Fig. 3-11). The system control system microcomputer contains random-access memory (RAM) and read-only memory (ROM). The RAM section stores up information for frequency channel change and customer settings such as current picture and volume range. This allows the operator to turn on the last viewed station with the same volume and picture adjustment when the receiver was turned off. The ROM section of memory stores information that is never changed.

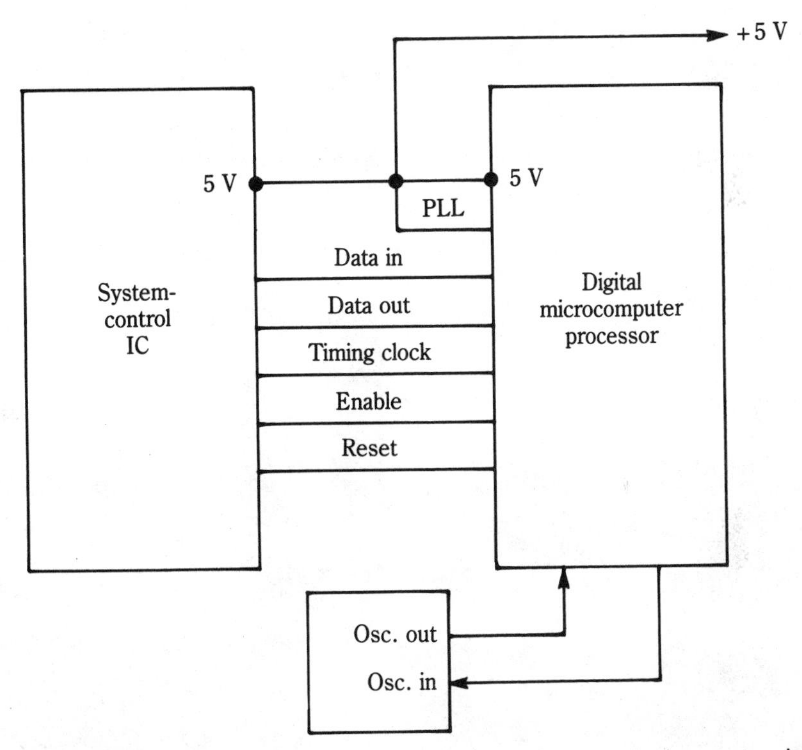

3-11 Block diagram of the system control reset and communications system in a higher-priced TV chassis.

The system control reset is to ensure the digital microcomputer starts at the same time in the program each time power is applied. System control communications controls the major functions within the chassis. The customer control interface circuits control the function operating from system control system. These controls consist of brightness, sharpness, contrast, color, tint, volume, balance, treble, bass, and power (Fig. 3-12). All these controls may be operated manually or upon the hand-held remote control unit (Fig. 3-13). You may find these digital control features in higher priced TV chassis.

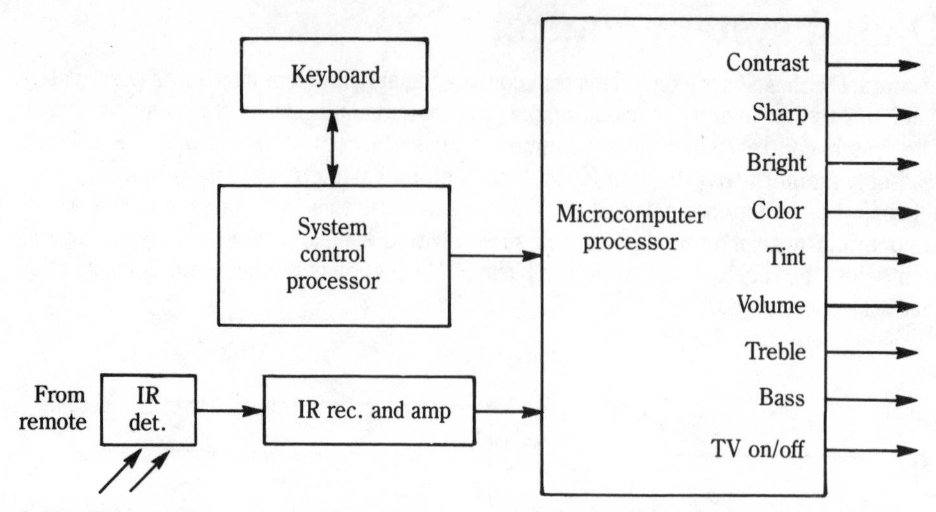

3-12 The system control processor and the remote circuits control the microcomputer processor IC, which activates the different TV functions.

3-13 A remote control can operate many features.

Comb filter

The comb filter is an electronic device to separate luminance (brightness) and chroma (color) video signals, which eliminates the cross-color or rainbow effect in the picture.

The signal from the video stages applies at the input of comb filter network and feeds to the luminance (brightness) amp and restoration circuits. A time clock signal from the luminance/chroma processor is applied to the comb filter charge-coupled device (CCD). The CCD is actually a time delay line. The comb filter device is only found in a few color TV chassis (Fig. 3-14).

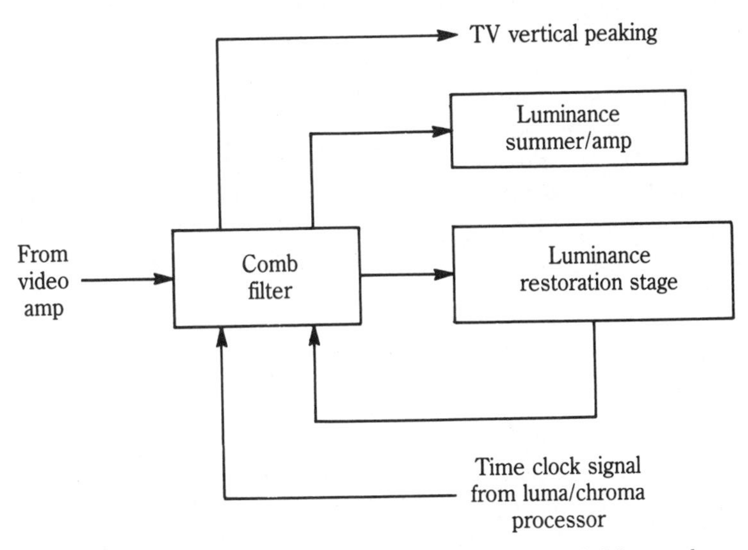

3-14 Block diagram of the comb filter that separates the brightness from the chroma signal to eliminate rainbows or cross-color effects in solid-state chassis.

Color picture tube

To better understand the construction and operation of a color picture tube, let's quickly review the features of a monochrome picture tube.

A monochrome TV picture tube consists of three main sections: the gun, the deflection system, and the screen. The gun produces a stream of electrons, which, with the aid of properly applied voltages, is focused to a sharp point on the face of the tube (the screen). The deflection system, again by virtue of correctly applied voltages of proper characteristics (horizontal and vertical sweep voltages), moves the beam to the desired points, in proper sequence, on the screen.

Now, if instead of a black-and-white phosphor (coating on the inside surface of the TV screen), we substituted a red phosphor (or a blue, or a green), the same exact picture tube just described would produce the same TV picture, but in red (or blue, or green), with all the various gradations in brightness as in the black-and-white picture. It is a bit of oversimplification, but nonetheless true, that the color TV picture tube is a combination of three tubes, a red tube, a blue tube and a green tube in one glass envelope; however, while this picture tube has three distinct guns, each emitting an electron beam for each color, it has but one screen, for all three colors. This accounts for the great complexity of color TV tubes. In order to understand some vital adjustments and corrections that are sometimes necessary on a color TV set it is first necessary to understand the structure and functioning of a color picture tube.

In contrast with the black-and-white picture tube screen, which has a continuous coating of phosphor and where the electron beam may be placed on any point on the screen, the color tube screen does not have a continuous coating but consists of a large number (hundreds of thousands) of color dots, and the electron beam must hit the dots only. Specifically, the screen is composed of three-dot groups; each group has a red, a blue, and a green dot closely spaced but not quite touching. A recent improvement in color picture tubes has been realized by putting a black border, a nonluminous coating around each color dot, to reduce unintentional blending of the colors. The dot arrangement is alternated in each group so that no two adjacent dots in any direction are of the same color. For better visualization, consider the dot groups as subminiature billiard balls, arranged in triangles (three in a group, almost touching) with no two balls in the triangle being the same. Figure 3-15 shows the dot layout on a portion of a color screen as just described.

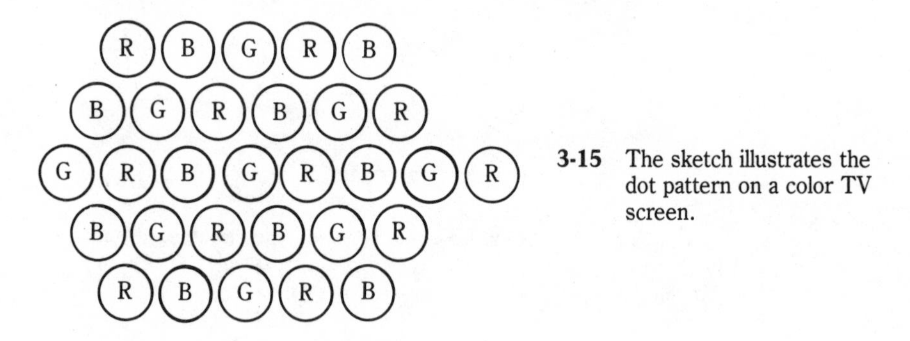

3-15 The sketch illustrates the dot pattern on a color TV screen.

Immediately behind the tricolor picture tube faceplate is a perforated metallic plate very precisely arranged and positioned so that each hole in the perforated plate lies in an exact position behind a group of color dots, a color triad, on the faceplate. When the plate is correctly positioned, the red beam goes through the hole and strikes the red dot only. Similarly, the blue and green beams passing through the same hole will strike the blue and green dots, respectively. This is a precision structural alignment and is independent of the characteristics of behavior of the TV set.

For the sake of clarity, let me state again that the three segments of the electron gun in the tricolor picture tube are identical. They are called red, blue, and green only because of their physical positioning with regard to the aperture mask (the plate behind the screen) and faceplate not because of any color difference between them. All three produce identical electron beams with no inherent color characteristics.

It should be apparent from the above description that for the proper excitation of the three colors the three electron beams originating at different positions in the neck of the tube (see Fig. 3-16) must be made to properly bend or *converge* so as to go through the holes in the aperture mask instead of striking the space between the holes. While this is approximately provided by virtue of tube design, it is not sufficiently accurate for a satisfactory color picture reproduction. To ensure this special convergence devices are

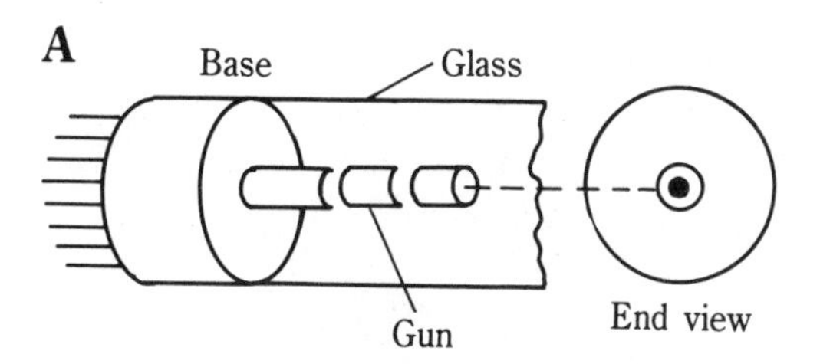

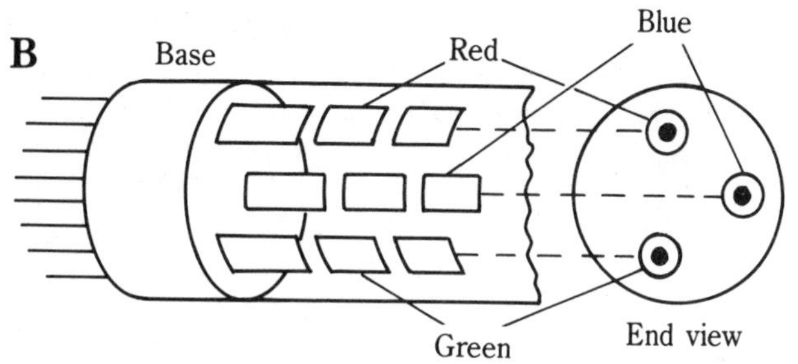

3-16 Sketches comparing the gun structure. (A) shows a black-and-white gun, and (B) illustrates a color picture tube gun assembly.

employed. There are two such sets of convergence adjustments. One consists of a set of magnets placed around the neck of the picture tube and adjustable for physical position. This set is called *static convergence* controls and is best left alone. The second set of convergence adjustments is of the rear-of-the-set shaft types. Three groups of this type of adjustment controls exist, one for each of the three primary colors.

While focusing is a function common to all TV picture tubes it is much more important in a color tube due to the strict compatibility requirements between the three electron beams, the hole in the aperture mask and the color triad just in front of each hole.

Block S of Fig. 3-4 indicates the convergence function just described. Although the convergence adjustment shafts are accessible at the rear of the receiver, these adjustments require extreme care as well as complex equipment in order to set them properly. Such is not the case with the focusing adjustments. With proper care (as I shall indicate under troubleshooting) this adjustment can be satisfactorily performed by a beginner.

Incidentally, at least one color TV manufacturer has built in a convergence tests accessory, making it relatively simple to carry out these adjustments when required. The trend in the industry seems to be toward inclusion of the simpler self-test or self-correct capabilities in the receiver. A common example is the demagnetization or degaussing coil which prevents and corrects color deterioration due to stray magnetic influences in the immediate vicinity of the color set. Prior to this the performance of a color TV set, even though it was in normal operator condition, was affected (sometimes seriously degraded) by the appliances or electric wiring in the immediate vicinity.

A third strictly color function mentioned earlier is the *color killer*, and it is described fully later. At this point, suffice it to say that the purpose of this is to make sure that no color appears on the screen during a black-and-white picture.

CRT socket components

There are many different components connected or mounted upon the CRT socket. Besides the focus cable, the color output transistors are mounted upon a separate PC board attached to the picture tube socket (Fig. 3-17). Some color receivers have the different color bias transistors which are directly coupled to the three color amp transistors. The red, green and blue color output transistors are mounted upon the PC board (Fig. 3-18).

Usually, the voltage supplied to the color amp transistors is from the horizontal output transformer circuits. The color output signal is tied directly to the cathode terminals of the picture tube. The same voltage found upon the cathode terminal is about the same on the collector terminals of the color amp transistors (+200 volts). Here is this RCA portable chassis, the focus and screen controls plug in after the CRT socket is in position (Fig. 3-19). When one color is missing or all of one color on picture tube, suspect a defective color amp circuit.

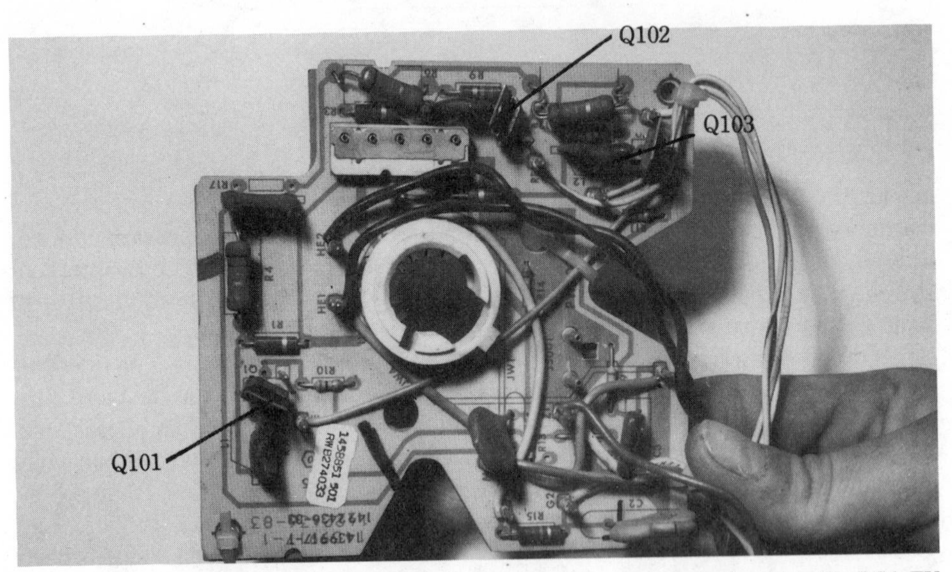

3-17 Notice the three color transistor amps mounted on the picture tube board in this RCA TV chassis.

Recent color tube improvements

Figure 3-15 illustrated the basic color tube using the three component colors in a triad arrangement, relying on the properly perforated plate or *shadow mask* behind the screen to converge the three component beams as required to produce the desired color. Even with a so-called border guard—a black, opaque area separating the color dots—and with improvements in convergence techniques, this construction left much to be desired from the viewpoint of color purity and picture fidelity. Subsequent to this, two significant improvements were made in color tubes so that now many TV sets have one or the other of these improved tubes.

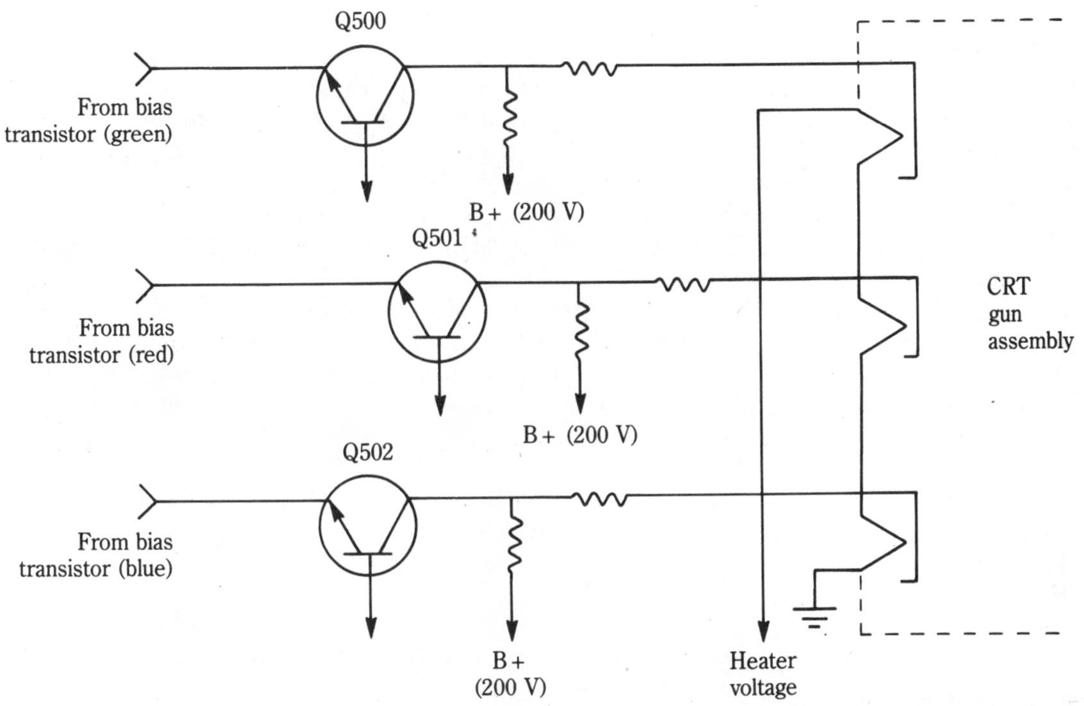

3-18 Block diagram of the color output amp transistors connect directly to the cathode elements of the picture tube mounted on CRT board.

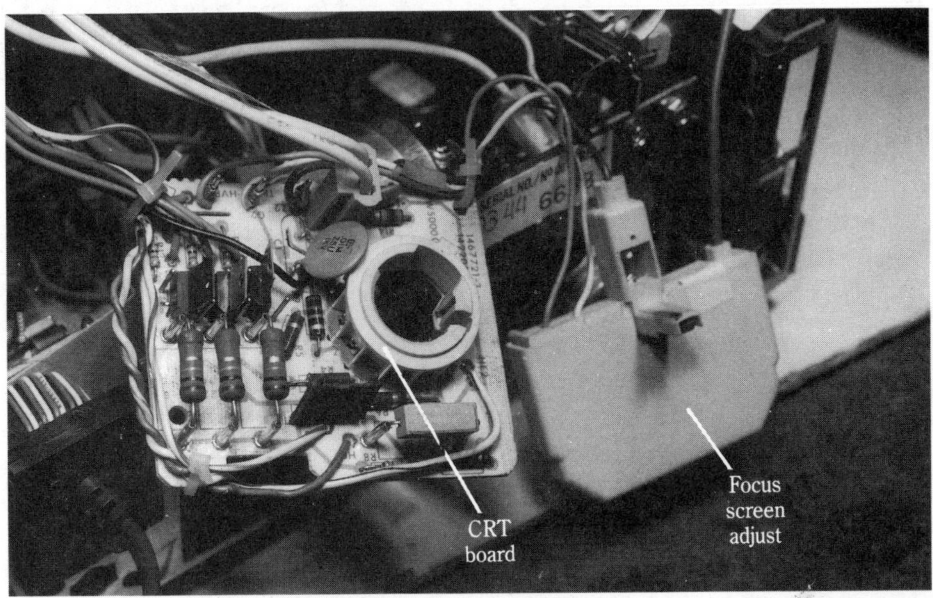

3-19 The focus and screen grid control assembly plugs directly into the CRT after the picture tube board socket is mounted on this RCA chassis.

The first of these was made a few years ago by one of the foreign manufacturers and still is used. This tube is known commercially as the Trinitron produced by Sony Corporation. This uses a three-beam gun, with a color faceplate consisting not of triple dots, but the vertical strips. These strips or stripes are also separated by opaque strips to prevent unintended color overrun or blending. The strips or stripes are continuous from top to bottom. Figure 3-20 shows in sketch form, the gun, focusing and faceplate structure of this tube. The Trinitron equivalent of the shadow mask, called the aperture grille, which has vertical slits instead of triad holes, is claimed to be more accurately producible and to provide a greater beam converage of the color screen. In a properly adjusted set of this type, that is, when the brightness, contrast, and color intensity are normal, it is possible to actually see each of the vertical color lines by getting close to the screen. The appearance resembles that of a fine cloth weave. In addition to the superior picture quality claimed for this type of tube are also the advantages of simplified convergence adjustments, as well as fewer adjustments compared to the basic original color tube described earlier.

The second type of improved color tube, and one found in more and more use by a number of manufacturers, also has a vertical slot structure, but, unlike the Trinitron continuous stripe arrangement, it consists of slots separated by opaque areas. In addition these types employ a three-gun in-line arrangement in contrast to the triad gun arrangement and the single-gun three-beam type. It is claimed that this type of shadow mask also provides for better color distribution and depth. Both of these types, the Trinitron and the slotted-aperture type, are superior to the old, three-dot type of aperture structure. In addition, and perhaps only incidentally, these tubes are also shorter because of the wider angle encompassed by the screen compared to earlier types. While this may not be of particular interest to the TV owner, other than perhaps the decreased depth of the cabinet, there are some definite advantages from a technical viewpoint also. While there still are a few sets with the three-dot matrix type of picture tube, they are said to have been improved with regard to focusing and color depth or brightness.

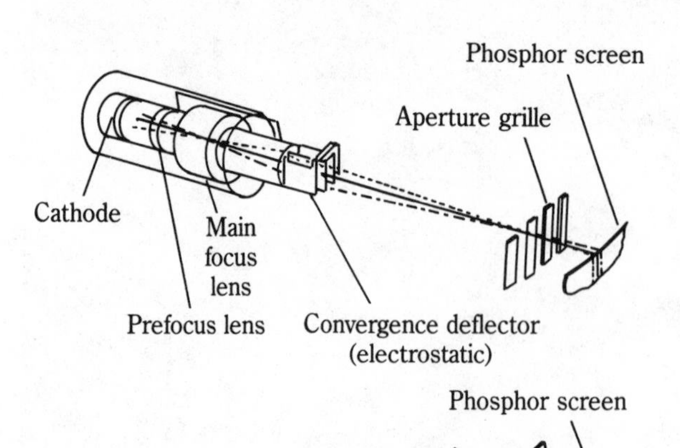

3-20 Sketches showing construction of the Sony Trinitron picture tube.
Sony Corporation

There is no simple way to identify the tube type from casual viewing by the observer. But all the manufacturers describe their products so as to leave no doubt about these features. Thus, Trinitron is a sure identification of the continuous-strip mask type, the in-line term identifies the slotted-mask tube (there may be other additional terms used in advertising), and the improved three-dot types are identified by their manufacturers as the mask-focus type, suggesting that each mask hole acts as a focusing lens for sharper, brighter images.

There is one other feature of some of these newer picture tubes worth mentioning, although technically of minor value. This is a mild light filter (picture tube seems dark when the set is off), of a smoky hue, intended as an antireflection device, so that the face of the tube will not act as a mirror and reflect objects in the room (Fig. 3-21).

Today, the color tube has several big improvements for the customer and service technician. The picture tube screens are much larger, have square corners, and do not need any type of convergence adjustments. RCA has developed a new color TV picture tube featuring a larger viewing area, flatter faceplate, and squarer corners than the current picture tubes.

3-21 The rear end of a CRT board and flyback section in an RCA 13-inch solid-state chassis.

Large picture tubes

Picture tubes are getting bigger every day. A few years ago, the 24-inch was the largest size with a 25-inch design around the corner. Since then the 26-, 27-, and now the 31-inch screens are being produced (Fig. 3-22). You may find the 27-inch quite popular and placed in portables and consoles.

The 27- and 31-inch screens have a flat front surface with almost square corners. The picture tube is measured diagonally from corner to corner. Very little picture distortion is in these larger glass picture tubes. Of course, the larger size tubes are a little more difficult to replace. It takes two people to lift, remove, and replace the larger picture tubes.

3-22 Notice how flat the picture tube looks, with square corners, in this 27-inch RCA console TV.

The defective CRT

How do you know the picture tube is defective or going bad? The defective picture tube will not light up, appears whitish around figures, blotchy pictures and distorted colors. The picture tube heaters may open and not light up at the end of the gun assembly (Fig. 3-23). When the picture appears dim with very little brightness, suspect a weak picture tube stream of electrons. The nickel-plated cathode around the heater or filament is worn off or bunches of electrons have built up on the emitting element.

The gassy picture tube may show blotchy areas with a close up view of a person's face. Usually, the above defects appear upon the black-and-white picture. Extreme distorted areas of color when the chassis turned on indicates a defective CRT. The picture

3-23 A defective CRT might not have any raster or picture due to an open filament in the gun assembly of a Sharp portable TV.

appearing dim at first and then improving as the set becomes warm, indicates a defective picture tube. Some of the above symptoms may also be caused by a defective component in the picture tube circuits. Most picture tube testers will indicate a defective CRT.

On-screen display

One of the recent displays added to the many TV circuits is the on-screen display (OSD). The display of numbers, letters and dots or dashes of the various operation controls are placed on the screen. On some TV screens, the contrast, color, tint, brightness, volume, treble, and bass bars or dashes, with correct function is shown upon the face of the picture tube. You can adjust these controls manually or with the remote control. The operation may be called menus or characters by several manufacturers.

By placing the function and graduation of bar dashes upon the screen, you can see where you like the best level color picture or sound. The dashes may go clear across the screen or when you select the best viewing picture for yourself, the color and brightness may be one-third the way up. This on-screen information is applied over the colored telecast picture (Fig. 3-24).

The system control and microcomputer processor select and control the on-screen display process. The on-screen display is fed from the microcomputer to a processing IC. The OSD signal is demodulated and connected into the R-Y, G-Y and B-Y CRT bias drive circuits (Fig. 3-25). A defective on-screen display may result in no or scrambled

3-24 On-screen display can be controlled by the remote or manually on a large TV screen.

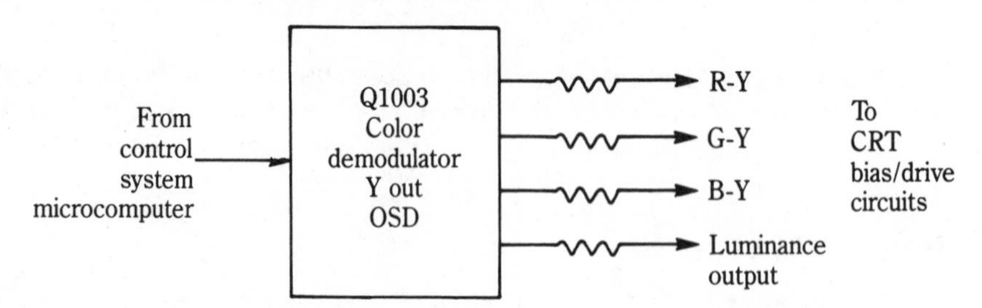

3-25 The on-screen display (OSD) menu is governed by the system control microcomputer via the OSD signal which is fed to Q1003. The signal then goes on to the R − Y, G − Y, and B − Y bias/drive circuits and to the picture tube.

on-screen display, no green character, a solid green screen, no luminance or brightness but the color is normal, and no black edge. Troubleshooting the OSC circuits takes some know-how and an oscilloscope.

Picture in a picture

You may be able to see more than one picture on the picture tube at the same time. The picture process allows one to watch the large picture with one or more small pictures over the larger picture or vice versa. You can zoom the small picture in size and freeze it at that point, in some TV chassis (Fig. 3-26).

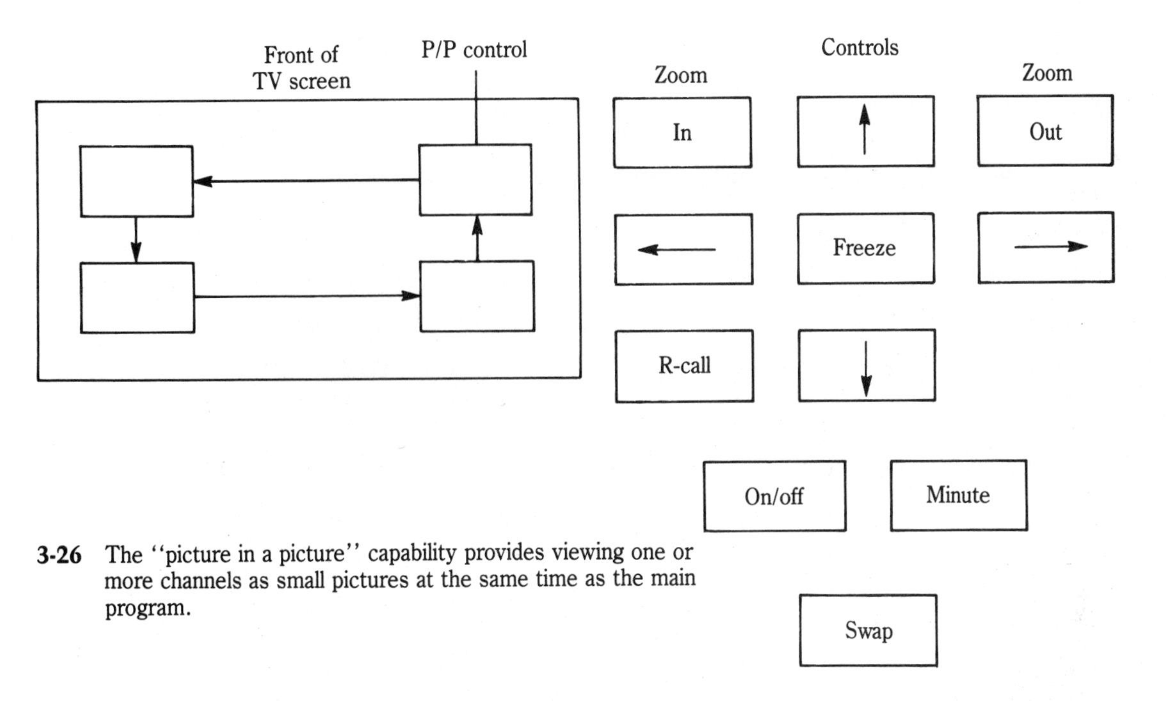

3-26 The "picture in a picture" capability provides viewing one or more channels as small pictures at the same time as the main program.

When the picture-in-a-picture function is selected, the small picture should appear in the lower right-hand corner of the screen. The on-screen display displays the big picture at the left and small pictures at the right. The small picture may be moved to any position on the screen by using the four direction arrow buttons. The picture will move until the arrow key or button is released. The sound is transmitted with the large or original picture upon the screen.

There are many different functions found with the picture in a picture process. These functions are controlled by a picture-in-a-picture module. Since many different signals are required and critical test equipment, the picture-in-a-picture module can be replaced or serviced properly by the electronic technician.

Big screen projection

There are several large curved screen TV receivers on the market with front and rear projection. Three separate small projection tubes with lenses are converged to the large projection screen. The control console with the projection tubes are mounted in front of the large curved screen. With rear projection, all projection equipment is inside the TV cabinet. Most TV sets with rear projection have large, flat viewing screens.

A disadvantage of large screen projection TV sets viewed in a normal living room setting was inadequate brightness and contrast. Today, projection TVs have improved brightness due to liquid cooling assemblies which are a part of each projection tube—red, green, and blue. A cavity holding a cooling solution permits each projection tube to operate at higher voltage potentials, resulting in brighter, sharper pictures (Fig. 3-27). Each tube has its own cooling assembly, lenses, and control circuits.

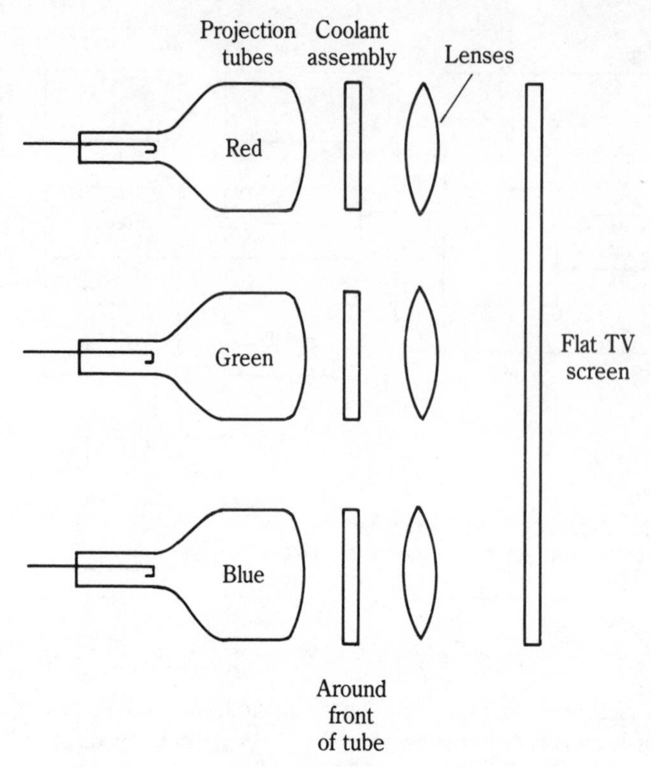

3-27 Liquid-cooled projection tubes in newer model protection TVs provide greater brightness. Wide-angle-lens assemblies result in flatter viewing screens.

Small black-and-white TV screens

Several years ago, the smallest TV screen was 5 inches. The small screen size has now (Fig. 3-28) developed into the mini television screens from a 3-inch to a 2.7-inch and down to a 2-inch portable TV that you can hold in your hand. The pocket-type portable operates from flashlight cells, while the large screen mini TV may be either ac or dc powered. You may find the small screen in a TV – AM/FM – cassette portable (Fig. 3-29).

Often, with small size screens, the high voltage may be around 1 kV or 1.5 kV at the anode connection. All circuits are solid-state with many IC components. In the early 3-inch portable TV the solid-state chassis was powered with a battery pack, which included several batteries or ac external power jack (Fig. 3-30). This transistorized chassis was built with plug-in type boards. The AM antenna is shown at the back and FM dipole antenna at the top (Fig. 3-31). The mini-screen TV components are packed and pressed close together, which demands a lot of patience, small, thin fingers, and a magnifying lamp to locate the tiny defective parts.

3-28 A 5-inch portable black-and-white TV with VHF low, VHF high, and UHF band tuner selection.

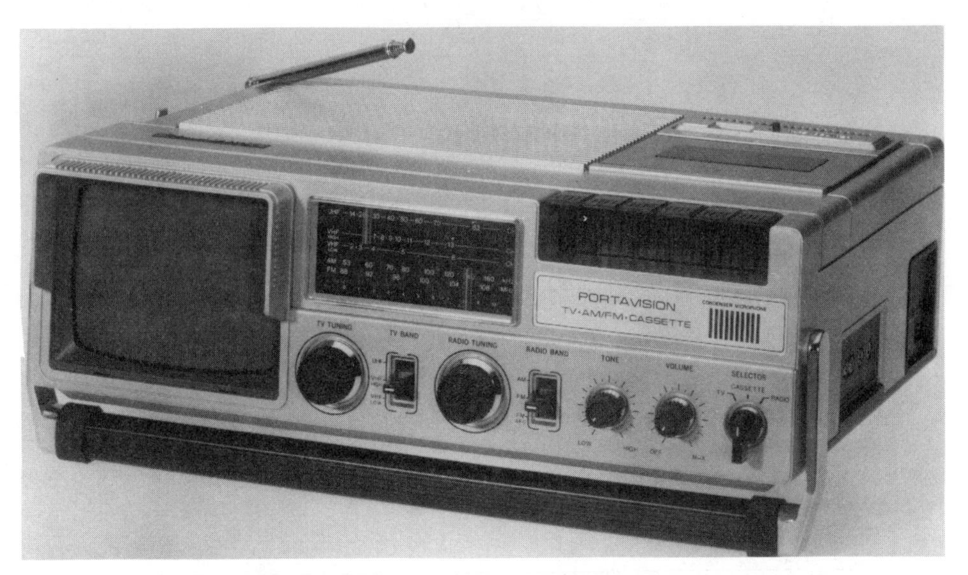

3-29 A small black-and-white TV in a TV – AM/FM cassette combination. Radio Shack

3-30 In one of the early small-screen TVs, the chassis operated with a battery pack or external ac-dc source.

3-31 The small transistorized chassis plugs into a multicontact socket and can be removed for easy service. The pen points to the small VHF transistorized tuner.

4
Color-only sections

THE BEST PLACE TO BEGIN AN EXAMINATION OF THOSE SECTIONS OF A COLOR receiver that deal solely with color reproduction is the video amplifier because that's where the differences begin. Figure 4-1 is a detailed block diagram of the color-only sections of a generalized TV set. It illustrates the functions, adjustments, and control locations of potential malfunctions that could be found in an actual TV set.

Beginning with block A, the video amplifier of the black-and-white portion of the set, we see that the color signal together with its timing (burst or sync) signal go to block E on the color side of the dashed dividing line. The dashed line from block A to block F is an alternate way for the color timing signal to be taken out of the video amplifier; thus, in some sets, two separate signals may go from the video amplifier to the color portion of the set—the picture information to block E and the burst information to block F. In more recent TV sets, the combined signal is amplified in the first color amplifier, then the burst separates and goes to the burst amplifier (block F) while the color (chroma) signal goes to the second color amplifier, block G. Here again some TV sets have only one color amplifier, so that block G would be missing. Whether a particular TV set has one or two *chroma* amplifiers, the front-of-the-set control marked color is located here and it adjusts the color content of the picture from all black-and-white (no color) to full color (and beyond, if turned to its extreme position).

Burst amplifier

The simplest way to describe the function of a burst amplifier (block F in Fig. 4-1) is to say it is exactly the same as the function of the sync amplifier (block H, Fig. 3-4). Input to the burst amplifiers is derived either directly from the video amplifier (block A, dashed arrow) or via additional amplification in the first chroma amplifier (block E, solid arrow). The output of the burst amplifier is ultimately applied to the 3.58 MHz oscillator (block J) for precise control (timing) of the oscillator (Fig. 4-2).

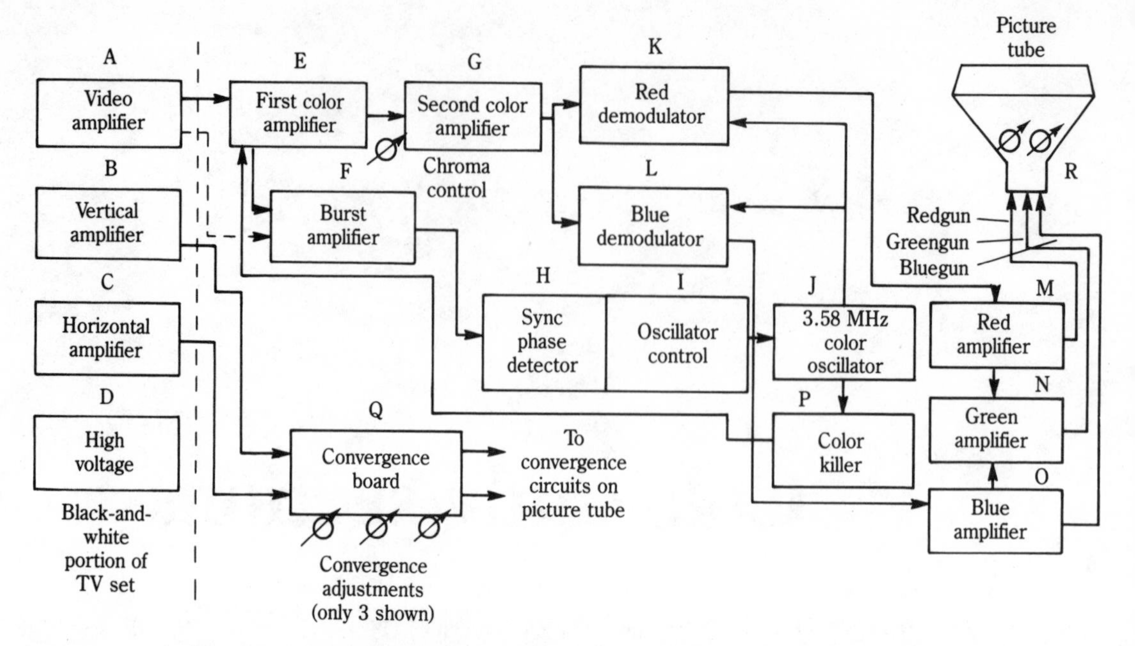

4-1 Detailed functional diagram of the color-only portions of a color receiver.

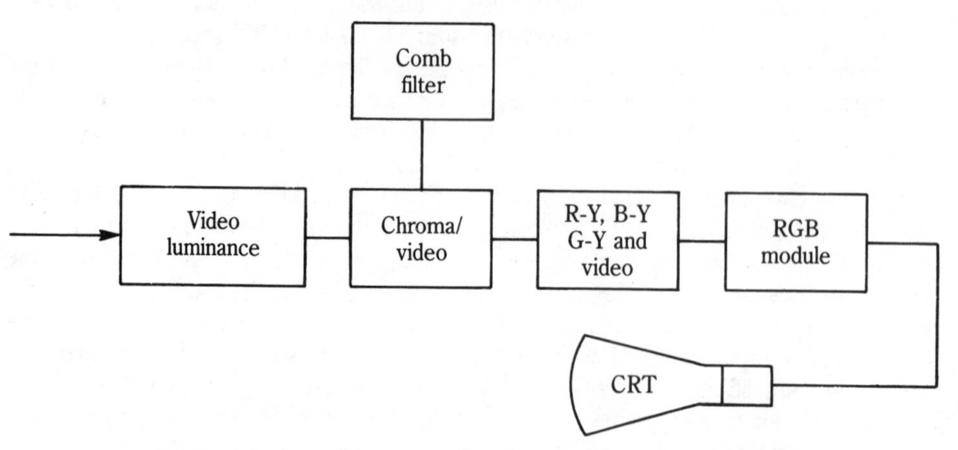

4-2 Block diagram of the color (chroma) circuits in a solid-state chassis.

Color sync and oscillator control

The color sync phase detector (block H) and oscillator control (block I) are grouped together for two reasons: first, both perform functions incident to controlling the accuracy of the color oscillator, and second, they may be considered, in a sense, optional. While some manufacturers use this function for oscillator control, others achieve the same result in a different way and do not incorporate these two in the TV receiver design. What is important to the TV set owner is the fact that when these functions are found in a set, they are links in a chain, so that if one or both should fail, the 3.58 MHz oscillator may wander sufficiently to cause some serious deficiencies in the color rendition and balances (by cutting off some of the color frequency components in the output signals).

As we shall soon see, the 3.58 MHz oscillator is one of the ingredients in the conversion of the chroma signal to the red, blue, and green portions of the final color picture. The balance between these three primary colors is, of course, essential for a normal color picture and can be achieved only when the 3.58 MHz oscillator signal is fed to the demodulators exactly in time with the corresponding signal in the transmitter. Control circuits and tubes in blocks H and I establish and maintain synchronization between the transmitter and the receiver, and a failure here invariably degrades if not altogether destroys the normal color picture.

Color oscillator

The 3.58 MHz precision oscillator (block J, Fig. 4-1) is a crystal-controlled oscillator, which is basically very accurate and stable even before being further controlled by the burst signal from the transmitter. An examination of the immediate vicinity will locate the crystal. It is a fragile, tissue-thin wafer of a synthetic, quartz-like material (the natural mineral quartz is seldom used nowadays) encased in a hermetically sealed plug-in unit, as illustrated in Fig. 4-3. Although the little box requires careful handling, it is relatively durable; however, occasionally it may become intermittent (stop functioning, then start again after the TV set is switched off and on) or it can fail completely. In either case, if the crystal malfunctions the color, but not the picture, disappears.

4-3 The crystal used to control the frequency of the 3.58 MHz oscillator is usually hermetically sealed in a metal container. Some are designed to plug in and others are soldered.

Solid-state 3.58 MHz color oscillator

Within the solid-state circuits, the 3.58 MHz color oscillator may be located in a transistor or IC component. When the IC chip was introduced, the color oscillator was included into the video-luminance-chroma IC (Fig. 4-4). Today, the chroma processing circuits are included with the luminance and chroma processor. You may find the chroma sections in one large module that is bolted or snapped into the chassis (Fig. 4-5).

The chroma signal is demodulated with a 3.58 MHz voltage-controlled oscillator (VCO). The frequency of the oscillator is controlled or stabilized by a fixed crystal (Fig. 4-6). The automatic phase control compares the phase of 3.58 MHz oscillator signal to the burst signal and gives an error signal. This error signal locks the phase of 3.58 MHz oscillator to the burst signal.

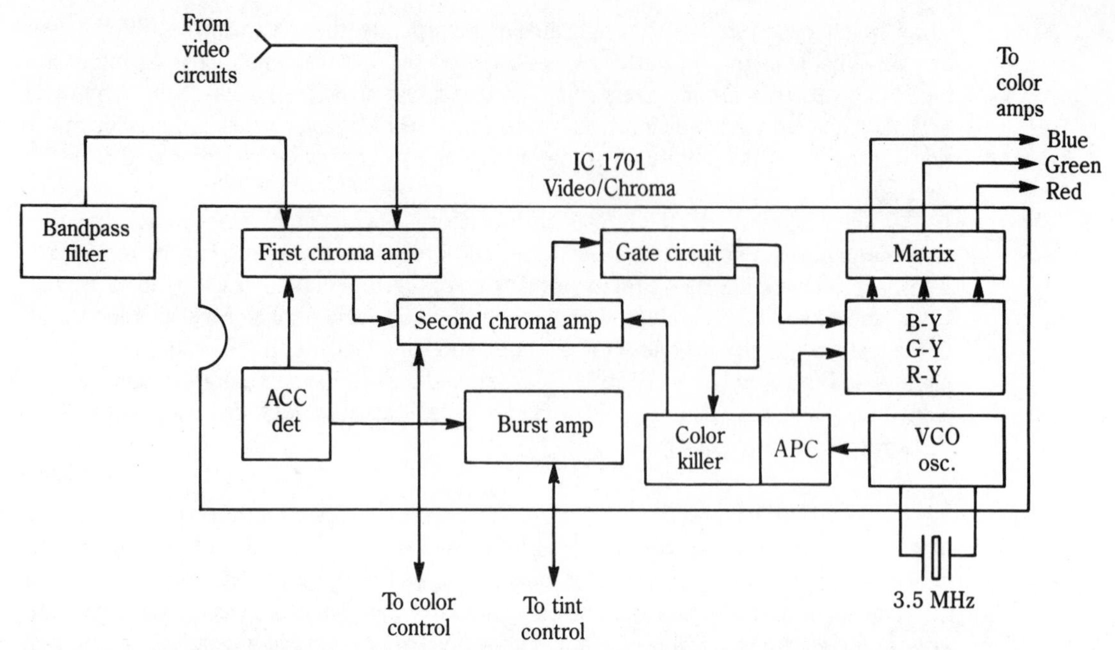

4-4 Block diagram of color circuits in the luminance/chroma IC processor of a solid-state chassis.

4-5 Here one large color board in a Montgomery Ward portable TV chassis contains all of the color circuits including color amps, which are usually found on the rear of the CRT board.

4-6 Location of a 3.58 MHz crystal oscillator component in a Sharp 19D82 portable color TV.

Color demodulators

Blocks K and L are the color demodulators. As shown in Fig. 4-1, the combined color signal comes from the (second) chroma (color) amplifier, while the second input is from the precision-controlled 3.58 MHz oscillator (or reference signal). Notice that there are only two demodulators, one for the blue signal, the other for the red. The green demodulator, as such, does not exist, and it is unnecessary. Since a combination of the primary colors, blue and red, will produce green, no separate demodulator is required. The green signal is derived by mixing a proportional amount of red and blue.

The function of the color demodulators, stated in nontechnical terms, is to extract the color information from the combined color-plus-carrier signal. The color demodulator is a detector. While two demodulators are shown, it does not necessarily mean that every TV set will have two tubes performing these functions; some may use two transistors for this purpose (in the most recent sets), while others may use a dual tube in a single envelope. Whichever the case, the function is the same.

In case of malfunction, a failure of a two-in-one tube will completely shut off all color. Should one section of such a dual tube fail (this is not unusual) or should one of the two demodulators fail in sets with two separate demodulators, two of the three colors will be missing. This is due to the fact, as mentioned earlier, that red and blue color components are combined to produce the green primary color as well as the hues of green.

Solid-state color demodulators

The outputs of the chroma limiter and phase detector are applied to the color modulator circuits. Sometimes the phase detector is called a *flesh* or *hue detector*. The output of the modulator is applied to a buffer stage with the generator 3.58 MHz voltage-controlled oscillator. The output of the buffer stage, which is a tint-controlled signal, is applied to the phase-shifting components.

The output signals from the phase-shifting circuits are applied to the chroma matrix amp. The chroma matrix amp produces R-Y, A-Y and B-Y signals combined with the luminance signal. R, G and B output signals are fed from the chroma processor circuits to the respective bias and color amplifiers (Fig. 4-7).

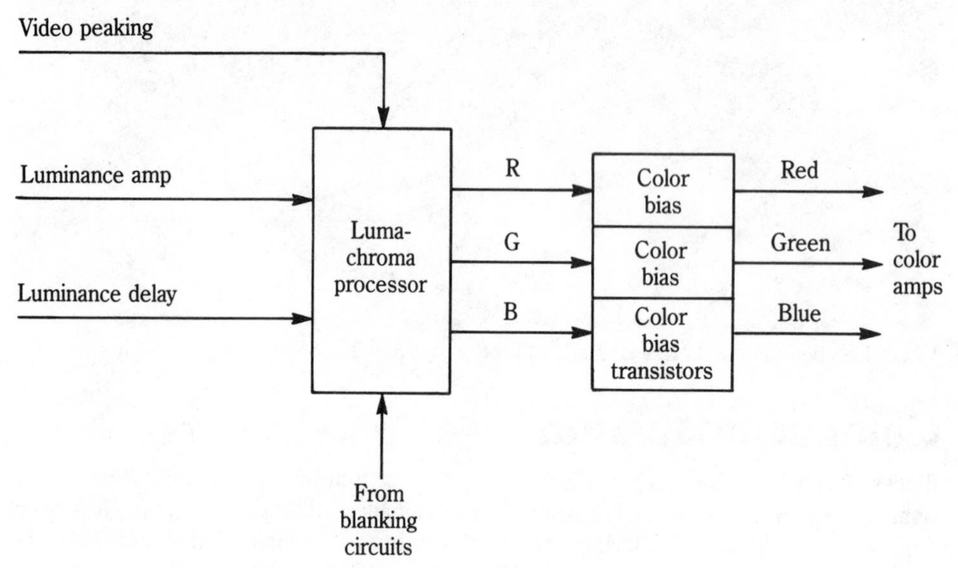

4-7 Block diagram of a large color-processing IC with R, G, and B color signals fed to color amp of a solid-state chassis.

Color amplifiers

Blocks M, N and O, the final color amplifiers for each of the primary colors, are similar to the video amplifier of a monochrome TV set. Since each color component signal has its own independent path from the demodulator to the picture tube, a failure in a particular color (or blend of that color) is unambiguously traced to the specific amplifier. As mentioned earlier, the block diagram in Fig. 4-1 is not of any particular manufacturer's TV set; hence, differences are possible.

You may find several transistors and ICs in the solid-state color chassis—electronically their function is the same as their tube counterpart. In fact, the entire color circuit function could be handled by one or two ICs. For example, in a Sharp model 13H22, the deflection video and color processor circuit are contained in a single IC type I801 (Fig. 4-8). A defective I801 may cause loss of color, poor color, or loss of brightness or picture.

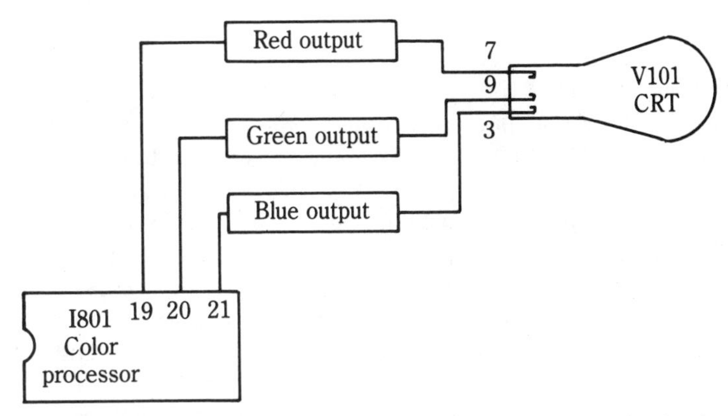

4-8 Only one large IC component is found in the color circuit of a Sharp 13H22 model TV. Replacing this IC can solve a variety of color or video problems.

Color killer

Block P represents the color killer, a violent title that means just what it says. The function of this section is to prevent any color from showing on the screen when a black-and-white picture is being received. To begin with, remember that a monochrome picture in a color TV set is produced from the correct combination of all three primary colors. In other words, a black-and-white picture on the screen comes not from no color, but from all colors in the exact proportion, just as white is not the absence of color but the presence of all colors.

Second, and this follows, that to transmit a black-and-white picture requires proper balance of all three primary color components. While balance would assure no coloration of a black-and-white picture, such balance cannot be taken for granted. Consequently, color fringes may appear on the color TV set when they are not supposed to appear, that is, when the picture is transmitted in monochrome.

It should be remembered that in all cases, whether monochrome or color pictures are transmitted, the picture tube receives a brightness signal which is actually the picture minus the color. Of course, when no color is transmitted this brightness signal is the picture while the chroma amplifiers have nothing to contribute. Similarly, when a color picture is received on a black-and-white set (in which there obviously are no color amplifiers) this brightness signal is all there is and it produces a black-and-white picture without any special adjustments or color killers.

Finally, it must be remembered that the color amplifiers as well as red, blue, and green guns of the picture tube operate on current (electron) beams, not on beams of colored light; thus, any current in the path of a chroma amplifier (which automatically feeds one of the three color guns) is capable of exciting the color dots on the picture tube screen regardless of whether or not color is being transmitted. To prevent this unwanted color the color killer function is required.

In operation, the color killer circuit keeps the chroma amplifier in a nonamplifying condition (or cutoff) until a color picture is received. At such time, a signal from the color sync (burst signal) disables the color killer, releasing the chroma amplifiers to perform their normal functions.

Improper adjustment of the color killer and fine tuning knob may result in loss of color or intermittent color in the picture. Sometimes the manual fine-tuning knob may be set clear to one side with the aft button (auto fine tune) in operation. When the aft button is switched off, the color may disappear. First, unlock the aft button and then readjust the fine-tuning knob for normal picture and sound. If there is no color in the picture, turn the color control to its maximum position. Then locate the color killer control (often located on the back chassis panel) and turn it to the extreme clockwise position and then back again counterclockwise (Fig. 4-9). Watch for the color to return. Some color killer controls give better color to the left or right side of rotation. Adjust both the color killer and color control (on front of TV) for the best color picture. Now, readjust the fine-tuning control and lock the color in with the aft and color monitor button.

4-9 Locate the color-killer control and turn it clockwise and counterclockwise to notice if the color returns.

Solid-state automatic chroma control (ACC)

The ACC circuits stabilize the chroma output signal against any fluctuation of the chroma burst signal. A voltage applied to pin 5 controls the gain of the first chroma amplifier (Fig. 4-10). When the burst signal at pin 1 and burst output signal is quite large, the voltage at pin 5 lowers, reducing the gain of the first chroma amplifier. R805 and C805 are the ACC filter network.

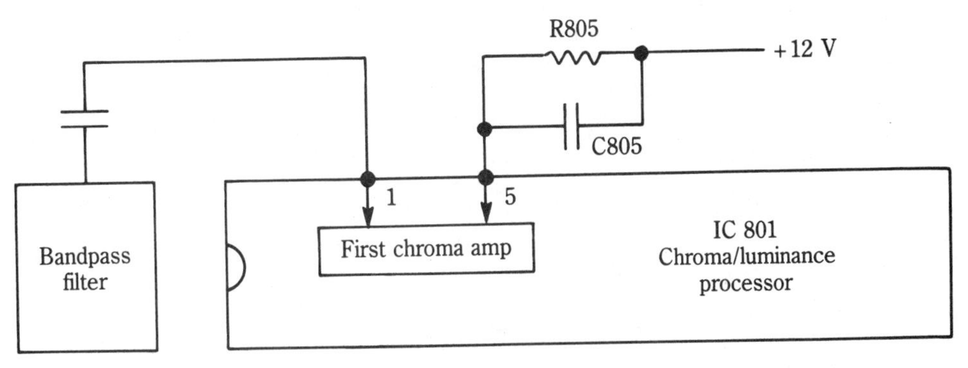

4-10 Block diagram of the solid-state automatic color control (ACC) circuits.

Solid-state color killer circuits

The color killer circuit is to prevent colored noise, which appears as colored snow, from appearing on the picture tube when a black-and-white telecast is received. In the early tube and transistor chassis, a separate color killer control was found upon the metal chassis (Fig. 4-11). Now the color killer circuits operate automatically inside the chroma processing. No color killer controls are found on the latest TV chassis.

When no burst signals are applied from the color processor, the color killer circuit output voltage lowers, making the gain of the second bandpass amplifier zero. When a burst signal is applied, the color killer circuit is off, resulting in a normal color picture.

4-11 The AGC color killer control adjustment in a Sears color portable chassis is accompanied by two different color/luminance IC components. Notice the ICs plug into an IC socket in this chassis.

Within the present-day color processing circuits, when a black-and-white signal with poor or weak color information is received, the color killer circuit activates, turning the second chroma amplifier off. Then, when a good color signal is picked up, the color killer is turned off and the output signal from the second chroma amplifier passes to the demodulator circuits (Fig. 4-12).

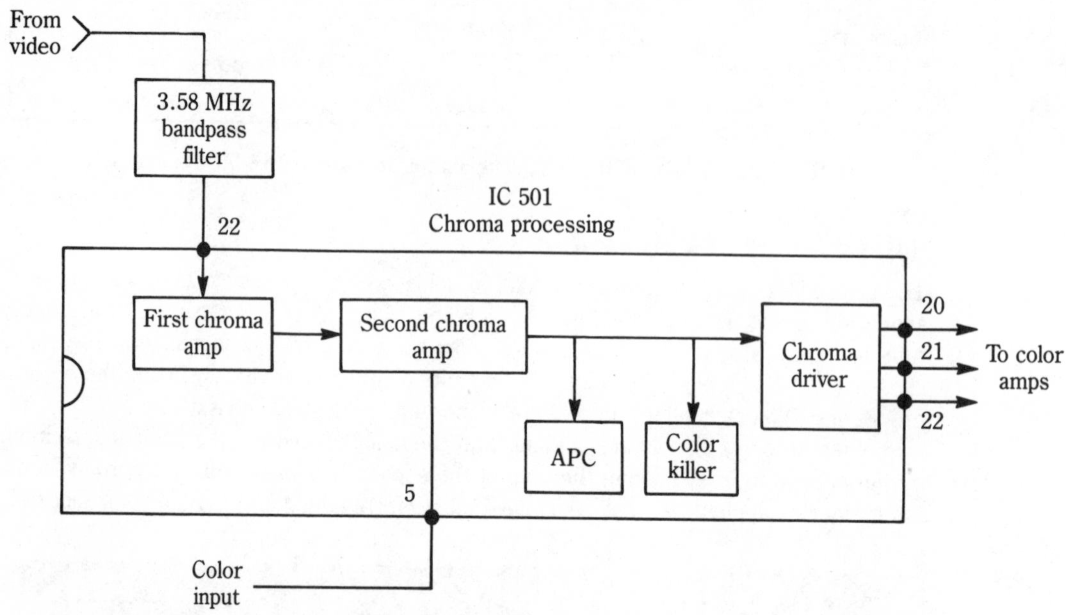

4-12 Block diagram of the automatic color killer control circuit in a present-day chroma processing IC.

Solid-state automatic phase control (APC)

The APC circuits regulate the subcarrier oscillator circuit to maintain proper oscillator frequency and phase to ensure correct color synchronization. The automatic phase control circuits detect the phase difference between burst signal from the burst amp and 3.58 MHz oscillator frequency.

The APC adjustment may be preset and field adjustment is not required. The pot may be sealed but can be broken for APC adjustment if readjustment is required.

Within the latest chroma processing circuits, the APC circuit provides automatic search function and is activated when the color killer is turned on (Fig. 4-13). The APC circuit is supplied to the variable control oscillator (VCO) and tint control circuits. The output of the tint control is applied to the chroma demodulator and R-Y, G-Y and B-Y signals. These color circuits may be located within one large front end RF/IF luminance/chroma IC processor (Fig. 4-14).

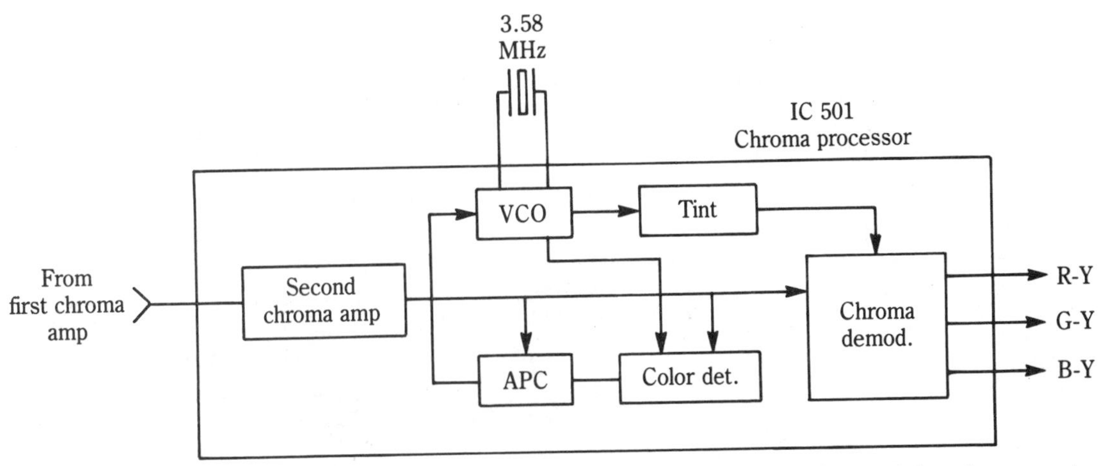

4-13 The APC circuit of today's TV chassis controls the correct color synchronization of the color-processing circuits.

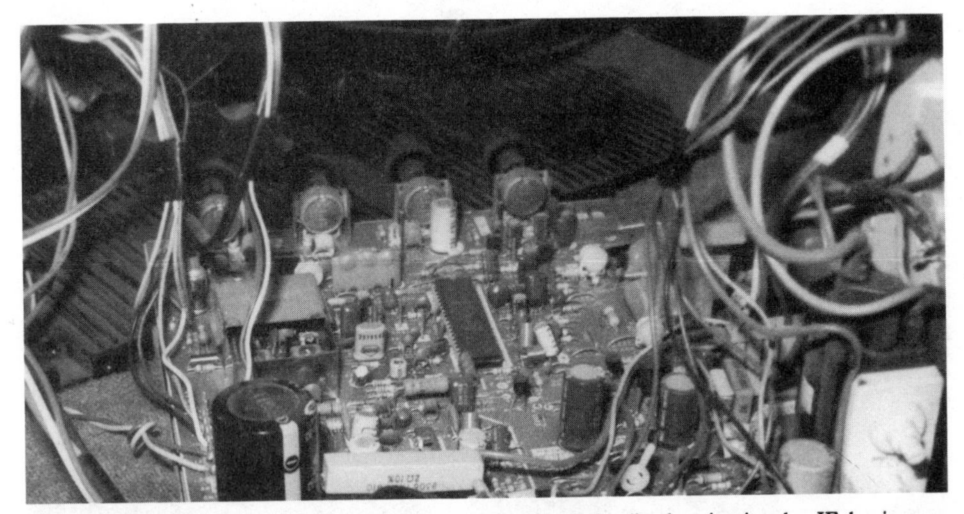

4-14 One large IC processor (center of board) might include all color circuits plus IF, luminance, and AGC circuits in later-model solid-state TV chassis.

IC chroma circuits

One large IC processor may include AGC, IF, AFT, IF sound, audio amp, sync separator, vertical and horizontal deflection, X-ray protector, and luma/chroma processing in the latest TV chassis. The IC color processing circuits consist of first and second color amps, APC, VCO, color killer detector, tint, and chroma demodulators (Fig. 4-15). The early solid-state chassis may contain the above circuits upon one large board (Fig. 4-16).

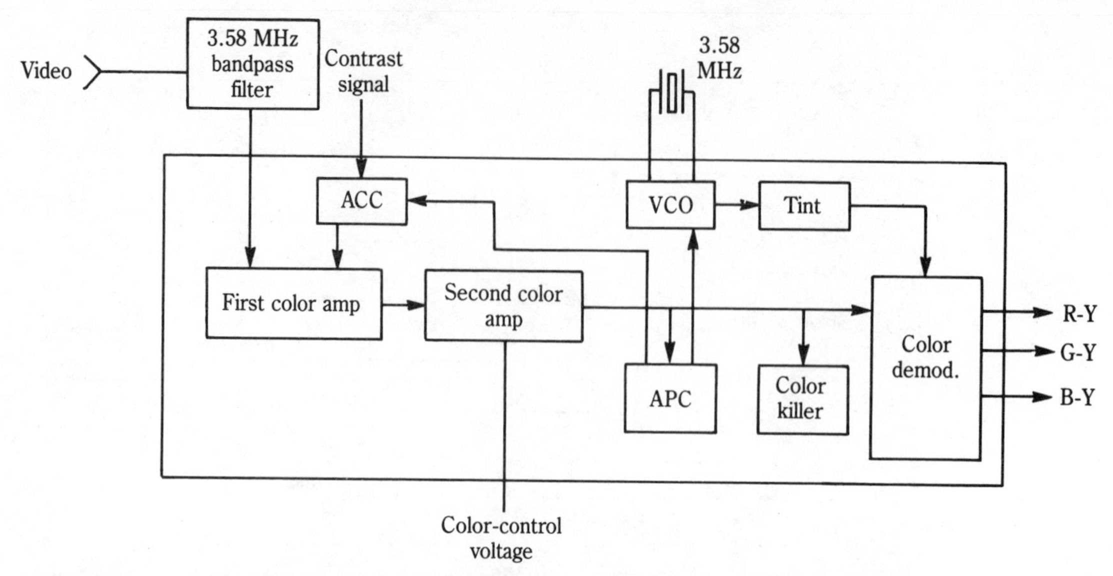

4-15 The chroma processing IC circuits might be located with other circuits.

4-16 The solid-state color circuits are mounted on a separate PC board along with the sound, luminance, and IF stages in a Sharp 19D80 portable.

The video signal is applied to the 3.58 MHz bandpass filter. The bandpass filter passes only color information. This color signal is applied to the first and second color amps. The gain of the first color amp is controlled by the automatic color control circuits

(ACC). The control voltage is applied to the ACC detector to track color and luma level.

The output of the first chroma amp is applied to the second color amp. The gain of the second color amp is controlled by the color control voltage. The color killer detector connects to the second color amp. When a black-and-white picture signal is received, the color killer circuit turns off the second color amp. But, when a clean color is received, the color killer is turned off letting the color signal pass to the color demodulator circuit.

The chroma signal is demodulated with the 3.58 MHz voltage-controlled oscillator (VCO). Then the automatic phase control (APC) compares the phase of the VCO and color burst signal, resulting in an error-phase signal. This error signal keeps the VCO locked to the burst signal.

The chroma demodulator signals are connected to the bias and color amp CRT drivers. The demodulated signals may be signal traced with a color-bar generator and scope. Many of the latest TV chassis have the color amps located on a separate CRT board (Fig. 4-17).

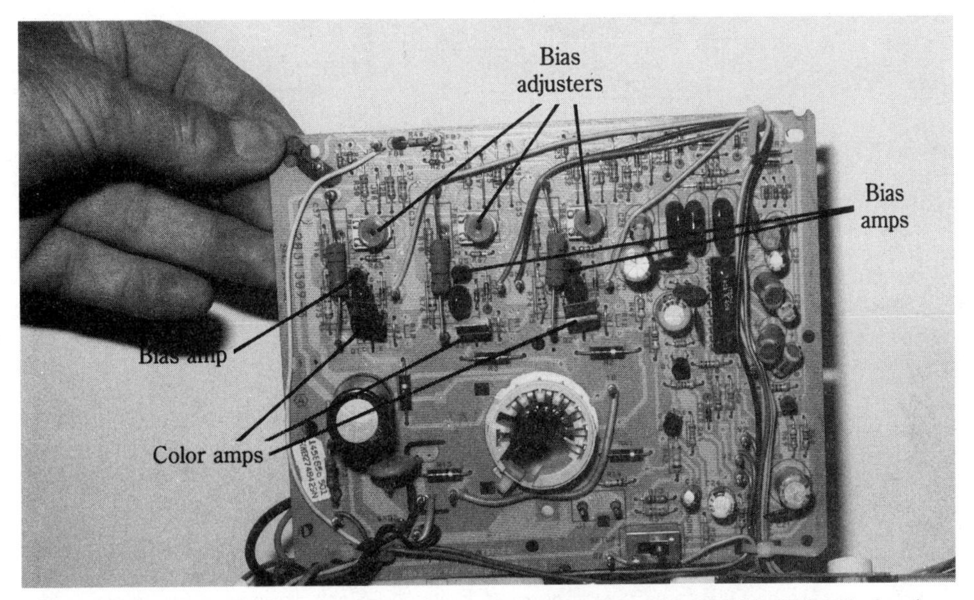

4-17 The color bias and amp drivers are on the CRT board in the RCA CTC 131 chassis.

Solid-state CRT bias and driver color circuits

The solid-state bias and driver amps may consist of separate transistors that apply luminance and color to the picture tube. The G-Y, R-Y and B-Y signals from the chroma processor are applied to each color bias transistor. The luminance signal is applied to the emitter terminals of the green, red, and blue bias transistors (Fig. 4-18). This allows the luminance information to be applied to the demodulator chroma signals.

The color amp drivers are operated in a common-base circuit. The green, blue, and red bias voltage is applied to the emitter and the output color drive signal from collector terminal applies directly to the respective color cathode in the color CRT gun assembly. Suspect the respective color driver amp transistor when one color is missing or the defective CRT. A leaky color amp may make the face of the picture tube have a dominant color.

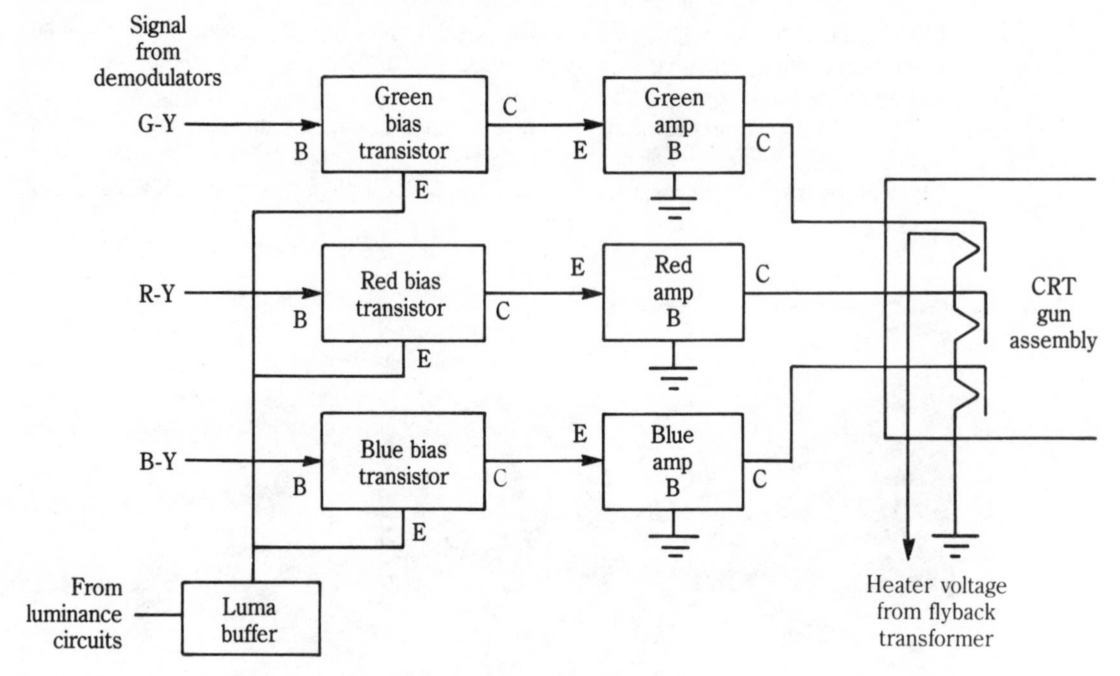

4-18 The bias and color-drive transistors provide color and video signals to the cathodes of each color gun assembly.

Convergence

Convergence (block Q) is mostly a problem with three-gun picture tubes. In the old-fashioned black-and-white picture tube, or for that matter, in the Trinitron color tube in some imported sets, the problem of convergence is greatly simplified. Convergence means aiming two or more beams so that they converge and meet at one and the same spot. As will be recalled from the description of the three-gun color tube the aperture mask behind the color screen has one pinhole behind each group of three color dots (triads). Not only must the three electron beams go through the same pinhole, but they must be so positioned and directed that the red beam, after passing through the aperture, strikes only the red color dot, and not one of the other two. Similarly, the blue or green gun must be so oriented that only the blue or the green dot is excited. It is the critical function of the convergence circuits and adjustments to ensure this if true color separation is to be realized.

It should be noted here that the basic compatibility between the three color guns, the single shadow mask aperture and the three color dots just in front of this aperture is ensured by the original construction of color tubes. This, however, does not obviate "live" in-the-receiver adjustments and corrections.

There are two types of convergence adjustments in current three-gun color TV sets. One is called static convergence, the other dynamic convergence.

Static convergence

Static convergence of the three beams is a mechanical adjustment of the positions of small magnets on the neck of the picture tube. Its purpose is to have the three beams converge at the center of the screen only. Stated simply, the purpose of static convergence is to ensure that the three guns, starting (necessarily) from three different positions at the back end of the tube, produce one and only one spot when pointed to the center of the screen. This is the correct starting point for dynamic convergence.

Dynamic convergence

The need for dynamic convergence stems from the fact that the three beams travel different (longer) distances when going to the far corners or edges of the tube than when going to the center of the screen. Figure 4-19 shows what a crosshatch pattern would look like as a result of the difference in the paths the beams travel. Furthermore, since the length of the path from the gun to the screen is different for each spot on the screen (the shortest being to the center, the longest to any of the corners) it is necessary to continuously adjust the convergence forces as the beams sweep the screen. This is accomplished by dynamic convergence, that is, continuous convergence correction during sweep.

In most color TV sets, all dynamic convergence circuits and adjustments are grouped together, usually on a discrete circuit board, often called the convergence board. While the required continuous correction takes place electronically, it is first necessary to preset a number of controls (two or more for each color) so that the automatic electronic correction will be just right. In other words, dynamic convergence adjustments are intended to preset initial conditions from which the automatic circuitry takes over.

Due to the complexity of convergence adjustments, the procedure may not be within the ability of the majority of beginners to perform. This is due not so much to the technical difficulty as it is to the need for some very specialized test equipment. There is at least one TV manufacturer, however, who incorporates such "specialized test equipment" into its TV sets and provides lucid instructions for convergence adjustments. But, even for the majority who will not be able to perform this function themselves, understanding the whys and hows will aid them in understanding what a professional TV servicer is doing.

In illustrating the phenomenon of nonconvergence, Fig. 4-19 shows a crosshatch pattern. In Fig. 4-19A, the lines are essentially straight to the very ends, while in Fig. 4-19B noticeable curvature appears and increases the farther the line moves from the center, both in the horizontal and vertical directions. In an actual color picture improper convergence will be accompanied by some color distortion and fuzzing near the ends of the lines. But for proper examination and correction, a test instrument capable of pro-

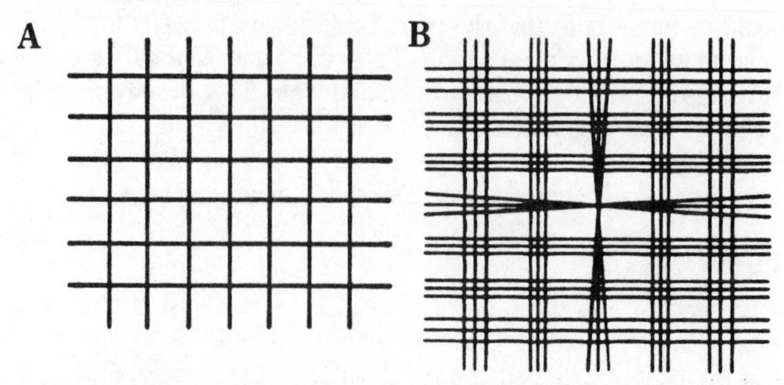

4-19 Crosshatch pattern A shows ideal conveyance both at center (static) and corners (dynamic). Pattern B shows good static conveyance but misconveyance at the edges.

ducing a crosshatch pattern on the screen, as in Fig. 4-19 is required. Such is the specialized test equipment referred to earlier.

The convergence procedure involves a sequence of adjustments of convergence board controls while observing a crosshatch pattern (from the test set) on the screen. Each color and each position (left, right, top and bottom) are individually adjusted until the overall crosshatch is as linear as possible. Some sets have two sets of controls, one for blue, the other for red and green combined.

From the viewpoint of the beginner, it should be added that convergence problems are neither frequent nor chronic. They may be required when a picture tube is replaced, but seldom otherwise; however, should the need be visually apparent, you know what is needed. We mentioned earlier that one of the color TV sets on the market has such a color signal generator built in. It should be added that the adjustment procedures for this set are quite specific, as they should be, because they serve a particular set and can point to each control by symbol and location, and outline step-by-step procedures for the complete convergence adjustments for this particular set only, of course.

Fixed static convergence

For some time now, improvements in picture tube design as well as in deflection techniques have made static convergence obsolete so to speak; thus, the starting point for convergence adjustments is the dynamic type—no static adjustments are required. In addition, dynamic convergence has been simplified. Since these changes are very much peculiar to each set, no general procedures can be given; the service information for each particular model should be followed for detailed procedures.

Color picture tube circuits

Block R in our typical color chassis (Fig. 4-1) is the color picture tube and its associated controls. There are a number of adjustments and controls, usually two for each color gun, although the red gun may sometimes have only one. There are other adjustments associated with the picture tube, but these are on the yoke and other assemblies on the tube neck, which we return to later.

At this point, we concern ourselves with the controls relating to the color aspects of the TV set only. The two groups of adjustments are the screen and drive adjustments. In most receivers, there is a screen adjustment for each color—red screen, blue screen and green screen; however, there may be only two drive adjustments, one for blue, the second for green. The design of the more recent color tubes and receivers often eliminates the need for the red drive adjustment.

In spite of the fact that these controls, whether six, or five, or even fewer in some sets, are marked with particular color names, their adjustments are best made for a balanced black-and-white picture. These adjustments are often referred to as the *gray-scale* adjustments, meaning that they are set so that the complete range or scale of light is reproduced on a monochrome scene. Once this is done, the color picture should require little, if any, touching up for the best color balance.

As to the basic need for these adjustments, this stems from the fact that the three-gun picture tube is in fact three fairly independent tubes with some variations in characteristics between them. Adjustments are intended to compensate for and equalize these differences, as well as for some differences in the three signals reaching these three guns. Furthermore, since the correct proportions of the three colors are required to achieve a pure black-and-white picture, correct adjustment for such a picture, therefore, implies that the color balance is correct.

Screen and drive adjustments, although interdependent, can nevertheless be made one set at a time. First, the receiver brightness control is set for a fairly middle-to-dim level with the color control to minimum. Next, each screen control is advanced to the point where a trace of color appears, then backed off to a point just beyond where this trace of color disappears. Next, the brightness control is advanced to what is considered normal brightness and the drive controls adjusted for a full gradation from dark to white on the screen, without any pitch black or glaring white patches showing. In other words the complete range of light to dark should exist. A second touchup of the screen controls may be required, depending on the end results.

If this is carefully carried out, there should be good color balance when a color picture is received, although some final touching up may be indicated if the color does not seem to be balanced. To the beginner, a balanced color picture is best described as what seems to the eye to be a most natural picture.

Perhaps it might help to indicate what is not a balanced color picture. If, for example, a color picture has a bluish hue (or greenish or reddish), regardless of whether it is sky, or skin, or grass, obviously there is an excess of blue (or green, or red, as the case may be) in the picture. It is actually possible to create such an artifically tinted picture by advancing one of the screen controls and observing the change.

A word of caution is in order, however. If you want to make the test, you can best ensure restoration of the color balance to its original state by observing, and perhaps marking, the position of the control about to be changed. Then it is only necessary to reset the control to its original position to obtain the original color balance.

Incidental to the discussion of color balance it is appropriate to comment on artificial settings of the color balance circuits for the sake of particular or special effects. It should be remembered that the color TV set, like a painter, starts with three primary colors, then mixes them in proper proportions to obtain any desired color; however, while the artist in order to achieve the exact hue or tint may have at his or her disposal many more than the three basic colors, not so the color TV tube. Unlike the artist's pallette, there are only three colors to start with. This necessarily limits the versatility of the color tube. While there are many other limiting factors, this is the major one.

For best overall results, the proportion of the three colors is such that all white or all blue or all green are easiest to reproduce, while gradations and shadings are not so easy to reproduce accurately. In practice, the eye is very tolerant on color shades, except on such tints as flesh tones. While it is possible to adjust the color balance of a set to favor flesh tones, such favoring is usually achieved at the expense of color balance; that is, the general background (or some not so intended objects) will take on the characteristic flesh tones. The clue to the problem lies in the favoring. If for technical (design transmission, etc.) reasons flesh tones do not look natural on a properly adjusted color receiver, misadjusting the color balance circuits to favor the flesh tones will invariably distort the color balance on most other tints. It is like misadjusting the tone control on a radio receiver or phonograph to obtain better bass (low-frequency) response by cutting off the high and medium frequencies essential to balanced sound.

Another form of color imbalance is poor purity, which produces tinted raster of various colors in a black-and-white picture and incorrect hues in a color picture. Most cases of poor purity are caused by the TV set's faceplate becoming magnetized from stray external magnetic fields. Stereo speakers adjacent to the TV cabinet, small radios or stereo units on top of the TV set, and turning off the vacuum cleaner next to the TV screen are some possible sources of external magnetic fields. In the majority of cases, poor purity can be corrected without removing the back cover of the TV set, by degaussing (a procedure using a coil designed to demagnetize) the TV faceplate. Figure 4-20 shows a cluster of magnets around the neck of the picture tube; they are the purity magnetic assembly. These round magnets are locked together and preset; do not adjust or move them unless you have the knowledge and test equipment to do so.

4-20 Do not adjust the fixed magnets located on the neck of the picture tube unless you have the knowledge and the correct test instruments.

5

Introduction to troubleshooting

IT IS NOT OUR INTENTION TO TELL A BEGINNER IN TV REPAIR WHAT TO DO WITHOUT also attempting to explain why something is to be done, or why a certain procedure should be followed. Unless you first understand what a tube, a transistor, or an assembly is supposed to do, basically, your chance of success in effecting a correction or repair is much smaller than it could be. That is also what I mean by potentially unsuccessful *fixes*.

While it is possible for you to look on a chart of symptoms and find the one that seems most like that in your TV set and follow the suggested remedy and actually get the desired result, the likelihood of such luck is rather small; however, your luck will improve in direct proportion to your familiarity with your set and your understanding of how it functions when operating normally. It is not impossible for you to look up a fault in the index, which will refer you to a certain page of this book, and by following the instructions end up with a *fixed* set. But your chances of continued success are better if you thoroughly acquaint yourself with the material in the first four chapters before tearing into your set.

Equipment

While it is not our intention to suggest that beginners equip themselves with a set of professional instruments and tools just to be able to do some simple repairs, there are a few necessary aids needed to undertake such repairs, in addition to the usual simple tools found in every home, such as screwdrivers, pliers, etc.

Multimeters

For checking continuity of antenna cables, speakers, tube filaments, and coils, select a low-priced digital multimeter (DMM) or volt-ohmmeter (VOM). You can find many uses for these small, low-priced test instruments around the house. Besides taking continuity tests, you can check resistance, voltage, and current measurements. With the latest DMMs, you can even check transistors, diodes, and capacitors (Fig. 5-1).

The typical DMM has a 0 to 1500 dc volt range, 0 to 100 ac volts, ac or dc current from 200 μA to 2 amps, resistance measurement of 200 to 20 megohms and a diode tester. The diode test may check a leaky or open diode or the transistor-junction tests. The transistor-junction test indicates if the transistor is leaky, open, or has a high internal resistance junction.

The typical low-priced VOM has a 0 to 10 kilohm resistance scale, 0 to 1200 volts ac and dc, and a dc current range of 60 μA to 300 mA. The VOM may load down the circuit being measured while the DMM does not. The DMM can measure a fraction of a volt while the VOM is not that accurate. So, if you are going to purchase a new test instrument, make it a pocket DMM.

The DMM with everything, may have besides the voltage and resistance measurements a frequency counter, capacitance tester, logic tester, transistor tests, diode tester, and audible continuity alarm. The frequency counter capability may extend up to 200 kHz with resolution of up to 1 Hz. The capacitance range may extend up to 20 microfarads. Bipolar transistors can be tested for H_{FE} gain from 0 to 1000. The audible signal can be used with the diode and continuity tests.

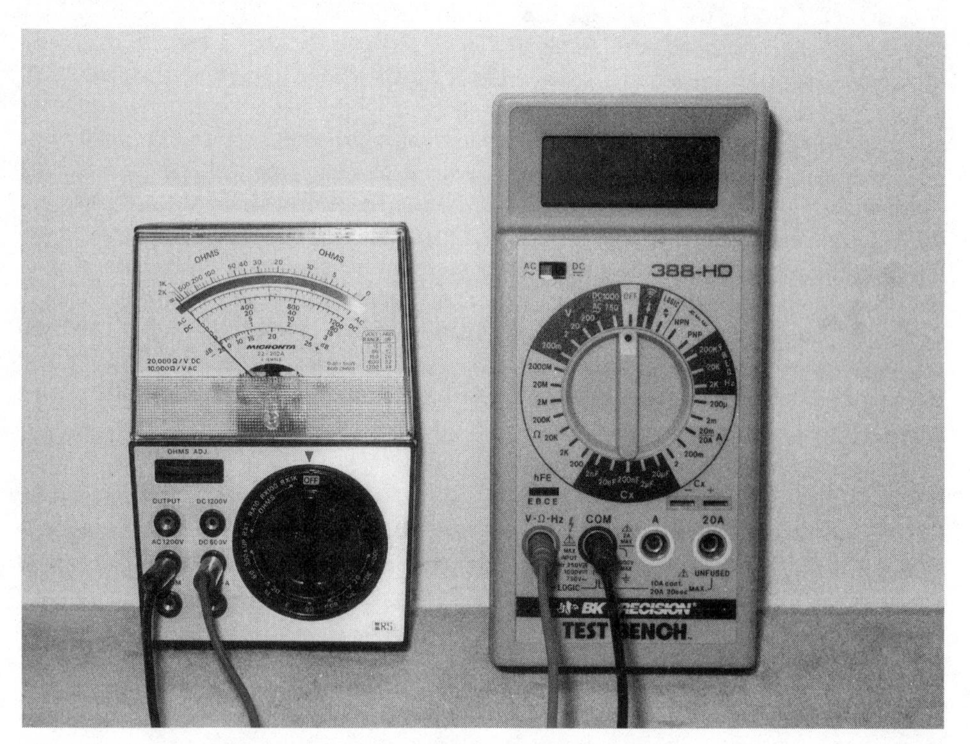

5-1 A multimeter, either analog (pictured at left) or digital (at right), is handy when making tests on the TV or around the house for making continuity tests.

Hand tools

The first of these aids is a small soldering tool, about 100-watt capacity, preferably one of the instant heating type commonly called a soldering gun. This is not to suggest that you immediately attempt to unsolder and solder parts of your TV set. It is rather a form of insurance in case of *accidents*. For example, if during a simple tube change or adjustment you unintentionally break an interconnecting lead (a number of these seem to be *floating* around the back of every TV set behind the cover), you should be able to repair it without having to run for help. With a soldering gun you need some solder—rosin core type only—never acid core or plumber's type, and preferably of the "thin wire" type. A half-pound spool of this stuff costs a couple of dollars or so in a radio supply house and will last almost a lifetime (unless you are in the business full time).

The second piece of equipment that is highly recommended is a complete set of tubes for your TV set (the picture tube, of course, not included). In hybrid sets (some transistors, some tubes) this advice still applies as far as the tubes in the set are concerned. The transistors are considered later as parts of board assemblies. A set of tubes involves some cost, of course, but actually it involves no additional expense, since defective tubes have to be replaced anyway. The great advantage in having a complete set of spare tubes is most obvious if you have a transformerless set (which are fairly popular) as we shall see presently, but it is no less convenient with transformer-operated receivers. Since the small tubes in a TV receiver will probably have to be replaced at least once during the life of the set, having a set of spares on hand actually involves no additional cost at all.

The third item of great help to any do-it-yourselfer is a schematic diagram or set of diagrams and other service data for your set. This can sometimes be obtained by writing to the manufacturer, giving the model of the set and the chassis number. This information is always printed or stamped on the back cover of the set or the back apron of the chassis. Sometimes the chassis identification is also given on the tube-transistor layout chart usually found inside the cabinet.

There is another and much more readily available source of service information. Instead of writing to the manufacturer, schematics, layouts, adjustment instructions and other helpful hints can be bought in almost any radio supply house. While most users of this book may not be familiar with schematic diagrams, these nevertheless are very useful because they give tube and transistor identification numbers, their functions as well as their relative physical locations in the chassis.

The last item of equipment is a substitute electric cord and plug, known as a *cheater cord*. The name is appropriate, since it enables its user to cheat or defeat the protective system designed into nearly all TV sets, namely prevention of the uninformed person from reaching into a potentially dangerous TV set with the power on. Let me add, however, that it is quite safe and permissible if you observe reasonable precaution. It is my intention to properly inform you regarding the dangers lurking behind that cover.

The safety interlock in the vast majority of TV sets consists of the following. The ac line cord is physically tethered to the protective back cover of the set. When the cover is removed the ac cord is automatically pulled out of a two-pin male plug on the TV chassis. To operate the set with the back removed requires a line cord identical to the one attached to the cover. This is the cheater cord, and it's available from radio supply houses for a dollar or so.

In a number of recent TV sets, the cord is not fastened to the back cover. Instead, it can be unplugged from the back of the set before the back is removed. If this is not done first, the cord will automatically disconnect from the chassis when the back cover is taken off. In either case removing the cover also disconnects the cord, and it takes a deliberate action to plug the cord back into the chassis without the cover. This still ensures that the TV set will not be accidentally exposed and operating; however, in this case, a separate cheater cord is not required.

Today, many TV chassis with adequate fuse protection have the power cord wired directly into the TV chassis. When working on one of these type chassis, remember to always pull the power cord before touching any component connected to ac or dc voltage. Extreme care should be exercised while working around a hot TV chassis.

Before proceeding with actual troubleshooting procedures, it is necessary to outline and explain a very useful and very simple procedure which is applicable to a large majority of TV repairs. This is the tube substitution method of TV troubleshooting. Earlier it was indicated how to carry out tube substitution the hard way—remove a suspected tube from its socket to check it on a tube tester. Of course, if you don't have a tube tester, that means going down to the neighborhood store displaying a do-it-yourself tube tester, checking the tube and buying a new one if necessary. If you have a complete set of tubes for your TV set, you can troubleshoot and repair-by-substitution much more easily and quickly.

Basic transistor troubleshooting

Locating and replacing transistors in the radios and TVs may be easy for some people and rather difficult for others. Simple diode and transistor tests in the power supply, horizontal circuits and audio stages may locate the possible defective part with the digital multimeter. Of course, digging into the video, sync, AGC, vertical, color and luminance circuits are quite complicated and should be left up to the electronic technician.

For example, if the audio stage is dead in the TV set, you can make several continuity tests with the DMM. Check the continuity of the speaker for a low ohm measurement (Fig. 5-2). No measurement indicates the speaker cone is open or a defective cable and soldered connection. The speaker is probably good with a 2- to 50-ohm measurement. The normal 8-ohm speaker may measure around 7.5 ohms while the 32-ohm voice measures around 30 ohms.

If the speaker is normal, check the coupling capacitor between speaker and solid-state circuit by shunting a good electrolytic (100 μF) across the suspected one (Fig. 5-3). Output transistor and audio IC components may be checked by taking voltage, resistance, and transistor tests with the DMM.

Besides taking voltage measurements on the transistor elements, a base bias voltage measurement may indicate if the transistor is normal. The base bias voltage is less than 1 volt and is taken between emitter and base terminals. The DMM is ideal for this measurement. For a silicon transistor, the normal voltage measurement is 0.6 volts and germanium transistors have a 0.3-volt reading (Fig. 5-4).

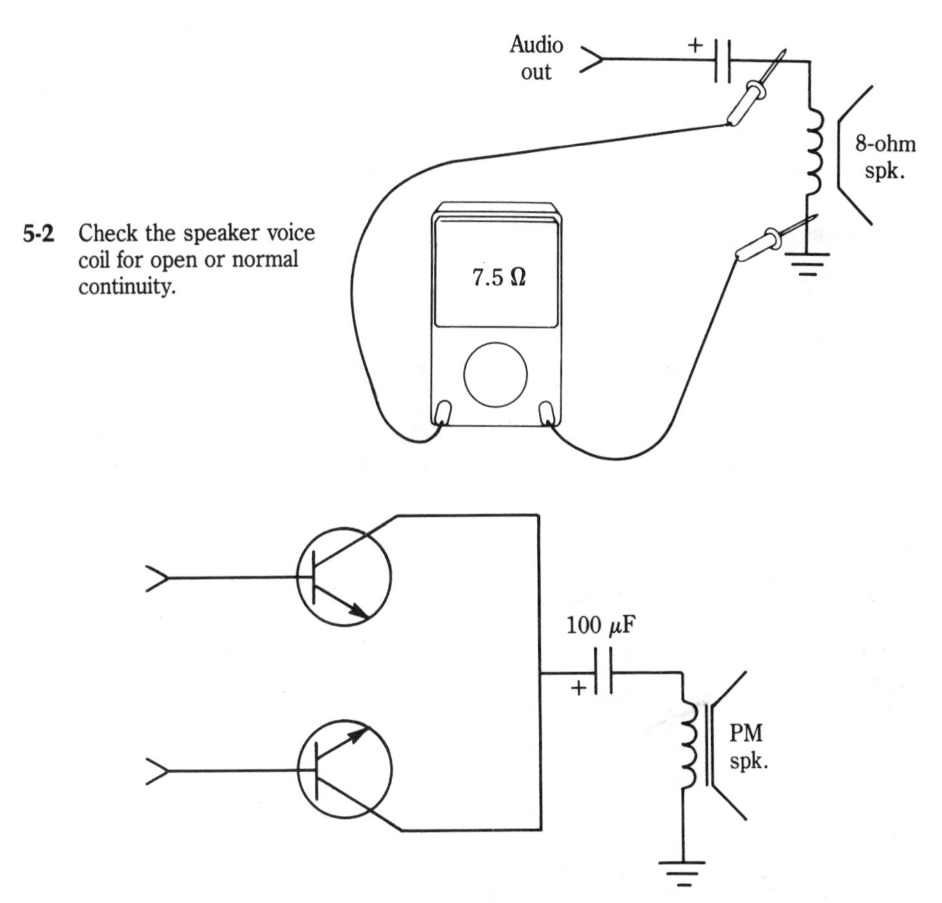

5-2 Check the speaker voice coil for open or normal continuity.

5-3 Defective transistors or coupling capacitors can be located with a DMM by taking voltage measurements.

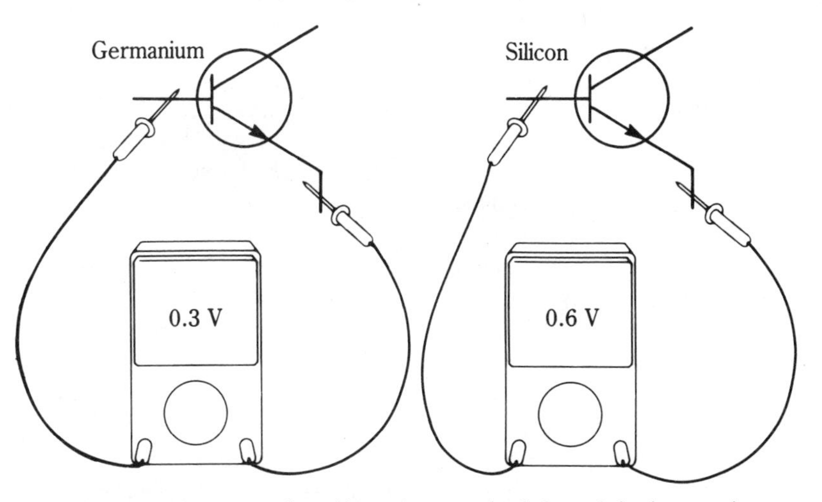

5-4 Measure base-to-emitter bias voltage to check for a defective transistor.

If the voltage measurement is zero between these two elements, the transistor is open or leaky. A higher voltage measurement between elements may indicate a high resistance junction inside the transistor. For more voltage and transistor junction tests, see chapters 10 and 11 (Fig. 5-5).

5-5 Transistors come in many different sizes and shapes in the TV chassis.

Where do you start?

In troubleshooting the tube, hybrid, or solid-state TV chassis, begin with the symptoms from the picture tube or speaker. No picture, raster, or sound may be caused by the power supply or horizontal circuits. First check the fuses for open or blown conditions. Next, measure the voltage across the large filter capacitor (Fig. 5-6). Does the tube light up? In a series tube chassis, one tube may be out, causing the whole string to go dark.

If the sound is normal with only a thin, bright horizontal line across the raster or screen, this indicates there is no or improper vertical sweep (Fig. 5-7). Vertical circuits are a little tricky to locate the defective component. But you can take transistor voltage, resistance, and transistor tests on each transistor in the vertical circuits.

The symptom may be no sound nor raster, but the tubes light up or voltage is measured at the low-voltage power supply. Since neither sound nor raster or sweep are indicated on the screen, check the horizontal circuits. Measure the voltage applied to the horizontal circuits. Do not check the voltage on the output transistors with any small meters.

5-6 Taking a voltage measurement across the large filter capacitor can indicate if the power supply is normal.

5-7 Nothing but a bright, thin, horizontal line on the screen indicates trouble in the vertical circuits.

Since the horizontal output transistor causes the most trouble in the TV chassis, you can check from the collector (shield) to chassis ground for a leaky output transistor (Fig. 5-8). The resistance with the DMM placed on the diode test should be somewhere between 0.450 to 0.550 ohms. If the resistance is lower, or less than 1 ohm, suspect a shorted output transistor or damper diode. Test the transistors in the circuit with the DMM.

If the picture is normal without any sound, go directly to the audio section. Most audio problems are developed in the audio output stages. Check the speaker continuity and output transistor tests. The sound stage may be normal in a no raster/no picture/no audio situation because in most chassis, the horizontal circuits must function before the audio stages can work.

5-8 A diode resistance measurement from the metal of output transistor (collector) to the chassis could indicate a leaky or open transistor or damper diode.

Always, start to check the TV chassis by a trouble symptom shown upon the screen or by the speaker. Feel components to see if they're overheating, such as transistors, resistors, capacitors and IC components. Smelling overheated or burned components may turn up a defective part. Don't forget to place the DMM in action to help locate the defective component. It's best to leave high-voltage and picture tube circuits alone, unless you have the know-how or test equipment to make these tests.

Tubes

The following section explains some basic tube troubleshooting techniques.

Parallel tube filaments

Figure 5-9 is a typical drawing of a parallel tube hookup. Only the heaters or filaments (the elements that glow when the set is switched on) are shown. Notice that every one of the tubes is connected independently to the source of current, most commonly 6.3 volts ac. An incidental clue to such a parallel tube hookup is the fact that the tube numbers (except the picture tube) all begin with the same number, usually 6 (such as 6AH8, 6GH8, 6BG6, etc.), but sometimes 12 (for example, 12AH6, etc.).

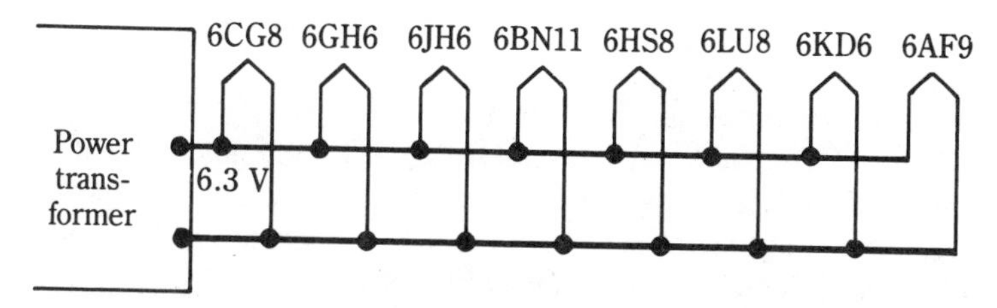

5-9 Diagram of tube filaments (or heaters) wired in parallel. The heavy lines in effect supply each tube individually. Thus, each heater operates independently.

Since each tube is connected to the source of the current, without depending on any other tubes to complete the path it becomes obvious that pulling any tube out of its socket will not prevent any of the remaining tubes from glowing or lighting; thus, when a particular tube is suspected, it can be pulled out, a new one of the same type plugged in, and the suspicion confirmed or disproved. In case of a burnout, a tube fails to light and can easily be spotted. In certain tubes, a positive visual observation or absence of light is not feasible, either due to the coating (the coating is a by-product of its being a vacuum tube and has no other significance) inside the glass, or because of some physical obstruction such as a metal sleeve (called a tube shield). But by carefully touching each tube with your bare hand, you can determine its condition by the warmth (sometimes a very high temperature) or coolness to the touch. By contrast, none of these tell-tale clues are available with a transformerless, series-string tube hookup. Other measures, therefore, are necessary.

Series tube filament

Figure 5-10 is a typical wiring diagram of heaters in a series hookup. Observe that the tubes follow each other, chain-link fashion. This is called *series wiring* (like the older or inexpensive Christmas-tree lights). As in the case of parallel hookup just described, the tube numbering is a reliable clue. In a series hookup there will be quite an assortment of tube numbers, such as 3AH8, 18BQ7, 14SH6, 12GD9, etc. Whenever any of these tubes is removed from its socket, or whenever one burns out, the circuit is opened and all tubes go out. There is absolutely no indication which of the tubes is the culprit, or for that matter, whether more than one is involved. While modern tube burnouts are not too frequent, this does not help when a burnout does occur.

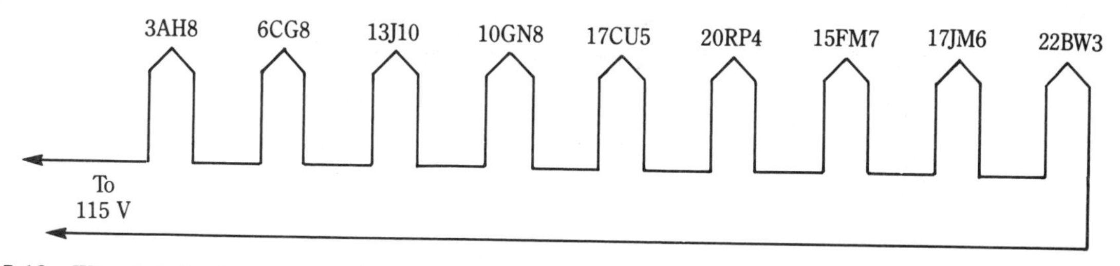

5-10 When tube heaters are wired in series, they depend on one another for power. If one goes out, the circuit is broken and they all go out.

Incidentally, it is the nature of filament circuits that burnouts are more likely to occur in a series-wired set where detection is more difficult. In the absence of a spare set of tubes the only practical procedure is to remove all tubes and check them in a store tube tester. With spare tubes on hand, one-at-a-time replacement can be made until the offender is found. Of course, it is just possible that the very last tube so tested is the defective one, but it is equally possible that the first, or second, or third tube pulled out is the culprit. The advantage of spares on hand is obvious.

In connection with series-wired tubes, it must be mentioned that there is one other possibility where all the tubes fail to light when none of them is burned out. This will happen when a protective device, called a fusible resistor or surge protector, burns out. As we mentioned earlier, tubes in a series-string circuit are more prone to burnout and shortened life than the same types of tubes in parallel-wired or transformer-type TV sets. To reduce this failure hazard, many manufacturers have incorporated in series-wired TV sets a limiting (or surge absorbing) resistor, wired in as if it were one additional tube in the string; however, since it is not as obvious as a tube and often much less accessible, it can be overlooked.

While some beginners may consider this task beyond either their present ability or ambition, others may be quite up to replacing the resistor themselves; therefore, here is a simple procedure for locating and replacing this resistor. Incidentally, this so-called fusible resistor is sold by radio supply houses, often in quantities of three or four in a package, for about a dollar or so. It is not a very critical part, and no precise TV set make and model need be known to buy one of these; a number of different makes and models use the same part. Figure 5-11 shows a simple diagram of the wiring from the ac line cord to the on-off switch to the surge resistor and tubes. This resistor is almost always located immediately after the on-off switch and in most sets is located in plain view and reasonably accessible.

In order to determine whether or not the resistor is burned out, disconnect the receiver from the wall outlet and remove the rear cover. Should it be required to remove

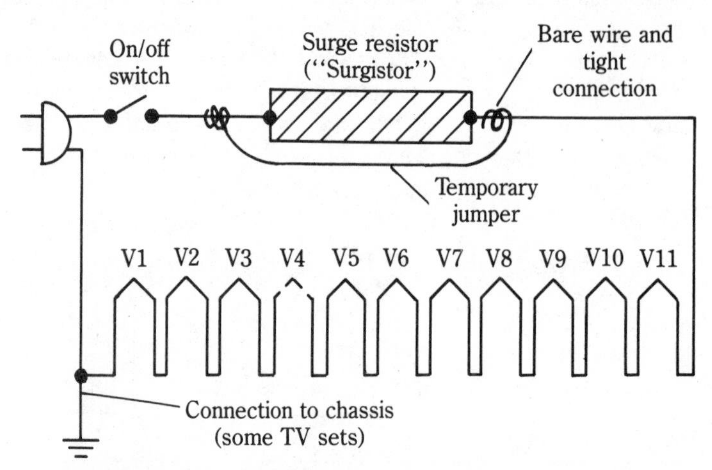

5-11 A typical series string filament circuit. The on-off switch must be on, and all tubes and the surge resistor must be good for all tubes to light up.

the chassis from the cabinet (this is usually not the case), the picture tube and the speaker are unplugged, leaving the chassis free.

Most service data clearly identifies what connectors exist between the chassis and the speaker and picture tube. Usually, there are two separate wires from the chassis to the speaker, each with a quick-disconnect device. Sometimes these two wires may interconnect through a two-pin plug. The picture tube usually has a multipin plug for the same purpose, and in some color TV sets, two such plugs. In addition, there is a high-voltage lead going to a snap button on the picture tube glass. This also must be disconnected, preferably a few minutes after the set has been switched off to allow time for the residual high-voltage to dissipate. Otherwise, an unpleasant jolt may be in store for the bare hand touching this lead. Incidentally, a large number of TV sets have a quick-disconnect on this lead at the chassis end, making it so much easier to handle. This applies to those receivers in which the picture tube is mounted to the cabinet. In some sets, however, especially some of the portable type, the picture tube is mounted on the chassis, and no disconnecting of any picture tube wires or plugs is required.

To determine whether the surge resistor is burned out, the professional TV servicer makes a continuity check with an ohmmeter; however, it is perfectly safe to *bridge* this resistor for test purposes (connecting the two ends together, as shown in Fig. 5-11), then observe whether or not the tubes light. If they do light, the resistor is burned out and should be replaced. If they do not light, one of them is burned out.

A flexible wire with alligator clips on each end is ideal to temporarily clip around a suspected surge resistor or thermistor. These clip on wires can be purchased or constructed with insulated flexible wire and two small alligator clips.

Series-parallel tube filaments

Before concluding our discussion of tube replacement in series strings, consider special series-parallel filaments found particularly in some older receivers. So far, to avoid complicating things, we have not mentioned the fact that in a series string (as in Fig. 5-11) the same current flows through all the tubes and that the tubes are purposely designed and chosen so that all tubes in the string, regardless of tube number, require the same current. In contrast, each tube in a parallel-filament TV set may (and sometimes does) carry a different current than its neighbor; they are independent of each other in this respect. Not so in the series string arrangement.

Let us again look at Fig. 5-11. When the string is complete, that is, all tubes light, the current is continuous and the same through the whole string. If, for example, tube V4 (shown dashed in Fig. 5-11) should burn out, all tubes will be out; however, consider a series string circuit of the type found in some of the older sets. Figure 5-12 shows a simplified version for the sake of explanation. Notice that V4 and V5 are, so to speak, together. Should one of these two tubes (say V4) burn out, all tubes will continue to light, and the second tube of the pair (V5) in this case will be overloaded and most likely will burn out or be permanently damaged. This type of hybrid tube hookup was employed by some manufacturers because there was not available a full choice of tubes for a simple series hookup as in Fig. 5-11.

In troubleshooting a TV set of this type, it is necessary to have the diagrams (mentioned earlier) showing the tube hookup. Before removing any tube for possible replacement, it is necessary to look at the tube heater hookup, which is always shown in the diagram. Any tubes connected like V1, V2, V3, V6, V7, or V9 in Fig. 5-12 may be removed, tested, and replaced without any fear of damage. Any tube hooked up in pairs, like V4 and V5, should be removed in pairs; that is, the set switched off, both tubes removed, each tube replaced in turn, then the set switched on to observe the effect of the substitution.

A final suggestion on hybrid tube wiring: in some (very few) obsolete sets, a tube or two may be paired not with another tube, as are V4 and V5 in Fig. 5-12, but with a heater element (called a resistor). The circuit might look like V8 in Fig. 5-12. Observe that should V4, V5, or V8 burn out or be removed from their sockets, the remaining tubes will still be able to light, but the heater element marked R or the second tube of the pair will be carrying all the current and may be damaged as a result. In such cases, the safe and proper procedure is to switch off the set, remove the tube or tubes, substitute others, then switch on the receiver and observe the effect of the tube replacement.

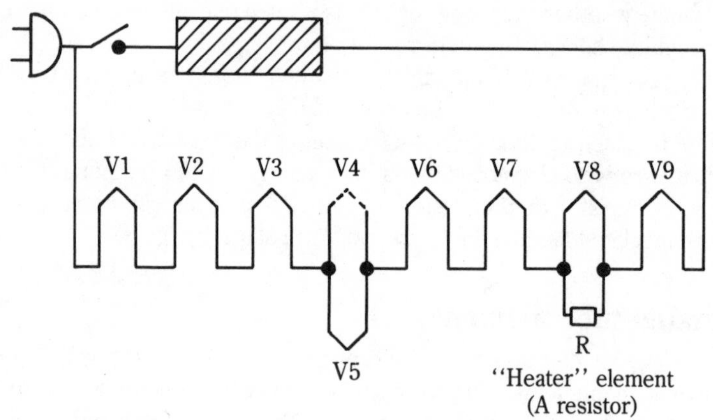

5-12 Some older TV sets have a series-parallel filament circuit, where two tubes V4 and V5 are connected in parallel or a resistor is placed across a tube filament.

Tube pulling

Since it is assumed that you have no experience in even the relatively simple task of tube changing, the following instructions should further simplify this task. In addition, these precautions will save time and prevent damage. More importantly, they may prevent potential (minor) injury as well as discouragement.

Because of the prevalent practice to cram the maximum amount of electronics into the minimum possible space, some tube locations are often less than ideal for removal or replacement. This requires extra care in removal and replacement of tube shields and tubes. The general purpose of this admonition is twofold. First, the removal of a tube must not affect or damage any adjacent part. Delicate components, fragile leads and potentially shock-producing (exposed) voltages may lurk in the immediate vicinity of the tubes. Plain care and common sense are all that is required to ensure that when the tube shield and tube are handled, other components in the vicinity are left undisturbed. Second, proper care must be exercised both in removal and replacement of tubes so that tube pins are not bent or broken, sockets are not damaged, or, what is less obvious, that socket connections are not loosened by careless or forceful handling. This is easier done in some cases than in others, as I shall show presently.

Octal tubes

Octal tubes generally are shown as the older generation. They are glass or even metal (rare) tubes with bakelite bases. Figure 5-13 shows the base and matching socket of an octal tube. Although they are called octal tubes, they may not have all eight pins. But whether the tubes have five, six, seven, or eight pins, they are all laid out on a symmetrical pattern of eight around a circle. The octal socket, by contrast, and for reasons of simplicity in manufacturing and stocking, has all eight pinholes, with connecting lugs underneath. Any unused lugs (as when a five-pin tube is to be plugged in) are simply left blank or used for anchoring or supporting other electronic parts in the vicinity.

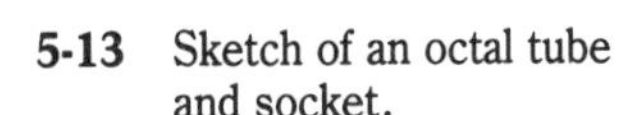

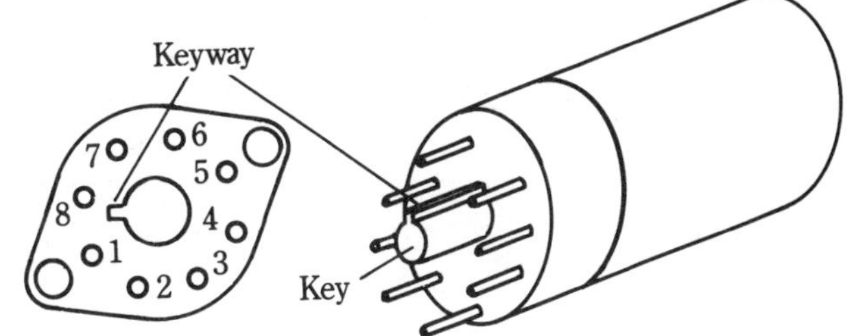

5-13 Sketch of an octal tube and socket.

Notice that the tube has a key (corresponding to the key on motor shafts and pulleys) and the socket has a keyway which must be positioned in one correct way only, otherwise the tube will not readily plug into the socket. Here is a good example of what is meant by plain care and common sense during tube removal and replacement. It is quite possible, with a little force, to wear damage the fiber or phenolic keyway in the socket and finally "get the tubes in" the wrong way. The proper procedure, and this can actually be done in the dark, that is, without seeing the keyway, is to hold the tube upright over the center of the socket. Then, and without any downward pushing whatsoever, the tube is moved slightly until just the tip of the rounded center pin barely enters the center hole of the socket. Now, slowly turn the tube until it drops a bit into the socket. (Up to now, only the rounded end of the tube center pin was pivoting in the hole.) When the tube drops, all the pins will begin to engage the corresponding socket holes. Gentle but firm downward pressure will fully seat the tube. An experimental trial with a tube and an accessible socket takes less time than it takes to describe it.

To remove the tube, a straight upward pull is used, taking care that your hand (with the tube in it) doesn't come up suddenly and hit something inside the TV set. Incidentally, a straight, perfectly vertical pull upward is not always easy, but with caution and judgment it is permissible, while pulling the tube up, to wiggle it slightly from side to side. The tube pins are strong enough and the socket pins flexible enough to permit doing this without causing any damage.

Tube shield handling

A metal sleeve over the tube, where used, is an essential part of the circuit; it is not ornamental or for mechanical protection. It is absolutely imperative that the shield be replaced and properly seated in its retaining base or clips after the tube has been replaced. Failing to replace a shield may not produce an immediate and obvious malfunction, but many a complaint of poor TV performance has been traced to the absence of shields as the result of failing to replace them after tube changes. In many TV sets, the shields on the tuner are soldered in place and are not completely removable. They are, however, collapsible to half of their normal height (by simply pushing down on the rim) so that the tube can be removed. After tube replacement, the shield is returned again to its original height.

All-glass tubes

These have been generally classified as miniature tubes and included, up to a few years ago, seven-pin and nine-pin tubes of rather small size. Since then, however, a new group of tubes includes types which often are as large as the largest base-type tubes. The method of handling is the same for all types so they are discussed as a group.

Seven-pin miniatures Pins of these tubes are made of relatively soft metal and will withstand a limited amount of bending and straightening; however, if any pins are bent in the process of plugging into the socket it can cause poor contact and resultant performance troubles. Figure 5-14 shows a bottom view of a seven-pin miniature tube

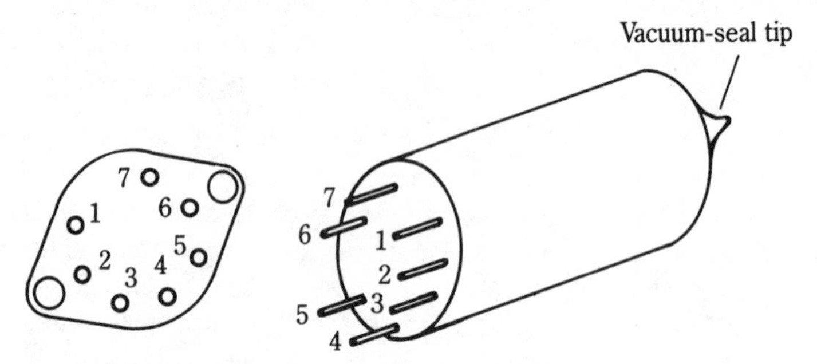

5-14 Drawing of a seven-pin, miniature, all-glass tube and its socket.

and its socket. Observe that unlike octal tubes, there is no key, no keyway, and no mechanical guide for correct insertion. The seven pins are symmetrically spaced on an eight-pin pattern, with one pin omitted. Therefore, more care is required in removal and especially in plugging in such a tube. In the latter case visual observation of the tube and socket is essential, especially the first time. As in the case of octal tubes it is advisable to experiment with a tube that is both accessible and visible. The proper procedure for plugging in such a tube is as follows:

- Align the tube so that the blank space on the tube faces the blank space of the socket. In other words, pins 1 and 7 (wide spacing) face holes 1 and 7 on the socket.

- With the tube resting on the socket (of its own weight), rotate the tube very slightly back and forth to get all seven pins to the point where they just begin to find the corresponding socket holes. This is best done by feeling rather than looking.
- Firmly push the tube straight down. If the pins are aligned the tube will move downward, although it will offer some resistance.
- Continue the downward pressure until the tube is fully seated. This, too, is done by feel, although it may also be seen.
- If the tube does not seem to begin moving downward when you first apply pressure remove it and carefully examine the pins. Due to possible improper alignment, one or more of the pins might have pushed against the socket material and become bent. Carefully straighten the pin and repeat the process.
- Now and then a gadget appears on the market which is intended to assist in plugging in these tubes. Most servicers, however, do not use them.
- As mentioned in the case of the octal tubes it is a good idea to practice removing and replacing a tube on a fully accessible socket where you can see exactly what happens. What has been said about tube shields before applies even more so here since most of the critical and sensitive functions in modern TV sets are performed by miniature tubes and the shield is an important part of the circuit.

Nine-pin miniatures Figure 5-15 shows the tube pin layout and socket of the nine-pin miniature tube type. What has been said about the seven-pin tube and socket applied equally here. Nine-pin tubes are somewhat larger in diameter, too.

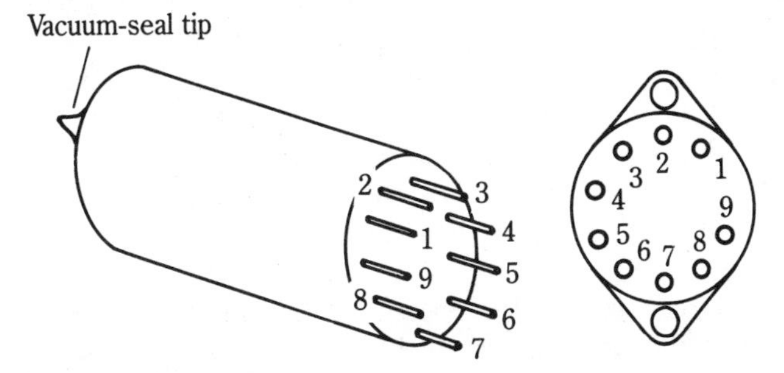

5-15 The nine-pin miniature tube and socket are similar to the seven-pin tubes.

Other all-glass tubes There are other types of all-glass tubes in use in TV sets, but these are basically no different than the two types just described. For example, a series of tubes known as Compactrons are considerably larger than the miniature types just described; in fact they are among some of the largest tubes found in a modern TV set. Their base structure differs somewhat from the miniatures, and a word of caution with regard to this difference is in order. All such types are, of course, evacuated the same as other types. But miniature tubes previously discussed have the bulb tip at the top, while Compactron types have this tip on the bottom right in the center of the pin circle as shown in Fig. 5-16.

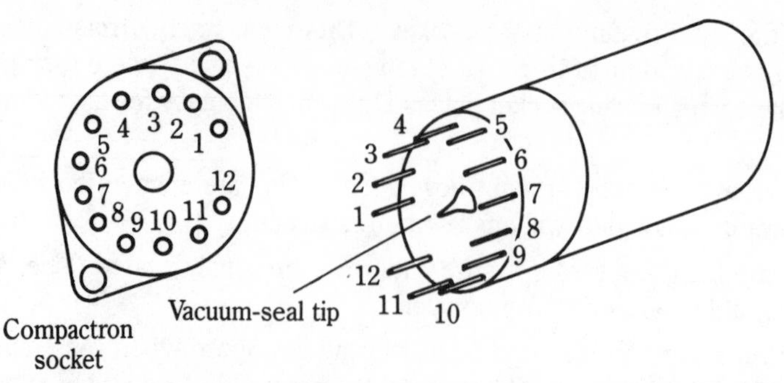

5-16 Some receivers use larger all-glass tubes, like this sketch of a compactron and its socket.

While handling this type of tube, care must be exercised that the glass tip is not accidentally broken. As shown in Fig. 5-16 the socket has a clearance hole for the tip. Nevertheless, it is possible to strike it against some object on the chassis. Breaking this tip will not cause any violent implosion but it will certainly destroy the tube.

Tubes with top caps

There are at least two tubes in a modern TV set that handle very high voltage (in the thousands). Because of this, one connection to such a tube is brought out to a metallic button (top cap) on top of the tube instead of to one of the base pins. Furthermore, in the interest of safety, the connecting top cap lead is often enclosed in a nonmetallic (bakelite, phenolic, etc.) outer shell. While the top cap button on the tube is electrically soldered to the appropriate internal tube element, the button is only cemented to the glass; quite often the cement separates from the glass, leaving only the soldered wire to hold the button. While this does not call for discarding the tube, it calls for special care in removal and replacement of the top cap lead.

To remove a connection from a loose button, first try (gently) to rotate the cap a bit. If it responds, gradually pull it straight upward while rotating it a bit. Under normal conditions, such rotation will be accompanied by a very slight scraping of metal against metal, indicating that the button on the tube is rigid while the cap is rubbing against the button. Since there is motion between the two metal surfaces, it is safe to pull the cap off; however, should this slight rotation be smooth and noiseless, it probably (even if not 100 percent of the time) means that both the cap and button are turning together because the button is no longer cemented to the glass. Further twisting is likely to break the connection to the tube element and thereby destroy the tube. In such a case, it should be possible, using both hands or perhaps a thin screwdriver blade, to hold the loose button on top of the tube while working the cap loose. If a subsequent test in a tube tester shows the tube to be in operating condition, the looseness of the cap may be ignored. If desired, any household cement may be used to attach the cap before plugging it back into the TV set.

Check the tube cap for corrosion (burned or green looking substance enclosed inside the cap area). Inspect the cap connecting wire for breaks or burned areas. The cap temperature of a horizontal output tube is quite high and may melt down the solder, producing a poor connection. If corrosion is found on the connecting wire tube caps of the high voltage or damper tubes, clean it off with a pocket knife or small wire brush. Failure to do so may result in a poor electrical connection.

Testing by substitution

Now that we have become acquainted with the mechanics of tube pulling and replacement, we can proceed to troubleshoot by tube substitution.

Remove the back cover with the line cord assembly or the line cord and the cover separately if they are not attached to each other. Connect the set to the ac wall outlet, using the cheater cord or the TV set cord if it is free.

Switch the TV set on. Observe, as far as possible, whether any tube fails to light. In case of the transformerless TV set, all or none will light. I shall outline the procedure for this type later. For a parallel-wired (transformer type) set only the burned out tube will not light. All others will light.

If it is difficult to see whether or not a tube lights, determine by the touch method which tube, if any, does not feel warm. If all tubes light or are warm to the touch, the defect is not due to a burnout, although a tube may still be defective. Be aware that some tubes operate at very high temperatures, sufficient to cause painful burns.

Determine from the troubleshooting procedures that follow which of the tubes may be involved. Remove the tube, following the procedures and caution notes. Substitute a good tube for the suspected one. In cases where more than one tube may be the cause, change only one tube at a time.

Switch on the TV set, allow a few minutes for warmup and stabilization, then observe the effect. If the tube was at fault, the set should now perform normally. If it does not, return to the section in the book describing the malfunction and continue the procedures given there.

Common-sense servicing

Many defects and defective components may be located with sight, small sounds, and touch procedures. Wiring connections may produce a burning smell or smoke rising from the various components. Water or a soft drink spilled down inside the portable TV cover may cause the pc wiring to burn. Cutting out a portion of the board and cleaning up the chassis may prevent further breakdown.

You might hear arcing around the picture tube anode or high voltage tripler unit, causing the fuse to open. Besides checking the speaker sounds, you may hear the high voltage rise with the expansion of the yoke winding, indicating high voltage is present without a raster on the screen. A high frequency noise coming from the flyback transformer may indicate the horizontal oscillator frequency is way off and needs adjustment or repair.

A close inspection of a cracked or burned resistor may locate a corresponding defective component. After replacing a defective tube or transistor, the circuits may still be dead because these components may cause voltage-dropping bias resistors to overheat. A filter capacitor with a white substance oozing from the pin connectors indicates a defective capacitor. A glowing red plate in a horizontal output tube may indicate a shorted tube, insufficient drive voltage, and defective damper circuits. Arcing signs around the flyback transformer may indicate high-voltage arcing (Fig. 5-17).

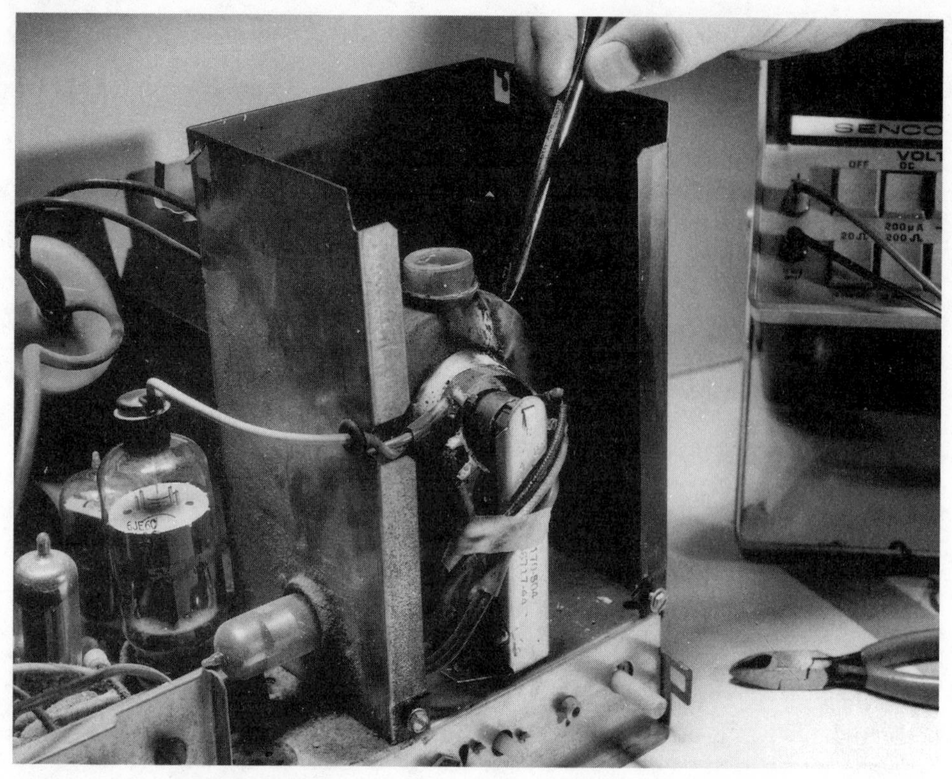

5-17 The pen points to evidence of high voltage arcover in the horizontal output transformer. The transformer must be replaced. Usually the horizontal output tube and damper tube then also require replacing.

A large-wattage resistor that is too hot to touch may indicate a shorted capacitor or tube. A power horizontal output transistor that feels too hot may be overloaded with a defective horizontal output transformer. A warm tripler unit may indicate excessive current. Remember, besides test equipment, common-sense servicing begins with sight, smell, sounds, and touch procedures.

Heat and coolant

The electronic technician may use heat or coolant to make a solid-state component act up in the TV chassis. Extreme heat may be applied by simply covering up the TV chassis with a blanket or coat. When the solid-state transistor or IC components reaches a certain degree of heat, it will break down. Most heat and coolant application is made with an intermittent TV chassis (Fig. 5-18).

Heat may be applied right upon the component with a light bulb, soldering iron, hair dryer or heat blower. Sometimes by holding the soldering iron or gun close to a capacitor or transistor, the extra heat causes it to act up. Several applications of heat with the hair dryer or blower may make the capacitor, transistor or IC become defective.

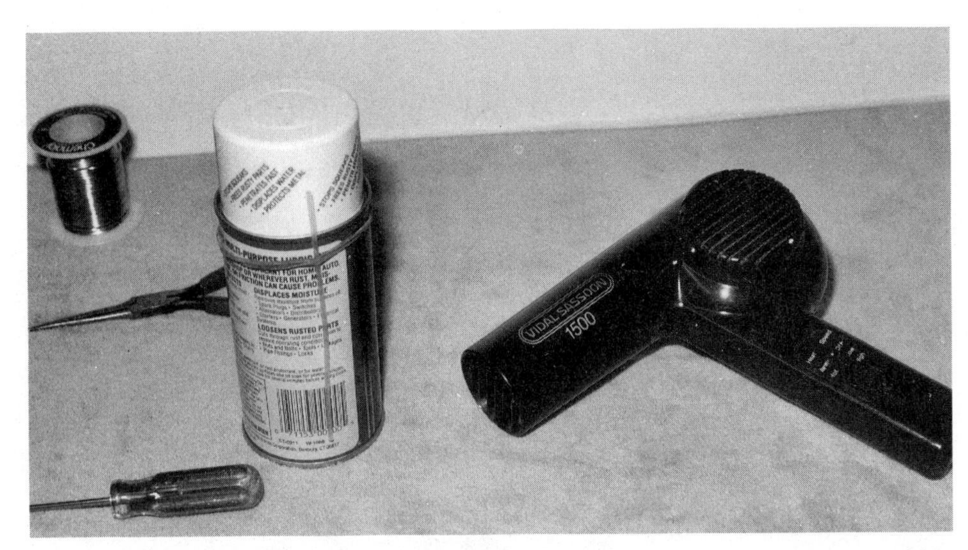

5-18 Use a hair dryer and a can of coolant to locate defective intermittent components or wiring.

Likewise, applying several coats of coolant upon the component may make it act up. Since transistor, capacitor, resistor and IC parts are made up of solid material, a spray of coolant may make the component go into the intermittent or normal state. Cold spray or coolant comes in a can like hair spray or paint with a plastic nozzle applicator.

You may find applying a coat of coolant and then blowing heat upon the suspected component will solve the most difficult intermittent condition. Be careful when applying heat to not melt components. Apply coolant right upon the suspected part. Sometimes you can locate cracked pc wiring or a poorly soldered connection with several coats of coolant.

Intermittent boards and poorly soldered connections can be located with the soldering iron. Try to locate the area where the intermittent occurs by taking voltage measurements or pushing up and down on the pc board. Then, solder all connections in that area (Fig. 5-19). Be real careful not to spill solder over the board or let two junctions be soldered together. Check the soldered junctions with a lighted magnifier or hand-held glass.

Typical solid-state troubleshooting charts

Here are five different troubleshooting charts. Remember to leave the most difficult repairs to the electronic technician. If in doubt, don't touch it (Figs. 5-20 through 5-24).

5-19 Locate a poorly soldered joint by resoldering each connection in an isolated area.

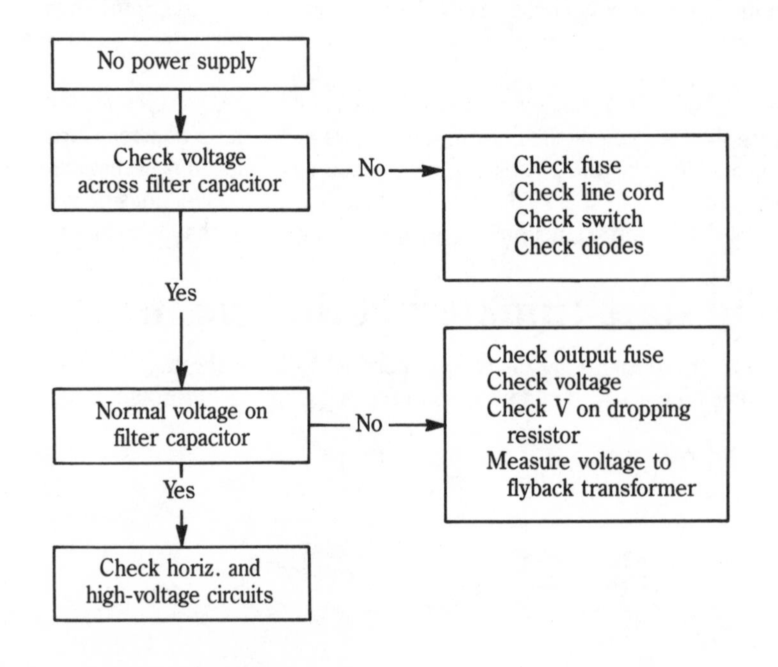

5-20 Flowchart for a solid-state TV with no sound, picture, or raster.

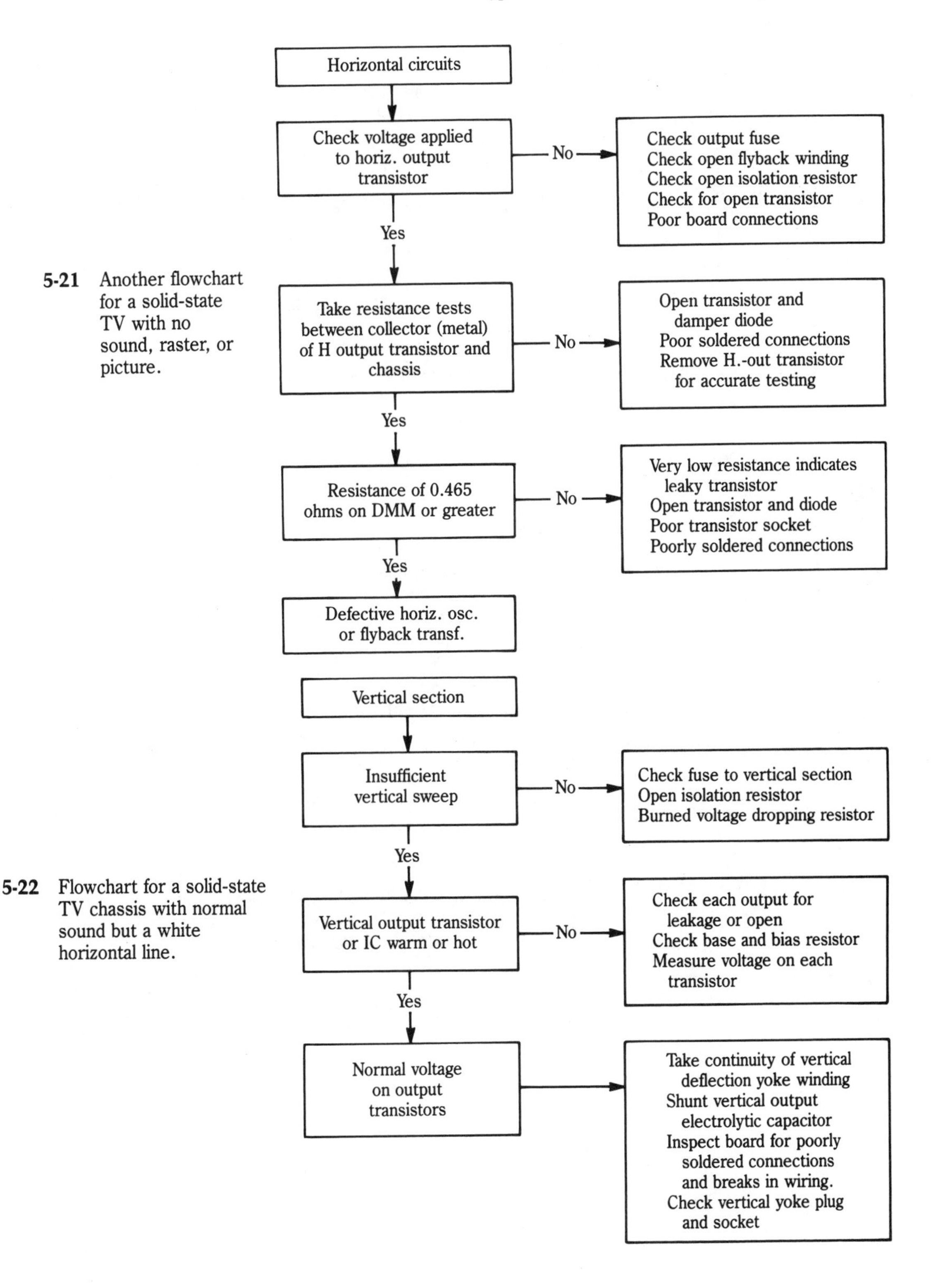

5-21 Another flowchart for a solid-state TV with no sound, raster, or picture.

5-22 Flowchart for a solid-state TV chassis with normal sound but a white horizontal line.

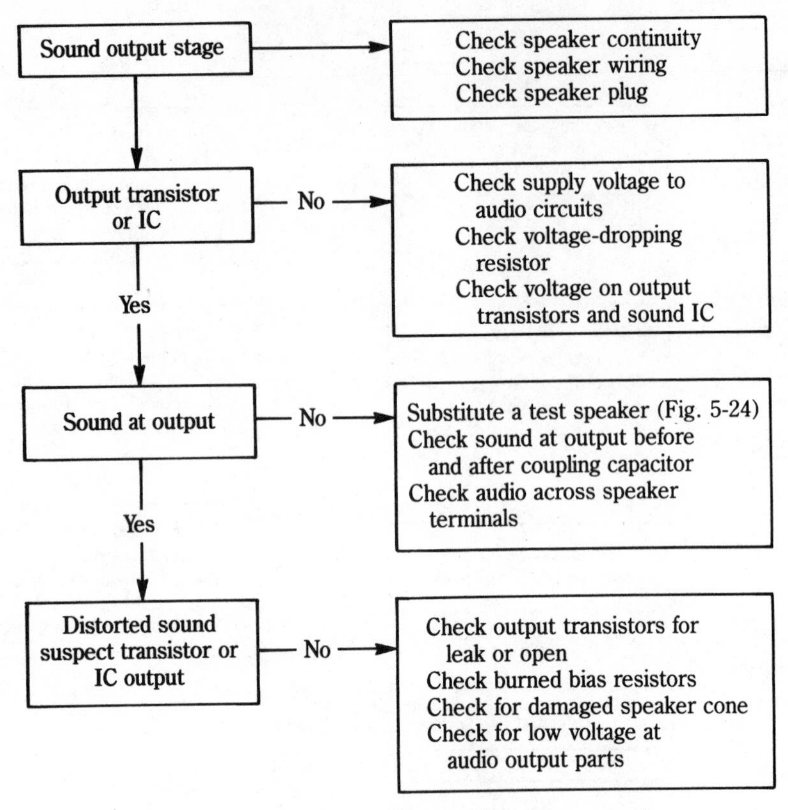

5-23 Flowchart for a solid-state TV with no sound.

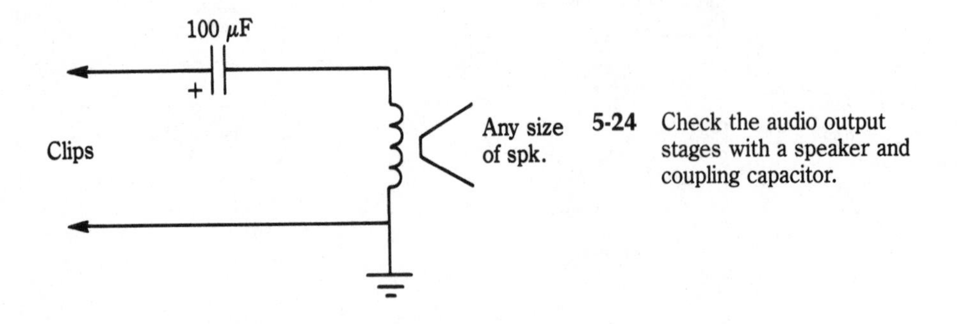

5-24 Check the audio output stages with a speaker and coupling capacitor.

6
Operating adjustments

AS WE CONTINUE TO EXPLORE TV TROUBLESHOOTING, THE NEXT LOGICAL STEP is a knowledge of the operating controls—not only knobs on the front of the receiver but those mysterious looking adjustments in the back.

Vertical size

The normal TV set has sufficient vertical size, or as it is commonly called *height*, to fill the screen completely with quite a little to spare. In other words, the back-of-the-set adjustment called vertical size (V-size) or height could make the picture as much as 30 percent higher than the screen height if turned far enough; however, normal design provides for a normal size picture with the control in approximately its midway position.

Another important consideration before adjusting the vertical size or height control is the avoidance of distortion. The normal picture format (or aspect ratio) by technical and NTSC standard is four units wide by three units high; that is, for every four inches of picture width there is a corresponding three inches of picture height. Under these conditions a circular emblem will look like a perfect circle, not egg shaped. In adjusting the vertical size or height care should be taken that the picture is not stretched to the point of distortion. This is best done with some circular object on the screen. Some stations still show a circular test pattern a few minutes before going on the air; however, it is not necessary to wait for a test pattern—many emblems and other commercial symbols shown between programs are circular and will serve for vertical size adjustments. Possible alternatives for height adjustments, therefore, are as follows:

- If the picture height is slightly below normal, adjust the vertical size or height control until the picture just covers the screen height. Check for nondistortion by observing the circularity of a suitable object on the screen, preferably a large one for ease of correct adjustments. For all adjustments where observation of the screen is necessary, a mirror is essential. The professional TV servicer has a special mirror on a stand for the purpose, but any mirror propped up in front of the screen so it can be viewed from behind the set is quite satisfactory.

- If the control setting for full-screen height is at or near its extreme position (little rotation remaining) the vertical amplifier stage is not operating properly. While it is just possible that a defective resistor, capacitor, or transformer, or even a voltage change, is responsible for the loss of height, the most likely and most frequent cause is a weak tube in the vertical circuit.

Referring to the typical functional block diagram in chapter 2, notice that there is a vertical oscillator tube and a vertical amplifier tube in every TV set. Sometimes these functions have different names for the same functions; thus vertical multivibrator is a synonym for oscillator.

V-output is quite commonly used for a shortened name of the vertical amplifier. What is significant is the fact that in the vast majority of modern TV sets, these two functions, vertical oscillator and vertical amplifier, are in one envelope and can be located easily by observing the number description, such as V6A and V6B, meaning parts A and B of tube V6. After locating and removing this two-in-one tube from its socket, if the TV set is switched on it should have normal sound and a simple bright horizontal line across the center of the picture tube. If, on the basis of probability, the pulled tube checks (in a tube tester) as weak or poor, replacement with a new tube will solve the problem but will almost always require adjustment of the *V-size* and perhaps also the *V-line* (linearity) control. The procedure is as follows:

1. Plug in the new tube and switch the TV set on. Wait a few minutes for warmup.
2. With a mirror positioned so you can see the picture from your position behind the set, reduce the V-size until the picture is too small to cover the screen both at the top and bottom; that is, until some blank screen shows.
3. Wait for a circular pattern or emblem to appear on the screen and observe its symmetry. It is quite easy to tell whether or not a circular shape is 100 percent symmetrical or not. Figure 6-1 is an example of patterns you may encounter during this adjustment.
4. If the circular pattern resembles Fig. 6-1A (the two halves symmetrical), adjust the V-size to increase the pattern or display until all four quarters are as identical as possible, or until the circle is no longer flattened at the top and bottom. This will produce the closest approach to a perfect circle. (There may be some minor irregularity due to horizontal and other imperfections, preventing the attainment of a 100 percent perfect circle, but if this deviation is very slight, no further correction is necessary.) The pattern should look pretty much like that in Fig. 6-1D.
5. If the final, most nearly circular pattern is obtained before the screen is fully covered in the vertical direction or, conversely, if you have to turn the V-size control to the point where the picture runs beyond the top and bottom edges of the screen in order to get a circle, then the width (horizontal size) adjustment is incorrect and will have to be corrected before completion of the vertical adjustment; however, at the moment we are proceeding on the assumption that only the vertical circuits require adjustment.

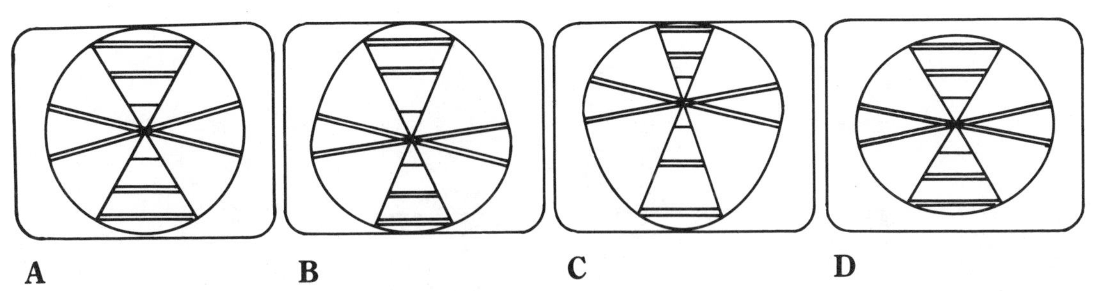

6-1 These drawings show three conditions of vertical nonlinearity. Using a circular emblem or insignia, (A) shows a symmetrical circle. At (B), the top half is stretched and the lower half compressed, where the nonlinearity at (C) is the exact opposite of the condition at (B). The circle at (D) is flat on both top and bottom.

Vertical linearity

If the reduced image on the screen resembles Fig. 6-1B or 6-1C, a second, closely related vertical adjustment must also be made. This control varies the vertical picture symmetry, usually called vertical linearity. While it is difficult to tell a nearly perfect shape by viewing an action scene on the screen, such is not the case for distorted proportion. For example, Fig. 6-1B corresponds to excessive stretching of the top of the picture; people seem to have very short legs and their heads seem to come to a point. The opposite extreme as in Fig. 6-1C produces long-legged, high-waisted people with very low foreheads. Correction of either requires the adjustment of the V-lin (vertical linearity) control, again using a circular pattern of some kind, plus the aid of a ruler and a grease pencil, as follows:

1. Mark a horizontal line across the screen midway between the top and bottom; use a ruler for a fairly accurate division of the height.
2. Adjust the V-size control, as previously, for a circle smaller than the full screen height, exposing blank space at the top and bottom.
3. Adjust the V-lin control until the upper and lower portions (above and below the painted line) are as nearly of the same height as you can tell with the naked eye (measurement is not necessary).

Vertical roll

In a normally functioning TV receiver, the picture will at times slide up or down one frame. This may happen when the set is first switched on, when channels are changed or when programs are switched at the transmitter. At all other times, the picture actually does not "want" to slip or roll as it is controlled by and synchronized with a signal from the transmitter (vertical sync, see chapter 2). If however, a TV set exhibits frequent rolling, requiring repeated adjustment of the V-hold control, the fault is most likely in the sync circuits.

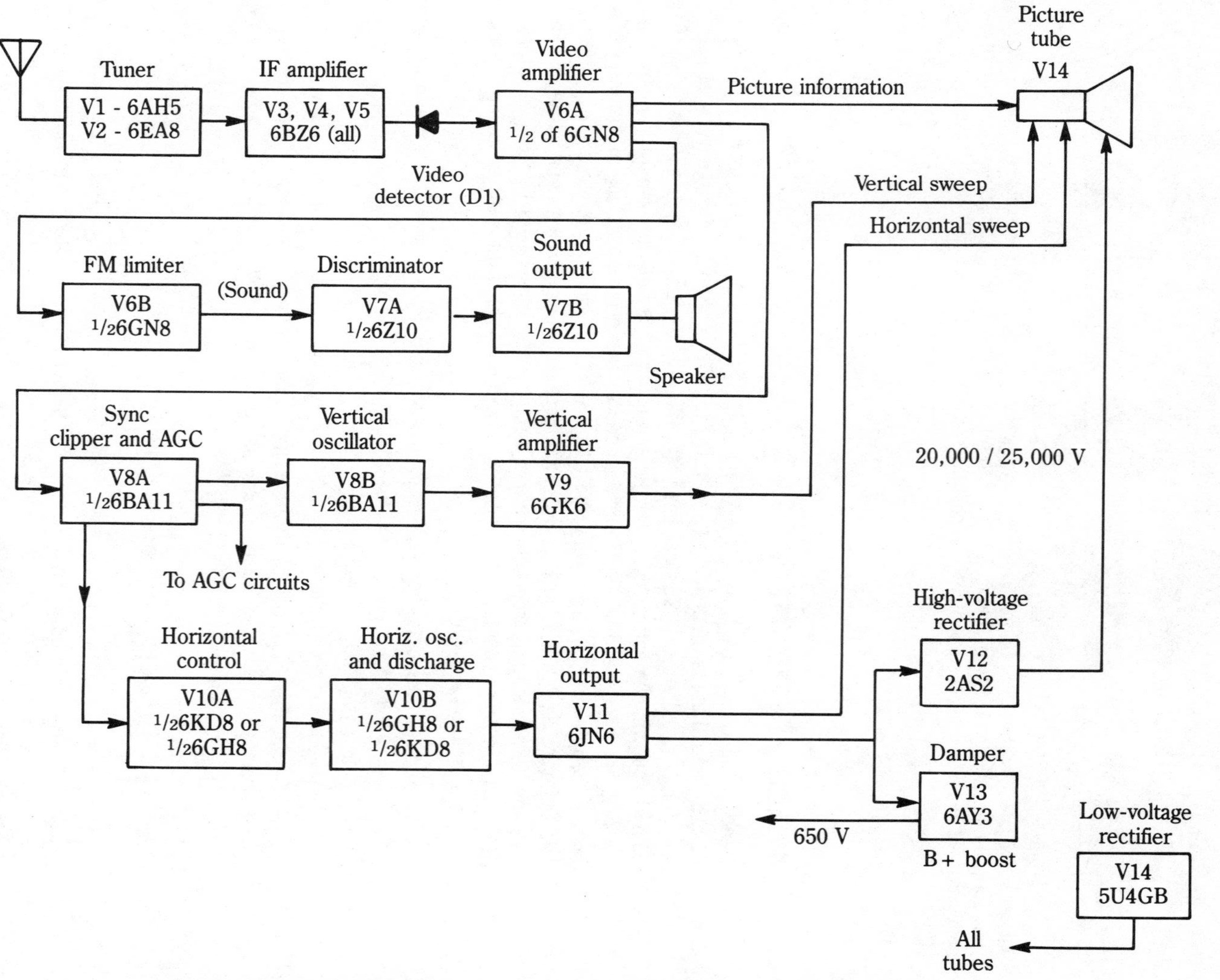

6-2 Block diagram of a typical black-and-white TV set with parallel-wired tube filaments.

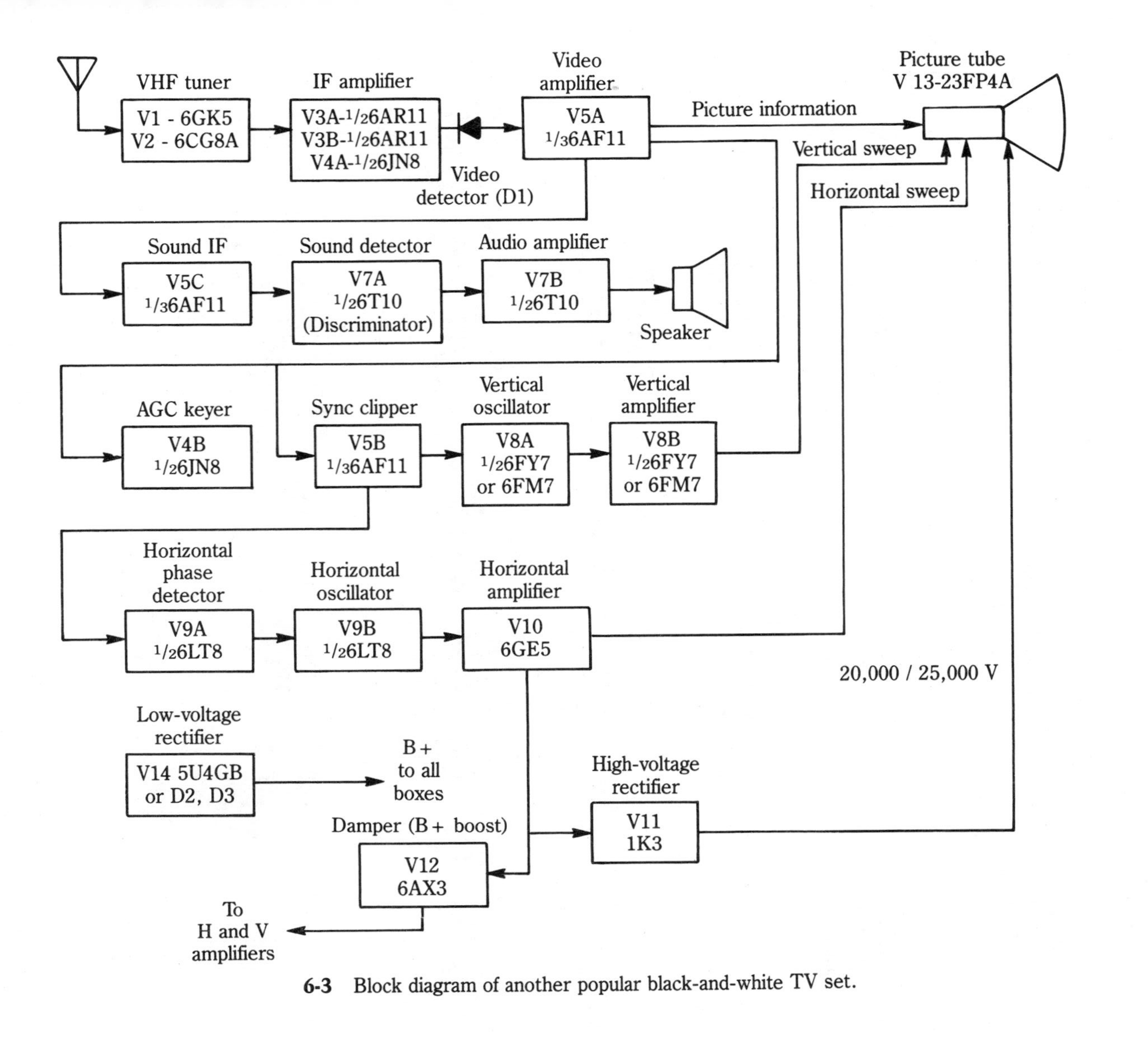

6-3 Block diagram of another popular black-and-white TV set.

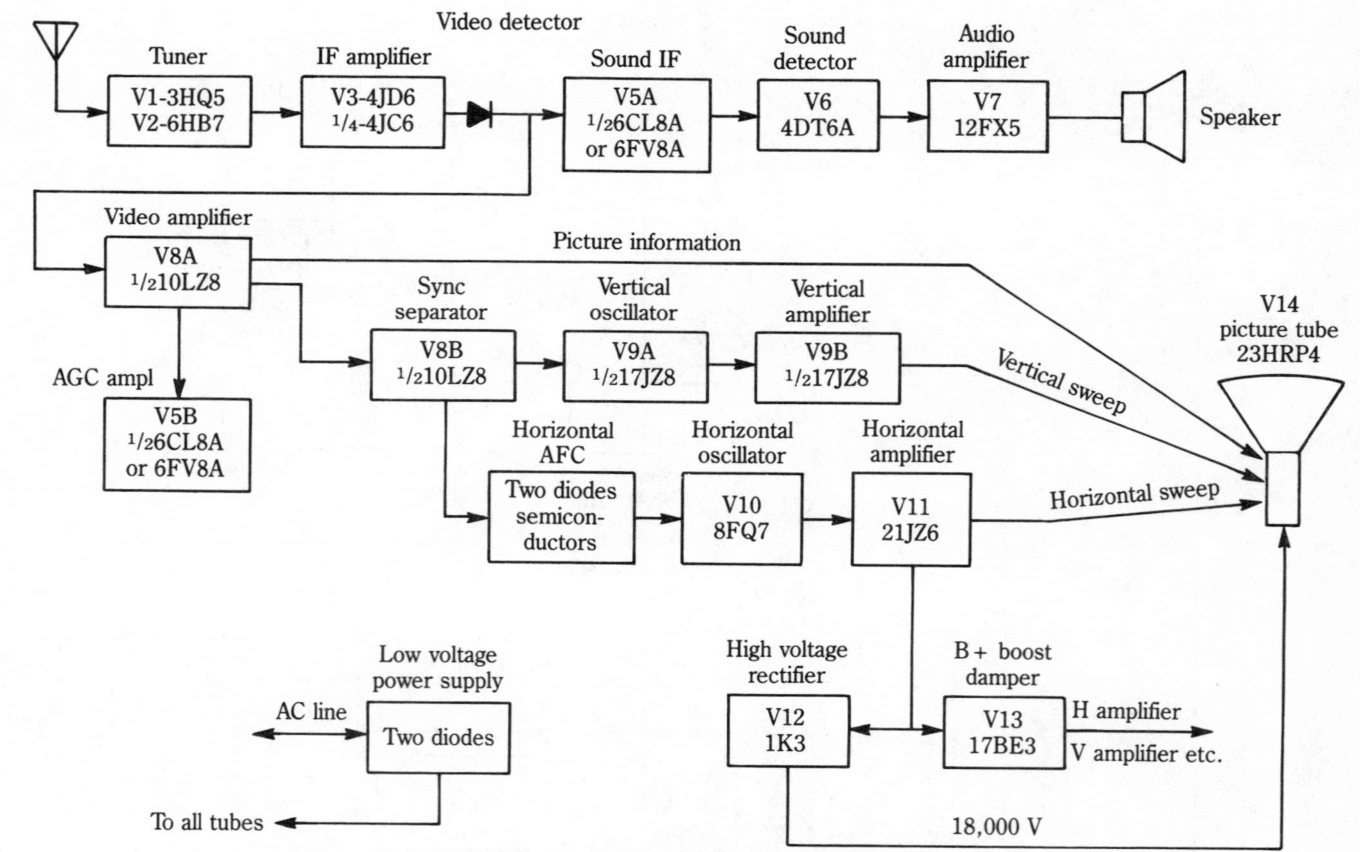

6-4 Block diagram of a black-and-white TV set with series-wired filaments.

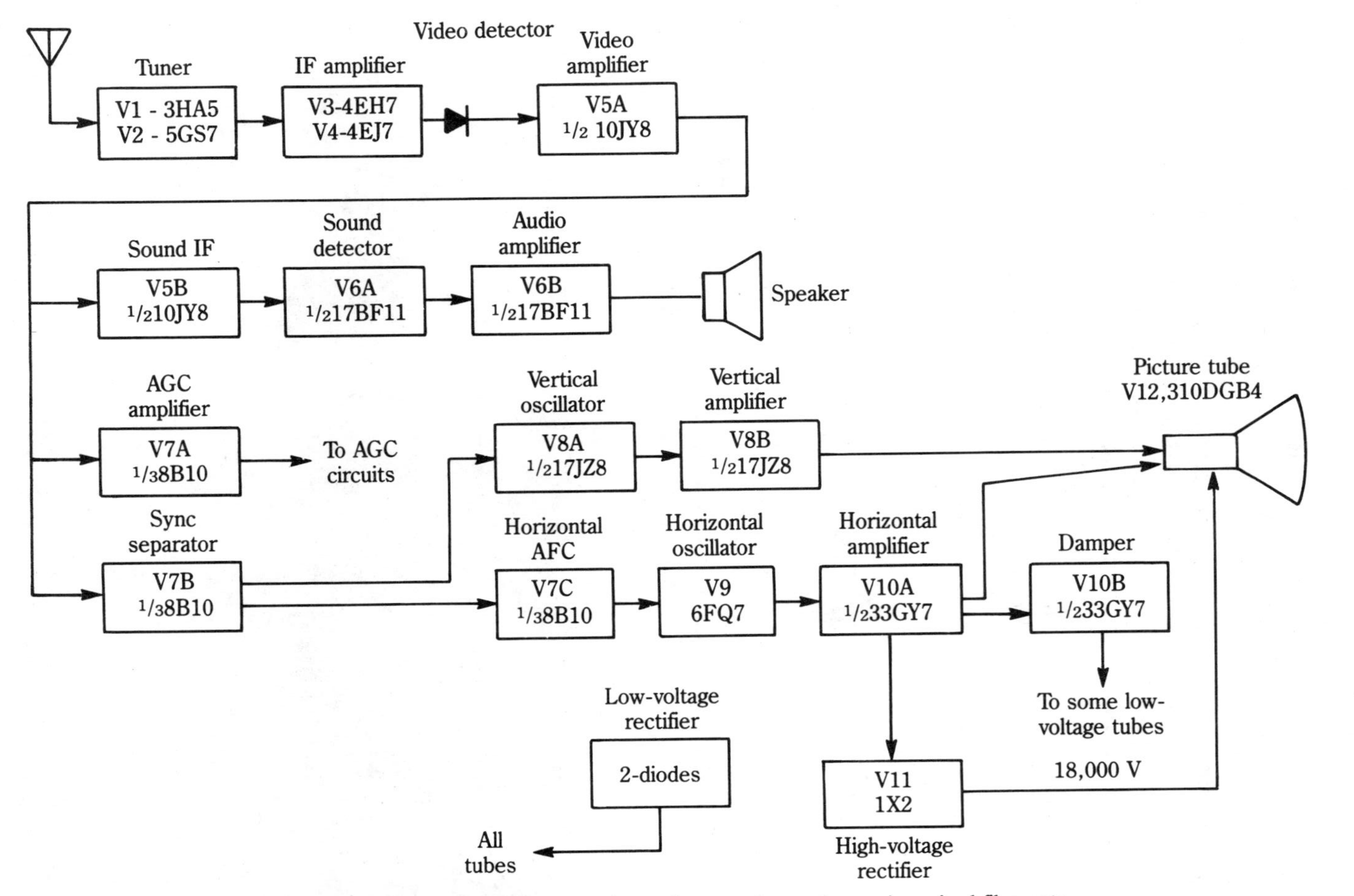

6-5 Block diagram of another transformerless receiver using series-wired filaments.

An examination of the typical block diagrams, and especially the diagram of the particular set in question, will reveal that there are such functions as sync amplifier, sync separator and sync clipper, or perhaps combinations of these functions, performed by one or more tubes all having the common term sync. A defect in one of these tubes is the most likely suspect causing vertical roll. As shown in Figs. 6-2, 6-3, 6-4, and 6-5, these are dual- or triple-function tubes, identified as sections A, B, or C (V5A, V5B, etc.). Each section of such tubes may be involved in separate unrelated functions so that removal of one tube actually disables two or three separate functions; thus, in Fig. 6-2, one-half of V8 (V8A) is a sync clipper amplifier, while the other half (V8B) is the vertical oscillator. With V8 out of its socket the picture will collapse to a thin horizontal line.

In most cases, a sync tube serves both vertical and horizontal sweep circuits. Removal of this tube will, therefore, disable both vertical and horizontal sync, and the picture will roll vertically and tear (zig-zag fashion) horizontally. As soon as a good tube is replaced in the socket, everything should return to normal, provided the tube was at fault.

Solid-state TV adjustments

Some of the early color TV chassis have the vertical, horizontal, contrast, brightness, color, and tint controls inside a hinged lid. Gradually, these controls moved to the rear of the chassis in portable TVs (Fig. 6-6). Today, in the solid-state color chassis both the vertical and horizontal controls are not found with voltage-controlled oscillators and fixed reference voltages. In fact, with the remote control TV, you may find all functions are controlled by the system control circuits.

6-6 The vertical, horizontal, brightness, and contrast controls are on rear chassis of a black-and-white TV set.

Black-and-white solid-state adjustments

The black-and-white portable TV adjustments are found in the rear apron. You might not find any width or horizontal hold adjustment. The brightness and contrast knobs stick out the back with a vertical hold control with a preset adjustment (Fig. 6-7). In the early transistor TV chassis, the vertical height and linearity controls were just inside the cover on the rear apron. The horizontal hold control turned a slug inside the horizontal oscillator control for horizontal lock-in (Fig. 6-8).

6-7 The brightness and contrast with preset vertical controls at the rear of a solid-state TV chassis.

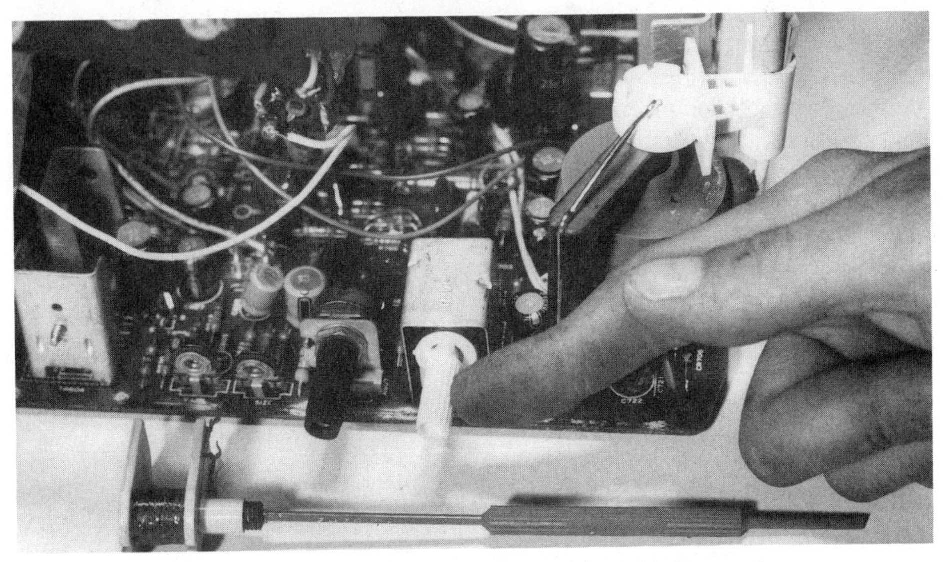

6-8 The finger points at a slug-tuned horizontal hold control.

In newer black-and-white chassis, you may find separate vertical oscillator, driver, and output transistor stages or IC components. The vertical oscillator, drive, and output circuits may be contained in one IC or combined with other circuits in one large IC (Fig. 6-9). Usually, the vertical IC has a long bar-type heatsink. Again, the vertical height and linearity controls are on the main chassis with the vertical and horizontal hold at the rear of the chassis.

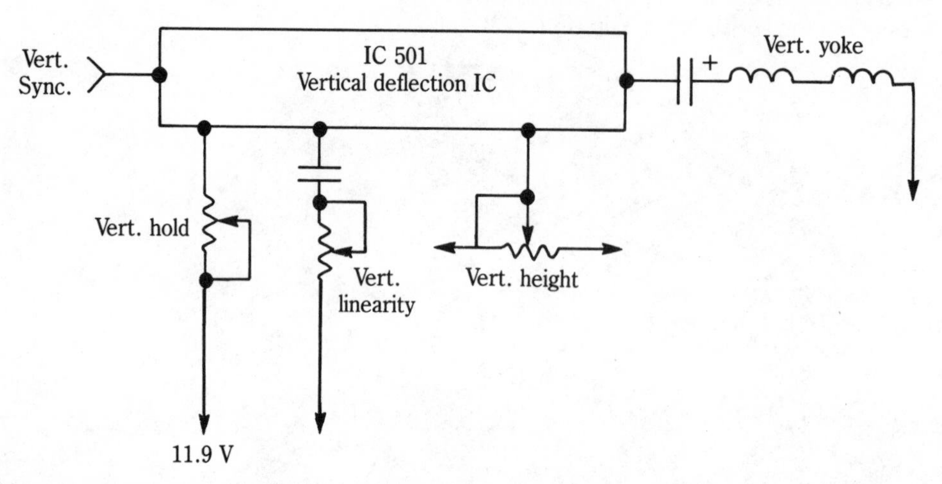

6-9 A single IC might contain all vertical deflection components in the latest solid-state TV chassis.

Color solid-state vertical adjustments

In yesterday's color chassis, the vertical circuits had vertical height, linearity, and hold controls. The vertical hold control was located at the rear of the chassis with the height and linearity having preset screwdriver type controls upon the chassis. Only the vertical hold control stuck out the back of the chassis (Fig. 6-10). The vertical oscillator and vertical amp were found in the deflection IC, which also contained the vertical sync, regulator, AFC detector, horizontal oscillator, flip-flop, sync gate, and X-ray protector (Fig. 6-11). In later vertical color circuits, the vertical centering, size, and hold controls operated from a sweep oscillator IC. Only the vertical hold control was found on the rear

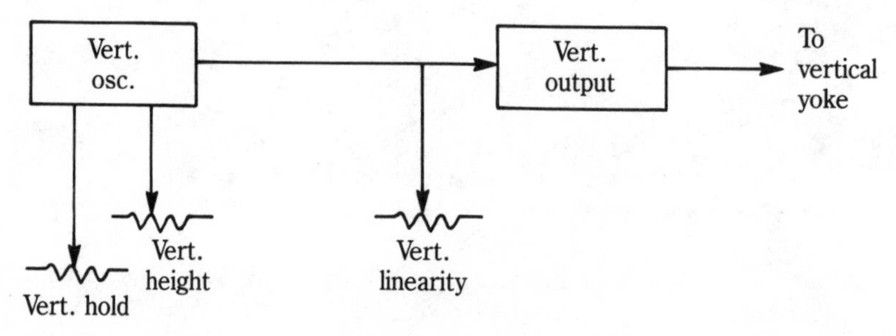

6-10 Vertical hold, height, and linearity controls are in the vertical block diagram.

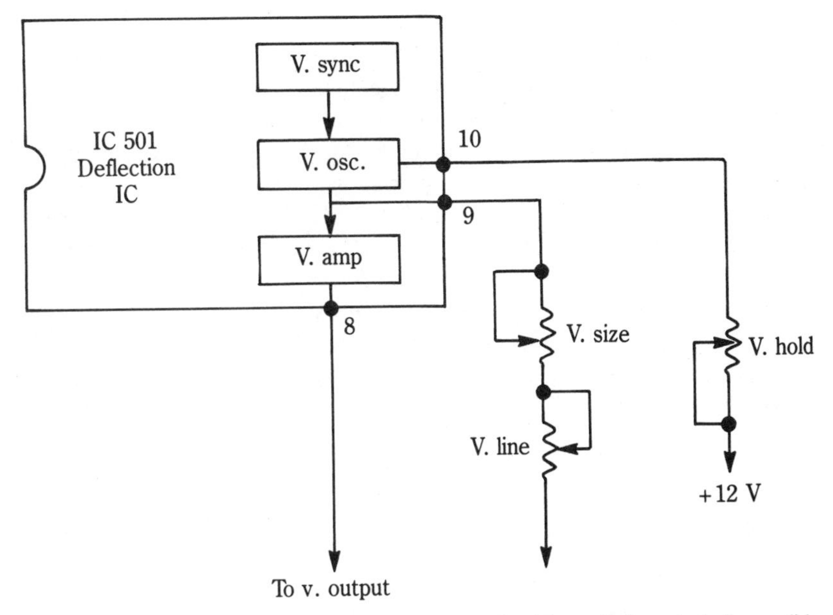

6-11 Block diagram of the vertical deflection IC with vertical controls in a solid-state TV chassis.

apron. In today's color TV chassis, the vertical and horizontal controls are unnecessary because of a countdown synchronization locked in circuits.

Sync failure

Weak or defective sync tubes are not the only potential causes for vertical roll. Referring again to the block diagrams, note that the signal to the sync amplifier or similar sync tube comes from the block marked VID. AMP. (video amplifier). Very often this stage, not the following sync circuits, is responsible. Should the video amplifier be either very poor or outright defective, a very poor sync signal will leave this tube; therefore, the input to the sync stages is inadequate.

Of course, in such a case the picture quality will also suffer, most often exhibiting a lack of highlights and contrast. While advancing the contrast control may result in some improvement to the picture, this adjustment may actually be detrimental to normal sync output; thus, chronic picture rolling in a set having a weak picture may be due to a subnormal video amplifier. Also, while a poor picture may be due to one or more substandard tubes anywhere from the antenna to the video amplifier, it is rather easy to tell which portion of the set is responsible. A weak, "thin" picture that is otherwise clean is almost certain to be due to a poor video amplifier, while substandard functioning in the TV tuner or in the IF amplifier is almost certain to produce a snowy, perhaps wavy, picture.

"Freak" sync trouble

There is one other potential case of sync problems worth mentioning here—interference from another TV station on the same channel. Standard FCC channel allocations are such that under normal conditions the geographic separation between the stations operating on the same channel preclude any likelihood of the distant station interfering with a local station on the same channel. Modern TV sets have, in addition, a certain amount of built-in immunity from this possible interference because they tend to favor the much stronger local station; however, during some not-too-frequent freakish atmosphere conditions, when distant signals tend to come in rather strongly, interference may result. This is due to the fact that although technical standards (frequencies, sync, sweeps, etc.) are the same throughout the United States, very minor and quite tolerable differences between those standards will cause severe interference with the local station sync and, consequently, with the picture as a whole; however, since these are very transient conditions, occurring on some stations only, nothing should be done about it. It is described here merely to acquaint you with the phenomenon, and caution you not to rush into making adjustments.

Horizontal size or width

Symptoms of insufficient width are quite obvious: the picture does not cover the full width of the screen and, quite likely, has shrunk more on one side than on the other. As in the case of vertical size, either tube deterioration or adjustments, or (most likely) both, are responsible. Referring to Fig. 6-2 we see that tube V10A, V10B, and V11 are involved. In Fig. 6-3, V9 and V10 are involved. In Fig. 6-4, V10 and V11 are involved. In Fig. 6-5, V9, V10A, and V10B are involved.

When insufficient width is the only complaint, the horizontal oscillator function block may not be involved; however, since adjustments are required after substituting a higher performing new tube for a worn out one, the horizontal oscillator block may also be involved since some of the adjustments are located there. Again referring to the typical block diagrams, other tubes and functions are shown, because they are intimately related to the horizontal amplifier.

In troubleshooting for insufficient width, first suspect the horizontal amplifier. Proceed as follows:

1. Switch off the receiver. Remove the back cover and ac cord. Use a cheater cord if necessary.
2. Carefully remove the top cap from the horizontal amplifier tube and remove the tube. This cap may be very tight or sometimes frozen. Observe the suggestions given in chapter 5 for removing the cap without damaging the tube.
3. Install a new tube and replace the top cap securely. This is a high-voltage point; a loose cap may cause arcing and even damage to the horizontal output transformer. If this tube is located inside a metal enclosure or cage be sure to secure the enclosure after tube replacement. Practically all of the components in this cage operate at extremely high voltages.
4. If the horizontal amplifier tube replacement does not correct the insufficient width one or two other tubes may be a fault. Looking again at the typical block diagrams, Figs. 6-2 to 6-5, we notice that each has a tube called a damper. Without going into the technicalities of the function of the damper, it is sufficient for

our purpose to know that this tube contributes in large measure to the normal operation of the vertical and horizontal amplifiers by providing them with a voltage boost. Should this tube be below par (and this does happen frequently) the horizontal amplifier will not provide sufficient output for full picture width. Again, caution must be exercised here since the damper tube often is located inside the high-voltage cage.

Another tube that may be responsible for insufficient width, also by virtue of providing insufficient voltage to the horizontal amplifier (as well as to all other tubes in the set) is the rectifier tube, where one is used. More and more TV sets nowadays no longer use vacuum tubes for rectifiers; semiconductor diodes, usually marked D (D1 and D2) on the diagram, are preferred for many reasons, not the least of which is a long, trouble-free life. In the previously mentioned diagrams, the rectifiers are marked V14 or 2 *diodes*. When a tube-type rectifier is used and if it is weak, the symptoms will be more than just insufficient width. The picture height may be subnormal, sound may be below par, picture may be dim and flat, etc. Replacement of semiconductor rectifiers is in our opinion beyond the scope of a beginner and is not recommended.

5. If it is obvious that the picture width is correct, no further adjustments are required; however, this is the proper time to adjust the set if, as is often the case, the picture is too wide, wrapping itself partly around the tube, causing cutoff of left and right edges. This can best be seen when some printing or writing is shown, such as at the opening of a motion picture. If it is apparent that the picture is too wide, proceed to steps 6 and 7.
6. If you have a set of diagrams or service information for the set examine it to find what type of width adjustment is provided. If no such data is at hand proceed to step 7.
7. Examine the back of the TV set with the cover removed. Compare it with the typical diagrams in Figs. 6-2 through 6-5. Notice that on these diagrams, three different kinds of width adjustments are indicated: pot, slug, and points. While all of these are meant to achieve the same purpose the procedure for each is different and must be understood before attempting to use any of them.

Pot This adjustment is of the volume control type and may be a knurled, round shaft like that inside the knobs of the ordinary radio, or it may be a slotted shaft designed for screwdriver adjustment. In either case, it is an adjustable potentiometer (pot for short) which can be rotated less than one full turn. It is the simplest to adjust and can easily be reset to its original (before adjustment) position. It is worth mentioning a second time that it is wise for the beginner to always observe and, where possible, record the position of an adjustment before changing it. In this way you can always go back to the original condition should you feel the adjustment makes things worse or is ineffective.

With a pot type of control, slowly rotate the shaft (using a mirror to watch the picture) until the width is less than full screen. If the picture extends more or less equally on both left and right edges, rotate the shaft in the opposite direction until the picture just barely extends beyond the edges of the screen. About 1/4- to 1/2-inch is all that should be *wrapped* on each side. The correct width setting is that which does not mask any part of a line of printed matter (the list of actors, etc. displayed at the beginning of a movie is most suitable).

Slug A slug-type width adjustment is found in many TV receivers and requires a different technique, as well as some precautionary measures. The word *slug* refers to a carbon-like rod that can be made to slide in or out of fiber sleeve on which a spool of wire is wound. The whole assembly is called a *width coil* or a *slug-tuned coil*. The tuning or adjustment of width is made by inserting the blade of a special tool (or midget screwdriver) into the protruding slot of a threaded screw-like shaft. Figure 6-12 is a simple sketch of such a coil.

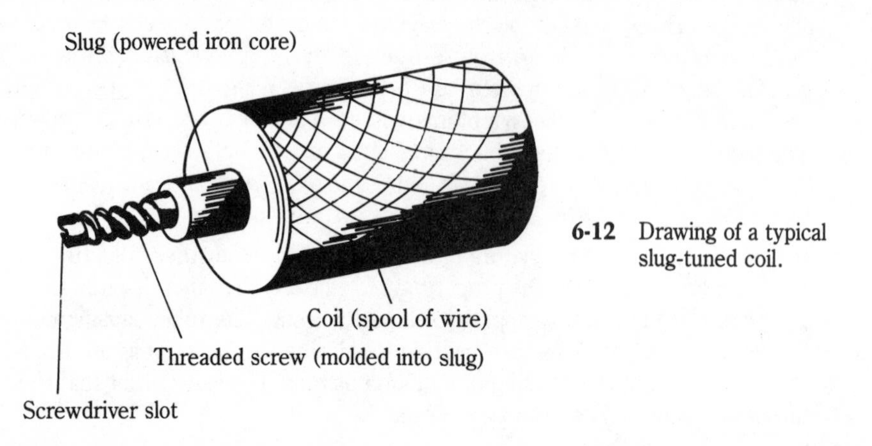

6-12 Drawing of a typical slug-tuned coil.

In contrast to the previous pot adjustment, the slug can be turned as many as eight or more times. In fact, there is no stop on this adjustment, hence the precaution: do not keep on turning a slug more than three or four turns in either direction. To do so may unscrew the slug completely, letting it fall out or, worse, fall into the dark recess of the TV chassis. Whichever way it falls, there follows a messy job of replacing it, certainly one to be avoided by the beginner. Fortunately, TV design is such that this slug is normally in an approximately midposition, allowing a minimum travel of five or six turns in or out—certainly enough for any width adjustment in the field. Normally most width coils are designed so that turning the slug out (counterclockwise from the screwdriver end) will increase the width, and vice versa. The procedure, therefore, is the same as the pot shaft rotation, except that here a few full turns may be required. The correct procedure for adjustment is as follows:

Rotate the slug clockwise until both left and right edges of the picture are visible; the picture is now too narrow. Rotate the slug in the other direction until a slight overlap or *wrap* is achieved on each side. Of course, too much overlap cuts off the picture.

Points In this method of width adjustment, a movable voltage lead can be attached to various voltage *points*. By changing the voltage applied to a portion of the horizontal amplifier, the width of the picture can be varied slightly. Usually there are three or four tie points to select. In some cases, not connecting the movable jumper lead to any voltage point provides a possible setting for the horizontal width. Since the movable jumper is usually insulated, this procedure can be made with the TV under power.

Other width adjustments In some TV receivers, a mechanical (magnetic) adjustment is provided for picture width control. No two of these are alike, and adjustment should not be attempted without some sort of service instructions. Two of the most common mechanical adjustments are the width sleeve and the width tabs.

The width sleeve is just that—metallic sleeve on the rear of the neck of the picture tube. This sleeve is easily adjusted by a combined sliding-rotation motion (it may sometimes stick to the glass until dislodged) until the correct width adjustment is obtained. Again I repeat the need for caution: there is no danger from either heat or high voltage in this procedure, but neither is there room for carelessness. Since the sleeve sliding along the neck of the picture tube also affects horizontal linearity, we shall return to this adjustment presently.

Width tabs function in much the same manner as the sleeve. The tabs have the same effect on width as a sliding sleeve. Since there can be a great variety of mechanical means of varying picture width, no standard, uniform procedure can be given, except as described above.

Horizontal linearity

As mentioned previously in connection with the procedure for horizontal width adjustment and correction, the horizontal linearity adjustment is most conveniently made at the same time. In some sets the two interact so that it is almost mandatory to do them together, unless a nonlinearity problem exists without any accompanying reduction in width. In that case it may be required to reduce the width in order to be able to see the nonlinearity.

Horizontal linearity can best be understood by referring to the discussion of vertical linearity. In simplest terms, *horizontal linearity* means that the left and the right halves of the picture are symmetrical. More specifically, and this applies to the vast majority of TV receivers, horizontal linearity means that a circle does not appear like an egg, and that the right side is neither stretched nor compressed. This is not always obvious when the picture is wrapped around the tube sides; therefore, nonlinearity is best shown when the width is reduced to a little less than the edges of the screen. As in the case of the vertical linearity, a circular object is most helpful in observing horizontal linearity, although this is not a must. It should also be kept in mind that although the standard aspect ratio (picture width versus height) is four to three—that is, the picture is a rectangle, not a square—a circle will appear (not an egg lying down) on a properly adjusted TV set.

Horizontal linearity adjustments are not the same on all TV sets; however, most sets have one of three types of adjustments. In earlier TV sets a potentiometer adjustment on the rear apron of the set provides correction for nonlinearity. In other sets a slug adjustment, identical in appearance with the width slug adjustment described earlier, is used. In still other TV sets, a metallic sleeve on the tube neck, by itself or in conjunction with a shaft adjustment, serves the same purpose.

Finally, some TV sets have no adjustment whatsoever, relying on the normal operation of the circuit (which was originally designed for proper linearity) to ensure linear horizontal configuration of the picture. It is important in the latter case to make sure that the horizontal oscillator and horizontal amplifier tubes are not significantly below normal.

Any tube that tests weak on a tube tester is a likely suspect for causing nonlinearity, even though the tube seems to work in the set. The step-by-step procedure for linearity correction is:

1. Make sure that the horizontal oscillator and amplifier tubes are good.
2. Reduce the picture width until both edges are visible. If either the linearity or centering (see "Centering") are incorrect the blank spaces may differ in width. With both edges clearly visible, examine the picture carefully to determine whether it is actually nonlinearity or improper centering. If after examining a suitable scene, preferably a circular object, it appears that both edges look symmetrical but the blank spaces are of different width, no linearity adjustment is made; the picture is centered according to the directions that follow. If, however, there is definite distortion on one edge (usually the right side), proceed to step 3.
3. Rotate the linearity adjustment shaft a small amount at a time, first to the right then to the left, and observe the effects on the picture. Figure 6-13 shows two opposites of nonlinearity. In Fig. 6-13B the picture should be made to shrink toward the center of the screen until symmetry is reestablished. In the case of Fig. 6-13A a certain amount of picture stretching is called for, until the picture most nearly resembles Fig. 6-13C. Most likely this will also affect width.
4. Adjust the width control, as described earlier, until the picture again just begins to wrap around the edges of the tube. It may be necessary to slightly adjust the width control during the linearity adjustment, but a final width touchup may still be required.
5. If the linearity adjustment is a slug type, proceed as for the slug-type width adjustment: in other words rotate the shaft clockwise or counterclockwise two or three turns until the effect is observed to determine if it is going the wrong way or not; then rotate it in the proper direction until the best linearity is achieved. The same precautions given for slug width adjustments apply here. If three or four turns seem to make no difference, discontinue the adjustment; the fault does not lie here. There are a number of components (resistor, capacitor or peaking coil) which, if defective, would cause horizontal nonlinearity, and no amount of adjustment can correct this. A professional servicer is then needed.
6. If the linearity adjustment is mechanical or a similar type (as verified either by the service data or as may be indicated on the interior of the TV set cabinet) the specific procedure should be followed. One popular method as previously mentioned is to slide the metal sleeve on the neck of the picture tube until the picture width is insufficient to cover the full screen width. Rotate the linearity adjustment shaft on the back apron of the TV chassis until the best linearity is obtained, following the general criteria outlined previously. When linearity has been achieved slide the sleeve on the picture tube neck until the picture fills the screen to a little beyond the edges, observing the effect if any on the linearity. These two adjustments (width and linearity) are somewhat interrelated; so it may be necessary to go through the procedure a second time until optimum linearity with proper width is obtained.

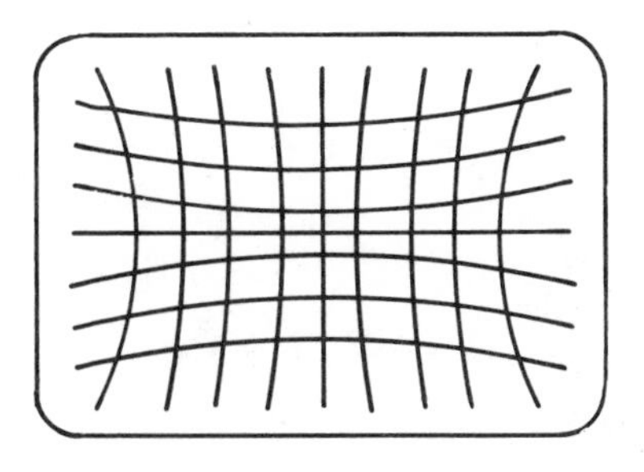

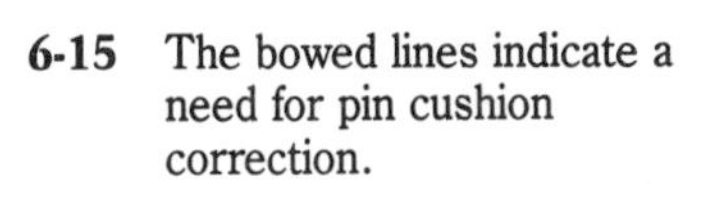

6-15 The bowed lines indicate a need for pin cushion correction.

Pincushion tabs or bars are fastened to the yoke structure and may be loosened for sliding, after which they are tightened again. As in other cases of nonlinearity described previously, these adjustments can best be made when edges of the picture are visible; hence as part of the other adjustments.

A note of caution is in order here. The tube and deflection yoke assembly are rigidly secured. The pincushion correction was made during installation of the original tube. It is usually not necessary nor advisable to make pincushion adjustments unless there is clear evidence of distortion or when the picture tube is replaced (by a professional TV servicer, of course).

In some TV sets, the pincushion correction is entirely automatic by virtue of a sensing, feedback and compensating circuit. Since this is usually a solid-state circuit and very reliable, there is little likelihood of malfunction during the life of the set. In case of failure, and after making sure that the distortion is not simple vertical or horizontal nonlinearity, replacement of the appropriate module board would be required. It should be noted, however, that even in some of the so-called automatic self-correcting circuits in some recent sets, there is a manual presetting adjustment, somewhat akin to the presetting of the vertical hold, from which the automatic correction circuitry takes over.

Solid-state side pincushioning circuits

In early transistorized and today's IC color chassis, both horizontal and vertical side pincushioning were corrected with a pincushion transformer. To compensate for pincushion distortion, the horizontal deflection current is modulated by a waveform during vertical scan period. The inductance generated in the secondary winding varies with the current applied in the primary (Fig. 6-16). The secondary coil winding is in series with the horizontal deflection yoke. The vertical output is applied to the primary winding. When the parabola waveform and dc current is applied to the primary winding, a linear current is applied to the secondary winding, eliminating side cushioning in the raster.

With today's 27-inch color chassis, the pincushion correction circuit may modulate the horizontal yoke current at a vertical rate to eliminate distortion. The pincushion transformer is in series with the horizontal yoke. The width changes with regulated B+ and modulated current through T502. As more B+ voltage appears across the yoke, the width increases.

The correction circuit develops a parabola waveform applied to the pincushion transformer (Fig. 6-17). Adjustment of the amplitude control determines how much vertical waveform is applied to error amp. The beam current input to Q401 allows pin correction

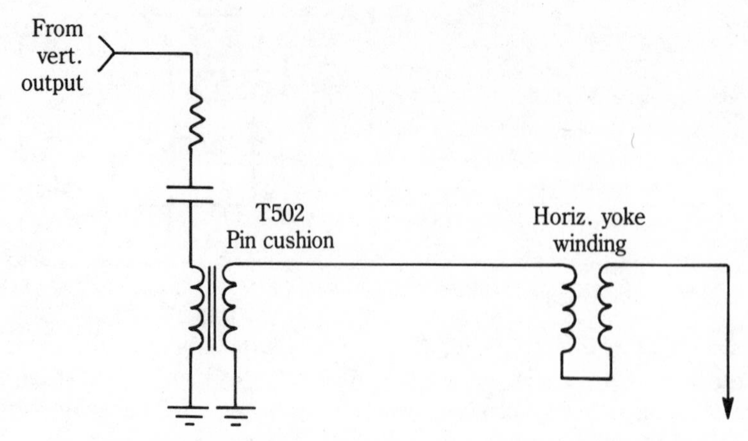

6-16 Block diagram of pin-cushion transformer connected to vertical and horizontal yoke winding.

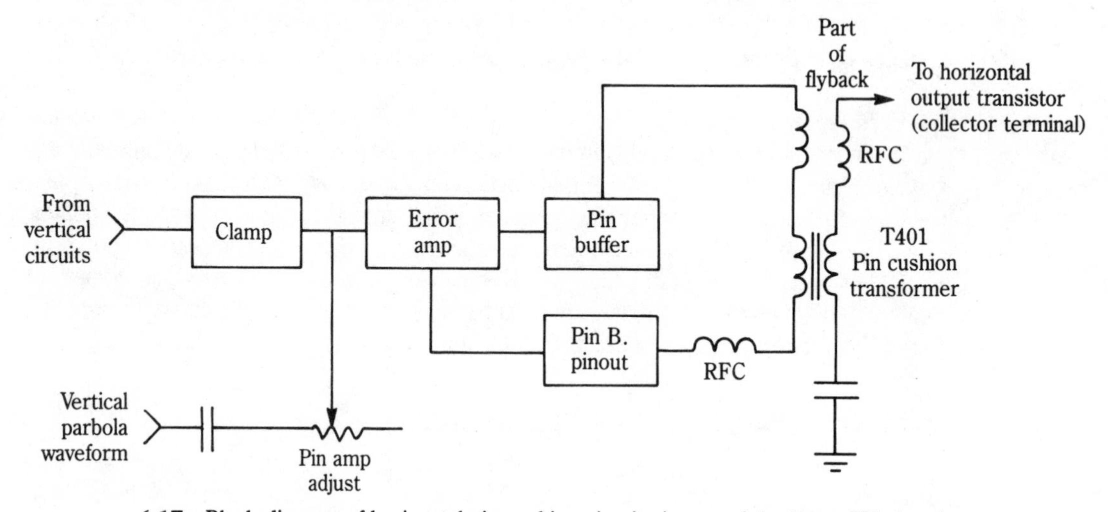

6-17 Block diagram of horizontal pin-cushion circuits in one of the latest TV chassis.

circuit to compensate for changes in width. The pincushion correction circuit interacts with both the power source and vertical circuit.

Centering

Another adjustment related to the various dimensional corrections just discussed is centering. The centering adjustments position the picture so it extends equally in all four directions. When the vertical size is adjusted, both the top and bottom edges should be reached simultaneously if the picture is centered; likewise, when the width is adjusted both sides should expand at the same time. In very old TV sets, but also in some recent color TV sets, two separate shaft adjustments were used, one marked V-cent, the other H-cent.

Centering adjustments are very simple; the controls have a total travel of less than 360 degrees (less than one full turn) and can safely be turned back and forth until proper picture centering is achieved. In some TV sets, the centering adjustments (if any) are mechanical/magnetic types, centering depending primarily on the original positioning and assembly of the deflection and focus components on the neck of the picture tube and only secondarily on after-the-fact adjustments.

The most popular type of centering consists of a pair of tabs, flat metal, magnetic strips, or bars, which can be rotated toward and away from each other until centering is achieved. One tab affects primarily the horizontal positioning, the other the vertical. Again, these are rather interdependent adjustments, and should, therefore, be made simultaneously. A variation of the latter type is the single tab centering device. It can be moved both horizontally and vertically (in fact, it can be made to move in various directions simply by pushing down on it and sliding it sideways at the same time). Still another centering adjustment is a thumbscrew on the picture tube neck components. This type requires only a clockwise or counterclockwise rotation to achieve centering.

Focusing

The focus adjustment varies the sharpness of picture detail. As in the case of the other adjustments, there are different ways of achieving the same end result, and different manufacturers use different methods; however, the procedures given here apply.

- *No adjustment*—There are a number of TV sets with what is called *fixed focus*; the picture tube is designed and constructed so that the picture is always in focus, although there may be a provision for correction (by a servicer) in the rare case when this type is out of focus.
- *Focus thumbscrew*—Some focus controls are in the form of a short, flexible steel cable with a thumbscrew at the end; the cable extends through the back cover of the TV set. Rotating the knurled end of the cable adjusts the sharpness of the picture. Sets with this type hopefully have long since been put to rest, but once in a while you'll see one.
- *Sleeve adjustment*—This focus correction device is in the form of a ring or sleeve on the neck of the picture tube. Sliding the ring or sleeve slowly along the neck of the tube adjusts the focus of the image. In those TV sets in which the width control is also a sleeve-type adjustment, it will be found that the width adjustment is very much forward on the picture tube neck, while the focus sleeve is very close to the rear end of the picture tube.
- *Shaft focus adjustment*—This is a control similar to a volume control on the panel of an ordinary radio receiver. This type of control is found on some (usually older) black-and-white TV sets and on practically all color TV sets where focusing is very much of a requirement.
- *Picture tube focus adjustment*—You may find the focus control and screen controls attached to a plastic assembly plugged into the rear of the picture tube socket. In many of the lastest RCA chassis, this plastic assembly must be unplugged before the picture tube socket can be removed (Fig. 6-18). Some of the focus controls

6-18 The focus and screen control might be located on the socket of a picture tube or fastened to the top of chassis in an RCA portable TV chassis.

are screwdriver adjustments, while others require a hexagon alignment tool to rotate the plastic control. When the focus control has little or no effect on the raster suspect a defective picture tube.

In all focusing adjustments, it is essential that a picture be watched at close range in order to obtain the sharpest focus possible. While any fine detail of the picture can be used for observing the effect of focusing adjustment, it is best to use the horizontal lines, preferably on a bright portion of a scene and focus for the thinnest possible lines with clear separation between them.

Arcing in CRT socket

Continuous arcing in the CRT socket or focus control may cause horizontal firing lines in the picture. Excessive dust collected in the spark gaps of the picture tube socket may cause the TV set to shut down. In the early color TV chassis, the focus pin and socket formed green arcing residue that would sometimes arc over. Sometimes the socket or just the focus pin was replaced.

Within the latest TV portables, the spark gaps in the CRT socket may arc over and cause intermittent arcing or chassis shutdown (Fig. 6-19). The raster may come up, flash, dim down, then shut down. The spark gaps are located on the lock color amp cathode, grid, screen and focus pin. Often, dust will filter through the air slots in the plastic cover and collect on the picture tube socket. Just blow the dust out or remove socket from CRT and give a couple of taps with a screwdriver handle to dislodge dust particles.

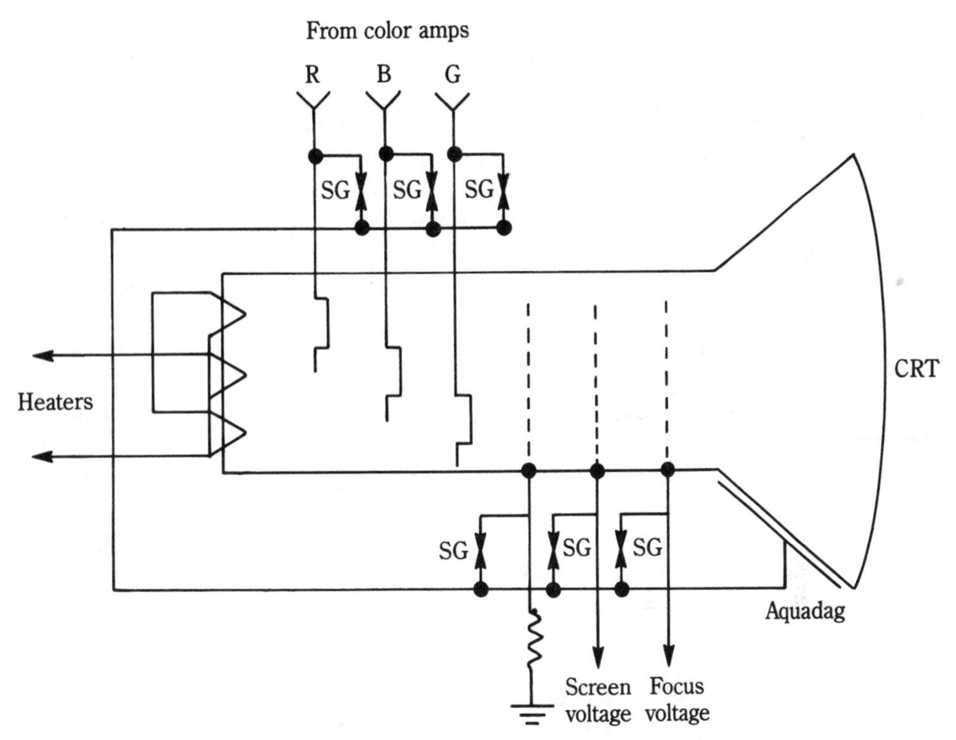

6-19 Excessive arcing of spark gaps could shut down the TV chassis if there are gaps in picture tube socket.

Ion trap

In some very old TV sets, there is an additional picture adjustment on the back of the TV tube neck—the ion trap magnet. Since this can seriously affect picture brightness, size, and picture tube life and since its position near the picture tube socket makes it vulnerable to accidental dislocation, it is important to consider its purpose and indicate the method of adjusting it if and when required.

The ion trap magnet is a form of bar magnet held on the picture tube neck by a flexible coil spring. Its purpose (in nontechnical language) is to ensure that the electron beam inside the neck of the tube is on course—moving properly along the axis of the tube. It is required because other electronic particles (ions) are incidentally generated inside the tube. Ions can damage a picture tube. The ion trap is thus part of a scheme of "separating the sheep from the goats"—getting rid of the ions, then making sure that the electrons move where they should be moving—through a number of small apertures toward the picture tube screen.

When adjusting the ion trap magnet, a combined sliding/rotating type of motion is used. The object is to get a dual result—full picture screen coverage at maximum brightness. It is actually possible, through misadjustment, to illuminate only a central (circular) portion of the TV screen and at a low brightness level.

Since ion trap adjustments can be detrimental to the picture tube, this procedure should be done carefully, repeated a few times, if necessary, until the best possible results are obtained. While a blank screen (no picture) is easiest to observe during this adjustment, there is no particular objection to do this with any type of program on the screen.

In connection with the adjustment of the ion trap magnet, a note of caution is required. An improperly adjusted ion trap magnet, in addition to producing a poor picture, can also damage the physical interior structure of the picture tube, and will ultimately result in poor focus and dark spots (burn spots) at the center of the tube screen. To avoid this, the magnet should be kept near the base end of the picture tube and adjustments made with the lowest possible brightness setting as quickly as possible.

For easiest adjustment and best results, it is recommended that the ion trap magnet be moved as far back (toward the tube base) along the neck as possible, consistent with obtaining the desired results.

The brightness and contrast controls must be kept low, so that slight changes are easiest to detect. After obtaining the best adjustment, it might be desirable to note or mark the position of the magnet both fore-and-aft as well as around the neck, with a marking pen or a small piece of masking tape; thus, should this adjustment accidentally be disturbed incident to other work in the back of the TV set, it is but necessary to move or turn the ion trap magnet to the marked location to restore the correct operation.

Ion traps haven't been used with new picture tubes in some years, but from time to time you may see an older set still operating with the original picture tube. New designs of picture tubes permit the deletions of the ion trap. If you own an old TV with a picture tube using an ion trap and purchase a replacement tube, you'll find that you can throw away the ion trap. That is, new tubes don't need the ion trap.

Magnet precautions

A warning pertaining to the effect of magnets, magnetic materials, and magnetic distortion on picture tubes is in order. While the subject of stray, unwanted magnetic effects on TV pictures came to the attention of the public only since the advent of color TV, the problem has always existed, even though to a much less degree in all TV receivers. We shall return to this subject in connection with the matter of *color purity* in color TV sets. At this time we shall concern ourselves only with black-and-white TV sets and how they can be affected by stray magnetism due to the careless handling of magnets and magnetic materials in certain areas of the TV set.

As explained earlier there are a number of adjustments, such as width, linearity, etc., which depend either on the use of a magnetic material, or on the modification of an existing magnetic structure on or near the picture tube. Since iron and steel are magnetic materials (even if they are not magnetized, that is, they do not seem to attract nails, paper clips, etc.), they may, when brought near an existing circuit, change that circuit, at least temporarily. Since screwdrivers, pliers, wrenches, etc., are almost always of iron or steel, their careless use on TV sets could adversely affect the TV set. For example, using a screwdriver blade to nudge or push an ion trap magnet or centering tabs is, therefore, not advised.

Similarly, using a pair of pliers to turn a tight thumbscrew, as used in some focus adjustments, can do more harm than good, particularly since the effect is not catastrophic but relatively minor. Of course, it is quite proper to use such tools on the chassis and other necessary areas in the cabinet, provided these areas are not too close to the tube neck, yoke assembly, focus magnet or the adjustments on the neck of the picture tube.

Other horizontal defects

Under the category of other horizontal defects comes such abnormalities as:

- *Tearing*—Picture breakup into horizontal strips shifted to left or right, somewhat like Fig. 2-12 (chapter 2).
- *Shifting*—What seems like a vertical split down the middle of the picture with the right-hand half of the left and vice versa, often accomplished by a very thin, faint picture (see Fig. 6-20).
- *Snaking*—The sides of the picture become wavy instead of straight up and down. This can also be accompanied at times by a rather thin, faint picture, although a normal picture may also be affected by this. See Fig. 6-21 for a typical example.
- *Streaking*—The picture seems normal in every way, except for dashed dark lines that seem to streak across the screen from left to right, somewhat like Fig. 6-22.

Referring to my description of the functions of a TV set, I said that there are such sections as sync (short for synchronization) circuits, whose function is to amplify the precise timing signals from the transmitting station and use them for horizontal and vertical frequency control, thereby keeping the picture steady on the screen. A failure or defect in any of the sync circuits will cause a loss of control with some such consequences as listed above. To localize the defect make two or three preliminary determinations.

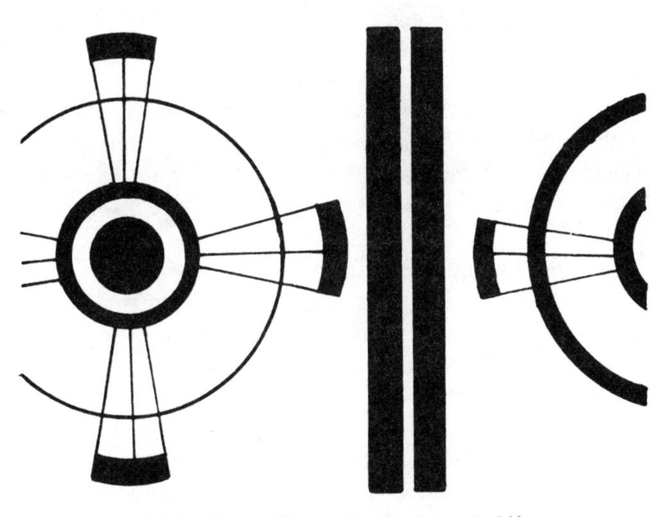

6-20 Sketch illustrating horizontal shift.

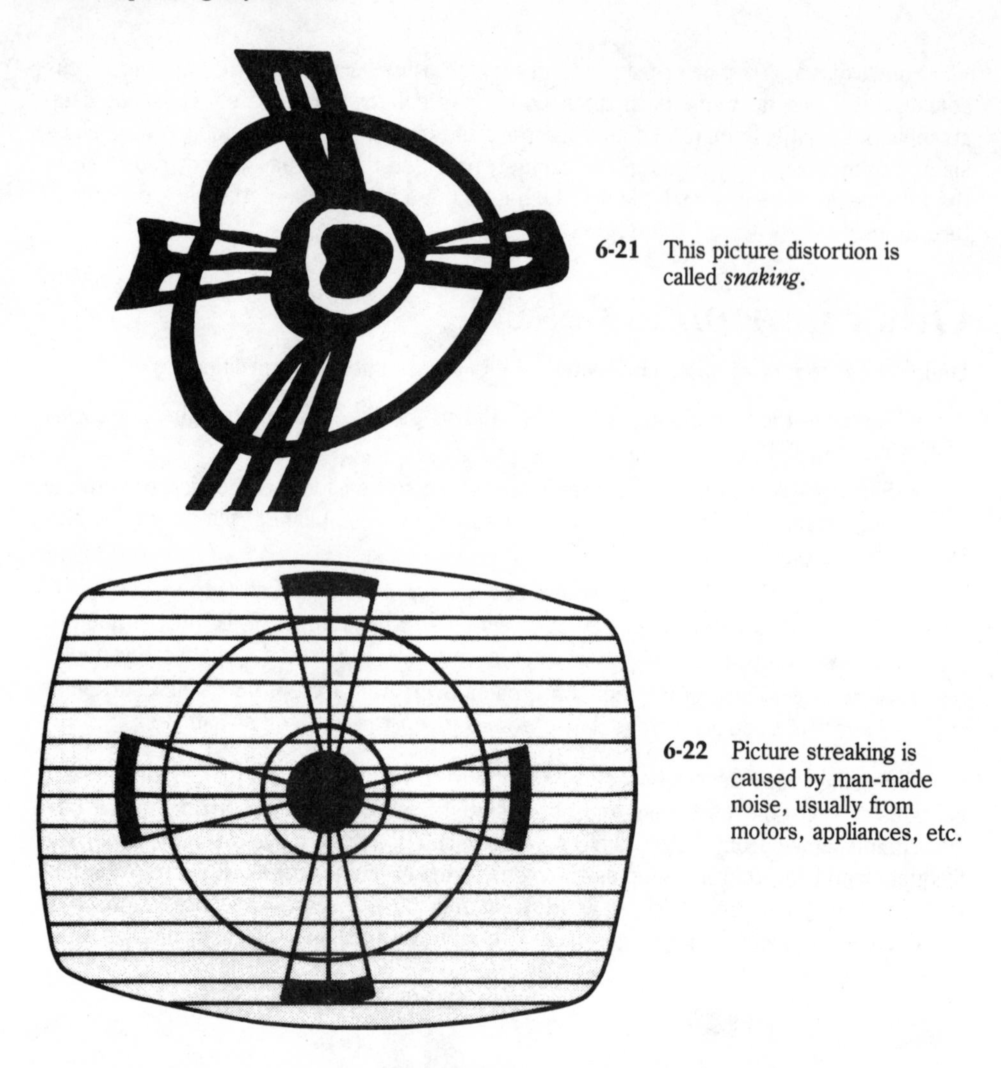

6-21 This picture distortion is called *snaking*.

6-22 Picture streaking is caused by man-made noise, usually from motors, appliances, etc.

First, is the picture generally weak, faint, or thin, even if the sound is relatively strong? The fault may be in one of the amplifier tubes (other than sync). In other words, any defective or old tube that causes a poor signal also causes a poor sync signal. While this condition affects both the horizontal and vertical picture stability, the vertical is far more susceptible.

Referring to Fig. 6-2, the offending tubes might be one or more of the following: V1, RF amplifier (on the tuner); V3, V4, or V5, IF amplifiers, and V6A, video amplifier. Replacing these tubes one at a time should locate the culprit. With the return of a normal contrast picture, the sync problem will automatically disappear. Sometimes in the normal course of aging a *number of tubes* may be responsible, each contributing partially to the trouble; thus, in a receiver with a set of tubes a few years old, more than one of the tubes listed may be responsible. Such tubes may test ''*weak-replace*'' or ''*?*'' on the tube tester scale. Replacing them one at a time will increase the performance of the signal stages, improving the sync and picture stability at the same time. If the picture contrast is normal, the fault is most likely in the sync circuit, either tubes or adjustments or both.

Notice that, for example, in Fig. 6-3 there is a tube marked "horizontal phase detector (V9A)," in addition to the horizontal oscillator (V9B) and horizontal amplifier (V10). Similarly in Fig. 6-5, V7C is called horizontal AFC, while in Fig. 6-2 tube V10A is marked horizontal control. In each of these cases, the tube referred to performs one type or another of horizontal sync function. A failure of any of the above mentioned tubes would invariably cause a loss of sync and some of the defects mentioned previously.

In those sets where a tube performs nothing more than a sync function, the defective tube is easy to identify: removing it from the set (in parallel-wired sets only) does not affect the already malfunctioning sync circuit. In most sets, however, the sync tube or the control tube, as identified above, is one-half and sometimes even one-third of a multifunction tube (two-in-one or three-in-one tube). In these cases removal of the tube will also disable other functions, often leaving just a bright horizontal line at the center of the screen; however, this is of no consequence. Replacement with a good tube should restore the picture as well as normal functions. Incidentally, when testing a twin or triple tube in a tube tester the tube should always be discarded if one section tests below normal, even if the others are perfect.

Overloading

Paradoxical as this may seem, too strong a signal can be just as detrimental to normal TV performance as too weak a signal. Specifically, overloading due to too strong a signal can cause snaking and picture tearing, and is usually caused by one or both of the following.

Contrast control Picture tearing can often result from excessive picture signal. Incredible as this may seem the technical structure of the average TV set is such that increasing the picture or contrast control to give a very hard or strong picture beyond that required for normal picture-viewing actually decreases the strength of the sync signal and often results in tearing. If this is the case, turning the same control toward a more moderate contrast setting should correct the tearing.

AGC pots In most TV sets manufactured in the past few years, there is a rear-chassis adjustment called AGC (automatic gain control). Its function, when properly set, is to place the receiver in such an operating condition that all stations in the receiving area will produce an acceptable picture with but an occasional adjustment of the front panel contrast control; however, misadjustment of the AGC pot can cause no end of TV woes, not the least of which is loss of picture stability (hum, buzz, and even a complete loss of picture). Since most TV sets have a tube whose function is to regulate AGC performance a defect in the tube could also be responsible. Because of these possibilities the following procedure applies to the AGC stage and the AGC pot setting.

AGC tube change With horizontal tear on the TV screen, remove the AGC tube and replace with a known, good tube. Touch up the channel tuning. If the picture becomes normal momentarily, switch to an adjacent channel, then return to the test channel. If the AGC tube was at fault, switching channels should not matter. If tear still exists, the AGC tube is not at fault.

AGC adjustment Assuming the AGC tube to be normal the adjustment is checked next. Using a moderately strong TV station, set the contrast (or picture) control about halfway, or a bit beyond (clockwise). Do not back off the contrast even if the picture seems too strong. While watching the picture in a mirror, advance (clockwise) the AGC adjustment on the back of the chassis until a buzz (or hum) is clearly heard. In some sets, advancing the control too far beyond this point may cause a complete loss of picture. Exercise moderation. Back off the AGC adjustment until all traces of the buzz disappear, or just slightly more than that. The AGC is now correctly set. Any malfunction of the horizontal sync will no longer be caused by the AGC setting, and other corrections must be attempted.

Horizontal hold, lock and range

While all three of these adjustments may not be found in any one set, all have a horizontal hold adjustment, and some have one or the other of the remaining two. Any one of these may be responsible for picture tearing or instability.

Horizontal hold Assuming good tubes, a picture may fail to hold if the horizontal hold control is at either extreme of its range. It is permissible for the picture to begin to show tendencies of tearing or instability when the control is at either end of its travel. As mentioned earlier some horizontal hold controls have less than a full turn of adjustment range. In that case a fraction of a turn is all that will be required for correction. In slug-tuned horizontal hold controls, two or three turns in one direction or the other should correct any instability due to this setting.

Horizontal range and lock These are always back-of-the set adjustments and are intended to set the range over which the automatic horizontal synchronization is effective. The proper setting is that which will produce no loss in horizontal stability when adjusting the fine-tuning control (if any) or when switching from station to station. Starting with a condition of instability or tearing, the horizontal range or horizontal lock control is adjusted very gradually until the tearing disappears (picture is normal). Then a test for stability is made by switching channels or by turning the front-panel hold control to near extremes (almost fully clockwise, then almost fully counterclockwise). In either case, there should not be a loss of sync.

Color solid-state horizontal adjustments

In early transistorized color chassis, the horizontal oscillator adjustment was rotating the ceramic metal core inside the horizontal oscillator coil. Within later horizontal circuits, they were located inside a deflection IC (Fig. 6-23). Instead of a horizontal oscillator coil, R607 varies the voltage applied to the internal horizontal circuits, locking in the picture. You may find one large IC processor which contains all the deflection circuits plus the sync separator, AGC and ABC detector, IF processing, and sand-castle generator.

No vertical or horizontal controls

You will not see any outside horizontal or vertical control in many of the new color TV chassis. Both the horizontal and vertical oscillator stages are synchronized internally with the incoming signal. The horizontal oscillator circuits may be found in one IC developing the deflection signals, sand-castle output pulse, coincidence voltage output and video IF processor (Fig. 6-24). The horizontal oscillator begins oscillating when voltage

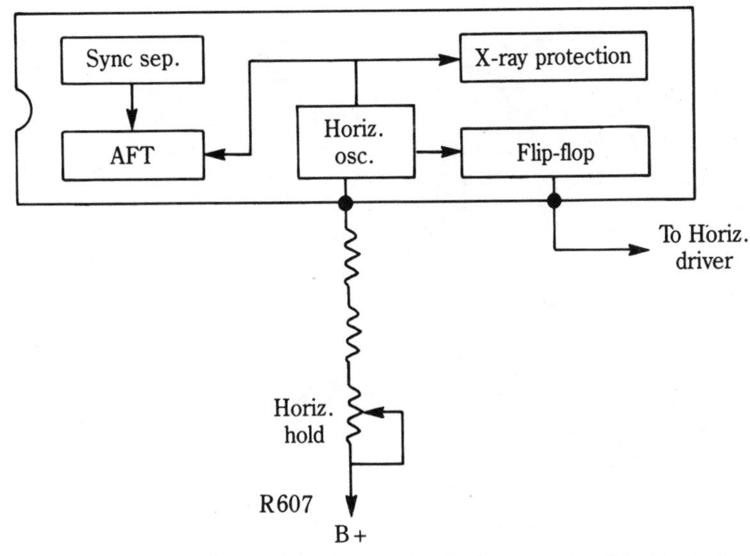

6-23 Block diagram of the horizontal circuits in part of an IC chip in the solid-state chassis.

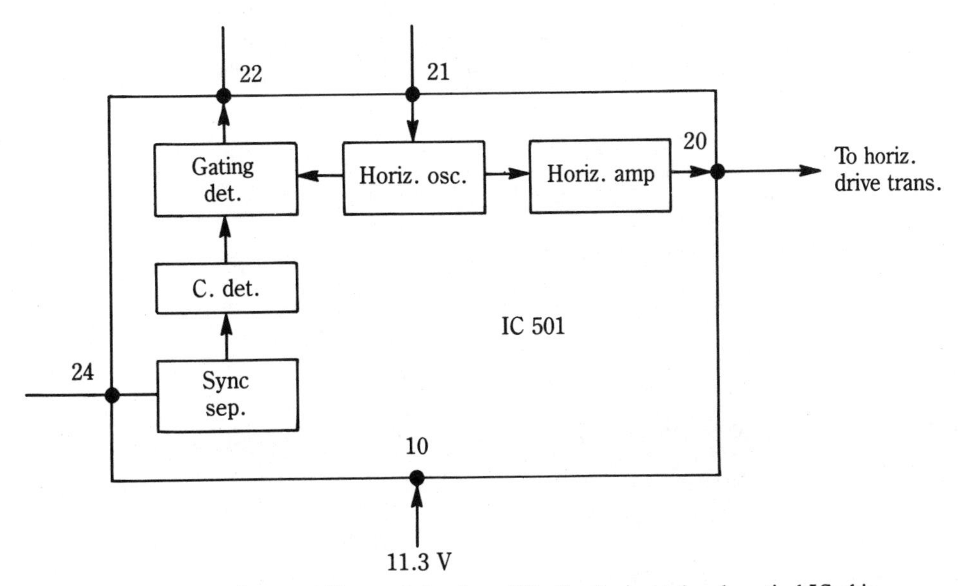

6-24 Block diagram of internal circuits within the horizontal and vertical IC chip.

is applied to pin 10. The horizontal oscillator is locked in with the incoming signal at pin 24. The sync separator circuit provides signal to the horizontal sync signals to the gating and phase one detector. The horizontal oscillator frequency is controlled by a dc voltage at pins 22 and 21. The horizontal oscillator drives the horizontal amp to horizontal driver stage.

The horizontal drive circuits without horizontal hold controls may consist of an IC processor which includes deflection circuits, AGC, IF amp, video detector, AFT, sound IF, audio processing, luma/chroma processing, sync separator, X-ray protector, horizontal AFC, VCO and horizontal countdown. The horizontal countdown signal generates the drive signal for the horizontal drive transistor.

The frequency of the horizontal oscillator is controlled by the VCO crystal (Fig. 6-25). The output of the VCO circuit is applied to the horizontal countdown circuit. The countdown circuit divides the signal to the proper frequency for the horizontal deflection circuit. The horizontal signal is applied from pin to a buffer and horizontal drive transistor. The crystal-controlled horizontal circuits provide stable operation without a horizontal hold control.

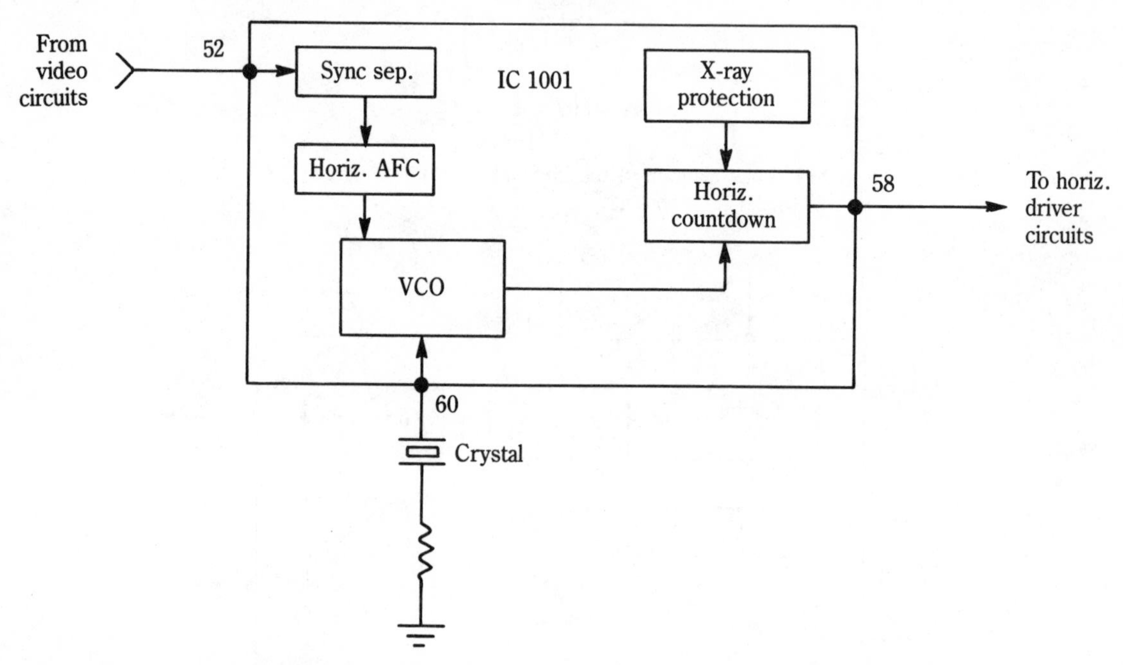

6-25 Block diagram of the internal circuits with horizontal crystal-control frequency of the latest RCA color TV chassis.

Horizontal shifting

This phenomenon is seldom present with a strong signal and it is likely to occur during freak reception; however, it can be caused by a substandard sync pulse, hence the procedure for sync tube changing or horizontal lock adjustment should be resorted to, if, upon careful thought, it is concluded that this is not a case of freak atmospheric conditions. Incidentally, a careful observation of the tube's behavior in a tester is of particular importance here. A *leaky* or *gassy* tube is quite likely to cause sync difficulties of every kind; so careful observation of the neon leakage test is imperative.

Horizontal snaking

Tube leakage is indicated by a neon glow, and even a level of leakage not detectable by most commercial tube testers, may be responsible for this type of picture defect. Tube substitution, regardless of the tube tester's verdict, is the only way of discovering such a defective tube. In addition all tubes in the horizontal group may be responsible, even if they appear to be less likely suspects than the sync tubes. An incidental clue to such an elusive tube defect (not shown on the tube tester because the tube doesn't operate as it does normally) is the fact that the snaking appears only after the set has been operating for a relatively long time, when the tubes and their surroundings have reached their highest temperatures.

Streaking

This display is unmistakable and it is caused by an electrical noise (commonly called static on the ordinary radio) of the manmade variety, and there is no simple way to eliminate it. It may appear either as long dashed lines traveling slowly across the screen, or as a random band or stream of snow, also moving across the screen in unison. If it is chronic, it may require relocation of the TV antenna by trial and error—a very costly remedy.

The noisy condition usually occurs from channel 6 on down. When channels 12 and 13 have excessive noise, you may assume the noisy system is close by. You can check for excessive noise coming into the power line of the house with a portable radio. If the noise is not noticeable at the power line entrance, suspect the noise is picked up by the outside antenna.

If the noise can be traced to its source (an arcing power company pole transformer, a heating system motor that sparks excessively, a similar workbench motor or some other defective electrical appliance), the proper procedure is to correct the condition at its point of origin. Should the source be a utility company device their cooperation will be required. It can usually be obtained provided a company representative is shown the evidence. There should be no difficulty in recognizing the malfunction and, if it is his company's responsibility, he will arrange for correction. Where any in-house appliance is causing the noise, correction will have to be made by a competent servicer familiar with such appliances (oil burner, for example). More on this subject in the next chapter.

Intermittent streaking and picture breakup

The tuner or front end can cause erratic, intermittent streaking, rolling, and tearing of the TV picture. This very critical and most sensitive of all sections of the TV receiver, black-and-white or color, is without question the most manipulated and most worn portion of the set. Regardless of make or model, the tuner contains a large number (up to 100 in some sets) of delicate switches or contacts, undergoing make-and-break operation whenever a station is changed. Since, as explained earlier, the precise sync-timing signals are an integral part of the TV picture, the signal interruption caused by switching from channel to channel also removes the sync pulses. For a moment the TV receiver is on its own.

In a normally functioning TV tuner, that is, one that is not worn out or otherwise defective, switching channels momentarily interrupts the timing signals, but they are restored almost instantly. During the switching interval a quick tear or roll may take place, or there may be a momentary diagonal zig-zag. One frame (picture) may slide up or down. In a worn tuner, and that includes virtually every tuner after a few years use, the switch contacts tend to become intermittent and erratic, causing some misbehavior of the kind just described.

With a color receiver malfunctions such as color confetti, absence of one or more colors, and especially color instability and shifting, can be due to poor contacts in the tuner. No special point is made of this fact under the color picture troubleshooting because the problem is not peculiar to color. In fact, it is much preferable (as is indicated in the introduction to color TV troubleshooting) to cut off the color when troubleshooting such color. When operation is restored to normal on monochrome, color performance will automatically have been corrected. In other words, this is a basic receiver problem, not a color problem.

There are two ways to solve a noisy or intermittent tuner problem: either thoroughly clean and lubricate it or have it overhauled by a specialist. A noisy tuner is so common in TV sets that a very practical and relatively inexpensive procedure has been established and is being followed by the repair industry. The old tuner is sent to a specialist overhaul shop where a factory-type overhaul is done on each tuner. The end results may be called as good as new. In view of the heavy use the tuner must endure, anything less than such a professional overhaul is a poor second choice, except where this service is not available or where a competent servicer is positive after examination that the overhaul is not required.

If you decide that cleaning and lubricating is the best course, you should begin, if you feel confident enough to tackle the job yourself, with a 100 percent detailed examination of the *before* conditions, including the method used to attach the tuner to the front panel or the chassis, and the exact positions, color, marking, etc., of each lead and cable from the tuner to the chassis. Most leads, in newer sets at least, are of the quick-disconnect type, but if one (especially wire braid or shielding) is soldered, carefully unsolder it, using no more heat than necessary. When you reassemble, you must exercise equal care in resoldering to the same spot. Be sure the soldered joint isn't a cold-solder joint (a seemingly rigid mechanical joint with a grainy appearance that will pull off if tugged hard or pried). It is very important to return all wires to the same locations as before.

After removing the tuner from the chassis, look at the contacts. Misalignment and looseness should be apparent now. In case of snap-in channel coils, remove one coil to observe the interior. Using a fine, soft brush (a small camel's hair artist's brush is best) clean the wiping contacts and the shaft bearing with a contact cleaner sold in radio supply stores. Do not use household grease solvents, especially carbon tetrachloride. Allow all surfaces to dry. Best cleaning can be done while repeatedly switching channels back and forth.

Tuner spray lubricant

There is a quicker, although not necessarily better, way of correcting noise and poor contacts in the tuner without resorting to the more tedius removal, cleaning, and lubricating procedure. It is not a cure, but a correction which may last for quite a long time. It is practiced by many TV servicers whenever it is decided not to remove or replace the tuner. This method involves using spray lubrication.

Electronic component supply houses sell a pressurized spray cleaner/lubricant, marketed under various names, usually described as a cure-all or similar solution. The procedure still requires removal of the TV set from the cabinet and removal of the protective shield from the tuner proper. A small amount (so as not to cause dripping) is sprayed onto the contacts of the tuner switch sections while the station selector is manipulated, as when changing stations, to enable the fluid to reach all contact surfaces. The solution dries quickly and is claimed to leave no objectionable residue, only a film of conducting lubricant. Incidentally, removal of the selector knob and spraying into the tuner, as suggested by one spray manufacturer, is hardly a satisfactory solution.

Tuner wash

There are many different types of cleaning chemicals available for cleaning the tuner contacts. In some wafer type tuners, excess grease attracts particles of dust, that become hard and result in a very dirty tuner. These tuners may not clean up with just any ordinary tuner cleaner. A large spray can equipped with additional plastic tubing and capable of delivering a heavy spray of cleaning fluid should be used to flush out the lumps of grease and dirt.

Place a drip cloth under the tuner and protect the chassis area from the washing liquid. Point the small plastic end right on each wafer switch contact assembly. Spray

6-26 Do not adjust any controls or slug adjustments to try and improve the color picture without the correct test equipment and knowledge.

both sides of the wafer switch to dislodge all greasy residue. Rotate the tuner switch assembly so each section can be washed out. After applying several coats, complete the tuner touch up by spraying the wafer switch contacts with regular spray lubricant.

Don't just turn any screw

Often, when the TV set acts up, a person has the tendency to turn a few controls to get the program back upon the screen. Readjustment of all external controls are placed conveniently for this purpose. Do not go inside and start turning a bunch of preset controls. There are many screws and preset adjustments inside the TV chassis that should not be touched unless you have the correct equipment for alignment and making certain tests (Fig. 6-26).

Leveling the picture

When the picture is not level, loosen yoke bolt with a 1/4-inch nut driver. This metal type screw is found on a metal clamp at the rear of the picture tube (Fig. 6-27). Just loosen screw and rotate the yoke to level. You should hold a mirror in front of the TV screen so you know when the picture is level. Now, take a look at the front of the TV set. Lock the metal screw with a nut driver.

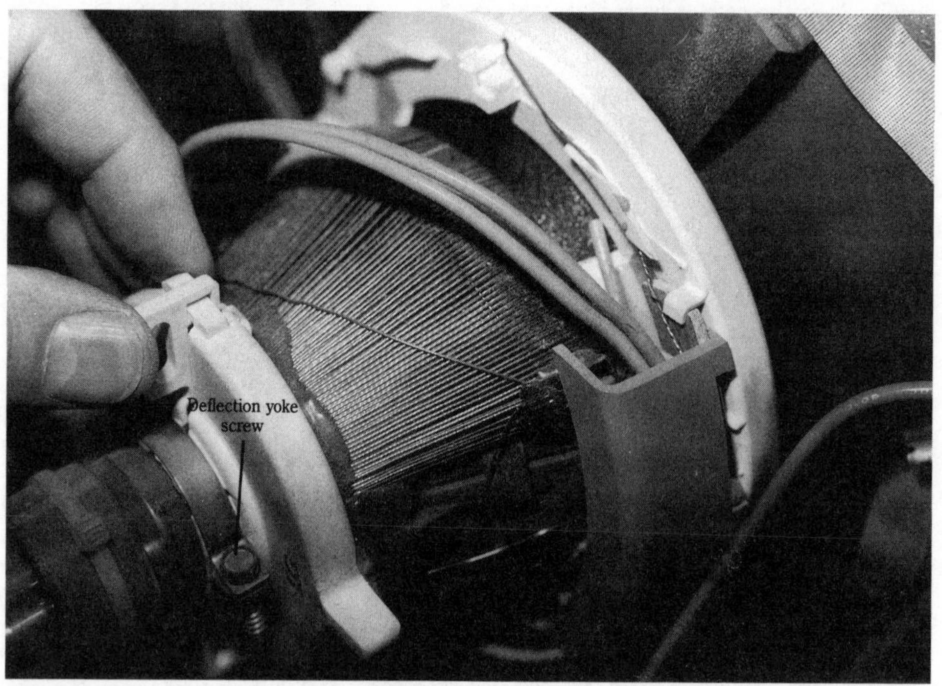

6-27 When the picture is not level, loosen up screw on back of deflection yoke. Turn yoke in the direction of side that is high. Use a mirror to get correct level adjustment. Back up and take a look at the picture from a distance.

7

Internal troubleshooting

CHAPTER 6 DISCUSSED WHAT MIGHT BE CALLED EXTERNAL TROUBLESHOOTING—the problems and solutions associated with the various controls and adjustments accessible from the outside of the TV chassis. This chapter and chapter 8 concentrate on defects and cures originating beyond these adjustments in the interior of the TV set. Since there is no absolutely clear dividing line between these origins and locations of defects, there will necessarily be an overlap between them; however, it is desirable to separate these, since the so-called external or adjustment type of defects are the most common and much more frequent.

Blooming and defocusing

This is a common phenomenon in older TV sets. A picture with blooming and defocusing problems exhibits insufficient brightness and contrast. Advancing either control, but particularly the brightness control, seems to have the effect of defocusing the picture, softening it as if a cloud were spread over the picture, fuzzing over the sharp details. The brightness may increase somewhat, but not enough before detail, and to a certain extent, shape is degraded. The fault lies in either or both of the following two areas.

High-voltage supply and picture tube

There is a considerable amount of interaction between the high-voltage power supply and the picture tube. In an old picture tube, the electron emission capability may be substantially below normal. Increasing the brightness setting further taxes an already depleted source of electrons, while at the same time it upsets the voltage distribution in the high-voltage (kilovolts) output. This is directly responsible for picture brightness. Similarly, the natural aging of some tubes in the horizontal deflection high-voltage system further acts to produce blooming and decrease electron beam acceleration.

Referring to the typical block diagrams in Figs. 6-2 through 6-5 in chapter 6, notice that each example has a high-voltage rectifier (V12 in Fig. 6-2, V11 in Fig. 6-3, V12 in Fig. 6-4 and V11 in Fig. 6-5), a damper (V13 in Fig. 6-2, V12 in Fig. 6-3, V13 in Fig. 6-4, and V10B in Fig. 6-5) and a horizontal amplifier (V11 in Fig. 6-2, V10 in Fig. 6-3, V11 in Fig. 6-4 and V10A in Fig. 6-5). In addition, in color TV sets there is also a high-voltage regulator which is covered in greater detail in a later chapter. Each of these tubes can be responsible for blooming in the following order of probability:

High-voltage rectifier

The high-voltage rectifier is almost certain to be one of the following tubes, regardless of the model number of the TV set: 1B3, 1K3, 2B3, 3AT2, 3A3, 1X2 and 2AS2. This tube is invariably inside the high-voltage cage or under a protective shield or barrier, as the case may be. As mentioned earlier, it is difficult to see whether or not this tube filament is glowing. This does not matter in this case since the tube is certain to be on when the picture is blooming; the suspicion is that the tube is substandard. Replacement with a new one is a sure test. The ordinary tube tester is less reliable in this case since it may show the tube to be weak or good which is not good enough in this case. Standard precautions in handling this tube during replacement should be followed, of course.

Damper tube

The damper tube may be indirectly responsible for blooming, but if the high voltage is substantially below normal, this tube may simultaneously affect vertical size, horizontal width, and the general performance of the set since this tube is an essential part of the overall power supply system. If the tube tests *fair* or even *fair to good*, it is not good enough and should be discarded in favor of a new one.

Damper diode

The damper diode in the solid-state TV chassis serves the same job as the damper tube. Of course, the damper diode often shorts or becomes leaky. In many cases, when the diode shorts out, the line fuse is blown. The damper diode does not act like the tube in the damper circuits for blooming conditions. Either the diode shorts or becomes leaky (Fig. 7-1).

The damper diode is found in the collector circuit of the horizontal output transistor. Within the latest TV chassis the damper diode may be included with the horizontal output transistor. The diode connects from collector to emitter terminal inside the transistor. When the fuse in the solid-state chassis keeps blowing, check for a leaky damper diode or horizontal output transistor. Simply measure the resistance with the diode check of a DMM. Measure from metal body of output transistor to chassis ground. A good diode and transistor will measure around 0.500 ohms in only one direction (Fig. 7-2). If a measurement in both directions from the transistor body and chassis is 100 ohms or less, suspect a leaky damper diode or horizontal output transistor.

Horizontal output tube

Since this tube and its circuits are responsible for producing the high-voltage supply for the picture tube, it can account for blooming because of reduction in this supply. While it

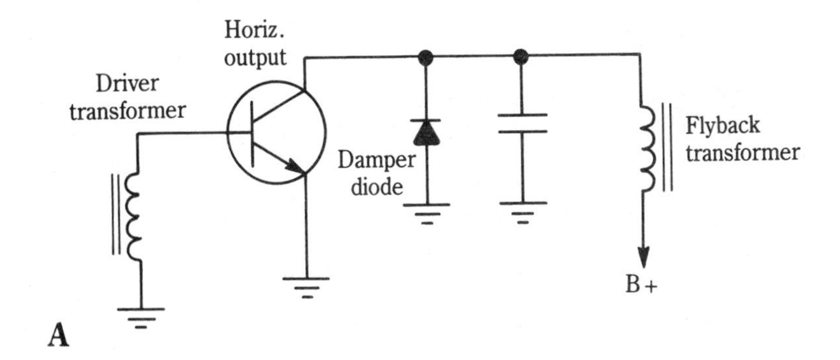

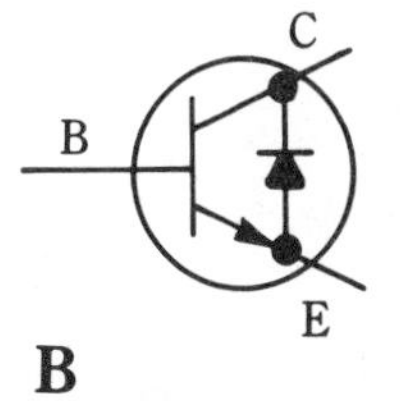

7-1 The damper diode outside of the horizontal output transistor (A) and inside transistor (B).

7-2 With the negative terminal on the metal body (collector) of the horizontal output transistor and the positive (red) probe at ground, a good diode or transistor should measure under 0.500 ohms.

was indicated earlier under decreased width in chapter 6 that a poor horizontal amplifier tube will cause reduced picture width, this is not always the case. Depending on individual TV set circuit details, reduced high voltage *prevents* a reduction in picture width since blooming and dimming also expands the picture size, both horizontally and vertically.

High-voltage regulator

Because of the more critical requirements of the color system, additional circuits and tubes are used to keep the high voltage as nearly constant as possible. This applies to both color and black-and-white reception. Looking at the color TV block diagrams in chapter 9, the regulators are V13 in Fig. 9-1 and V14 in Figs. 9-2 and 9-3. On color reception, blooming may evidence itself as color smearing, fuzzing and what is sometimes called color instability. In troubleshooting such a TV set, in addition to following the suggestion of turning down the chroma control to remove all color, a check after replacement of a suspected regulator tube should include turning up the chroma control to see how the remedial action has affected color reception.

Picture tube problems

When no symptoms other than blooming and patchy, chalky white results from advancing the brightness and contrast controls, the picture tube is the most likely culprit. When too much brightness or contrast is raised, notice the fuzzy white, blotchy picture with a close up of a person's face. Often, this condition is caused by a weak picture tube gun assembly (Fig. 7-3). If the color will bleed in sections, suspect the CRT of weak and gassy conditions. Also, a very weak picture tube results in not enough brightness.

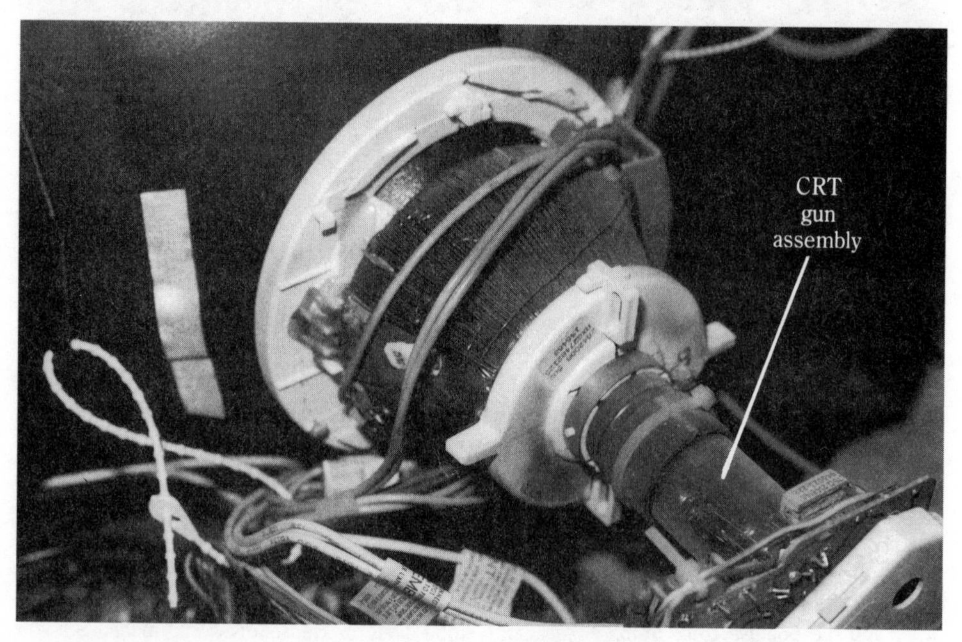

7-3 A weak picture tube gun assembly can cause weak, dim, or botchy color picture.

A picture tube rejuvenator or booster will extend the remaining life of the tube by a number of months. Care must be exercised to obtain the correct type of booster (for series or for parallel tube filament sets). Actually, the higher filament voltage results in more electrons flowing from the cathode and bouncing against the color beads on the front of the picture tube.

For three-gun color tubes, boosters have become available and may be used, provided the instructions are carefully followed. Since the color tube consists of three separate electron-emitting structures, some boosters and rejuvenators have a simple provision for boosting one of the three, depending on which of the color guns is deficient, by externally adjusting the booster before used.

Picture tube rejuvenation

Improper heater voltage or no voltage may produce a weak or entirely black screen. Of course, improper high or low voltage applied to the picture tube elements may produce a weak or dead raster. Sometimes by peering down towards the picture tube socket you may see all three heaters light up in a color picture tube.

The color picture tube has three separate gun assemblies, including heater and cathode elements (Fig. 7-4). The filament, like any tube, is enclosed inside a nickel-plated cathode that emits electrons. When the cathode is used up or is bombarded with layers of ions, the cathode will emit fewer electrons. If the heater voltage is raised or the cathode is cleaned off of ions, the gun assembly will appear almost good as new for a few months.

The electronic technician may have a color picture tube rejuvenator that restores the picture tube. The rejuvenator is enclosed in the same test instrument as the tube tester. This instrument will strip or clean off the ions from the cathode element, letting thousands of electrons to emit and strike the color beads on the glass front of the picture tube. The ions are a result of electrons bombarding the screen and falling back upon the cathode, decreasing the flow of electrons.

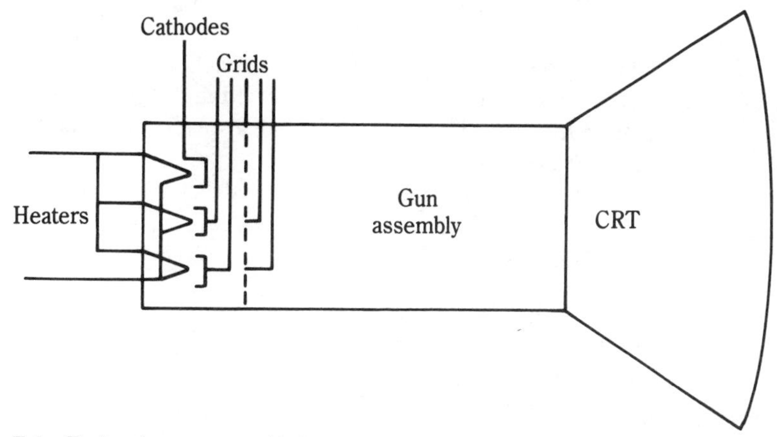

7-4 Each color gun assembly has a separate set of elements: heaters, cathodes, and grid terminals.

A high voltage is applied to the cathode element from the picture tube rejuvenator when the heater voltage is raised removing or cleaning up the cathode surface (Fig. 7-5). You can see the flashes of arcing and light around the cathode element when the rejuvenation occurs. Each color gun is cleaned up and rejuvenated. Sometimes all three guns will clean up restoring the picture tube, almost like new. Other times, one or two guns will not come up at all. Usually, there is a separate charge for rejuvenating a color picture tube. Sometimes the rejuvenated picture tube will last the life of the TV set. Other times, only six months or more.

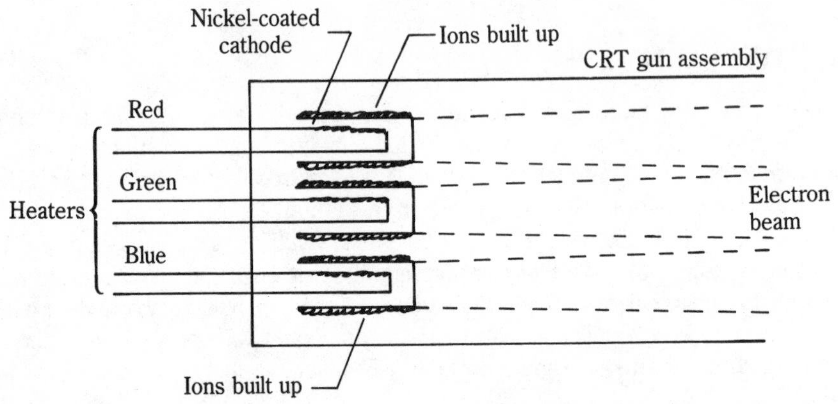

7-5 A picture rejuvenator test instrument removes ions built up on the cathode and filament elements.

Dim and blooming when set is cold

This behavior is exactly the same as described previously except that the picture gradually seems to achieve normal brightness and contrast after a good warm-up period (a quarter-hour or more). The temporary condition is almost certain to be due to weak tubes. While a number of tubes can be responsible for this condition, the picture tube is the most likely suspect. The procedure, therefore, starts from this point; however, since it is impractical to remove the picture tube and test it like a small tube, a suitable booster is employed to obtain the same result.

Attach the booster to the picture tube, following instructions given in chapter 8. Allow the tube to operate with the booster for a few days. If the trouble disappears, the picture tube is at fault. Three remedies are possible:

- Leave the booster on permanently. It will prolong the life of the dying tube.
- Have the tube rejuvenated. Approximately the same results will be obtained but the booster will not have to be used.
- Have the picture tube replaced. Under normal conditions, between 80 percent and 90 percent of the tube's useful life has already been realized, so the cost of a new tube is quite justified if the TV set is otherwise performing satisfactorily.

If the booster does not materially improve the performance, a check of one or two other tubes in the set should be made. This includes the high-voltage rectifier and the damper tube, in that order.

If none of the above checks locate the cause there is the possibility that a component (most likely a filter capacitor) in the high voltage or B+ boost circuit is responsible. This, however, is not a task for a beginner.

Hum and buzz

These two names, hum and buzz, actually stand for two different symptoms; however, since it may sometimes be difficult for an inexperienced ear to clearly distinguish between the two, and further, since the two often appear together, I shall describe them together, but suggest how to identify their sources which may or may not be the same in all cases.

Hum

Hum can best be described as a smooth, steady low-pitched note that is the standard characteristic of household (60 hertz) ac. It can usually be heard on almost any tube-type radio with the volume control fully lowered if you place your ear close to the loudspeaker. It is almost always independent of the station being received, the weather, volume control setting, etc. In fact, it is the sound of 60-hertz ac when audible. In most tube receivers (and sometimes in transistorized sets when not operating on internal batteries), the hum is due to a small residue of this ac sound that is not filtered (Fig. 7-6).

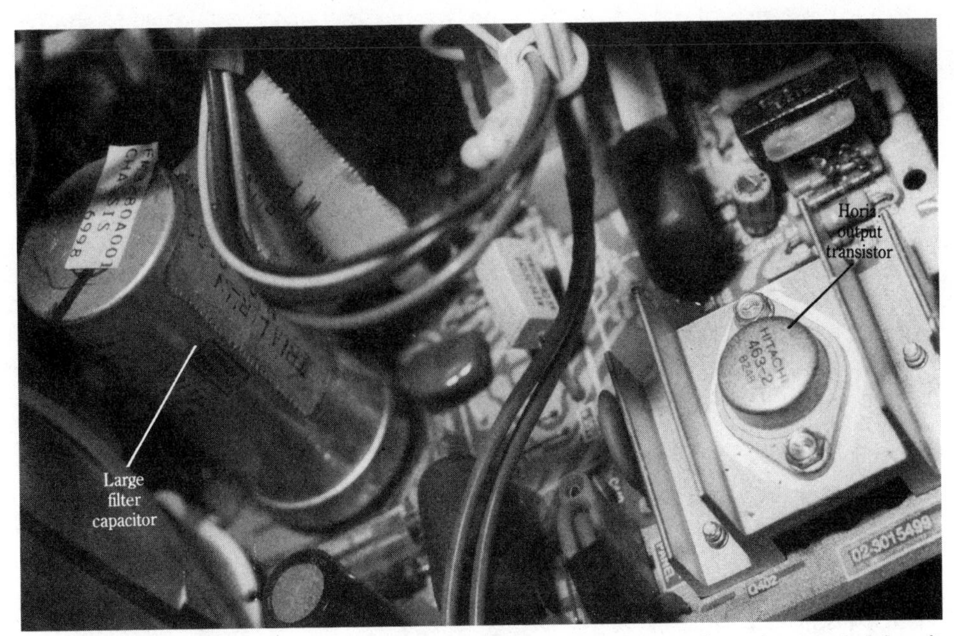

7-6 A dried out or defective filter capacitor can cause hum in the sound or appear as dark bars in the raster.

In severe cases, the appearance of hum signifies one of three possibilities: First and most likely is that one or more of the filter capacitors in the receiver (radio or TV) has deteriorated with age and heat so that it does not function at maximum effectiveness. Since there is no simple visual indication or identification of the defective component, the services of an experienced technician are required. A second although less frequent cause of hum is a defective tube which has developed a *partial short* between the ac input and one of the functional elements in the tube. In technical parlance, this is usually called heater-to-cathode leakage. This can be analyzed, located and remedied by following a simple test procedure.

Turn the volume up or down. If this seems to make no difference in the hum level, turn the volume fully off. If no change in the hum results (there probably will be no change) the fault most likely lies on one of the filter capacitors. Incidentally, this type of defect seldom happens suddenly, it is a gradual deterioration, not a catastrophic failure.

If the hum level changes with changes in volume a tube is quite likely to be the cause. Furthermore, such a defect is peculiar to tubes only; transistor circuits are not subject to this difficulty, although a filter capacitor failure as described could produce hum in a transistor receiver as well. To confirm your suspicion, change to a different station and adjust the fine tuning. Any resultant changes in the hum level further point to a defective tube, and the isolation to a specific tube is carried out in one of two ways, depending on whether the TV set has a parallel or series heater string.

In the case of parallel wiring, remove the sound discriminator tube. This is the first tube in the TV receiver which handles nothing but sound (audio). If the hum persists with this tube out of the socket, the fault lies beyond this point, closer to the speaker. This includes the first audio amplifier (if any) and the final audio amplifier. To find out which of these two tubes is responsible, replace the discriminator tube and remove the first audio tube. If the hum still persists, the final audio amplifier is most likely responsible. Replace it with a new tube.

If removing the discriminator tube stops the hum, either that tube or one of the tubes preceding it is at fault. If replacing the discriminator tube does not cure the hum, the fault lies in a preceding stage. In that case, remove the first IF tube. The logical sequence here is the same as just outlined for the audio sections; thus, if the hum disappears with the first IF tube out of,its socket, either that tube itself or one of the tubes from there to the antenna is responsible, including the RF amplifier (V1 in most sets) and the converter (V2). Remove the RF amplifier. If the hum stops, this tube is at fault. If not, the next tube in the tuner (converter V2) is defective.

If pulling the first IF amplifier does not stop the hum, the fault lies in either of the remaining two IF amplifiers. Remove and replace the second IF amplifier. If the hum stops, try a new tube. If not, the third IF amplifier is defective.

A final note on this hum problem. Earlier it was stated that while a constant hum regardless of the channel selector position or adjustments is most likely due to a defective filter component, and that a hum that responds to a change in station or volume control setting is probably due to a defective (although still functioning) tube, it must be pointed out that a tube-caused hum problem may produce effects other than audible hum. This includes snaking or weaving, or other image distortion on the screen; so be sure to observe any abnormal displays on the screen.

Removing tubes to trace a hum source is a convenient method for sets with parallel-wired filaments. But what about those sets with series filaments? Since removal of one tube (with rare exceptions) disables all others, the only alternative is outright substitution all the way until the offender is found. Some time can nevertheless be saved here by proceeding as before from the audio output tubes back toward the antenna.

A general precaution in all tube-caused hum problems: after plugging in a new tube, if the hum seems to have gone, wait at least a few minutes before considering the problem solved. Many cases of tube hum (heater-to-cathode leakage) develop only after a thorough warm-up period and disappear after some cooling. Since removal and replacement of a tube lets all other tubes cool down, it is necessary to wait until all the (suspected) tubes again reach their operating temperature. Of course, this does not apply to parallel filaments, where all but the removed tube maintain their operating temperature.

Buzz

By buzz, I mean a sound characteristically peculiar to TV sets (except the very ancient ones having a split-sound system), and originating in the TV circuitry incident to normal operation but due to one or more misadjustments. This buzz is similar to the hum described earlier, and is of the same pitch (60 hertz) except for a raspy, buzzy quality instead of the relatively smooth tone of the hum. There is also another type of buzz, not related to the electrical operation of the TV set—a mechanical buzz. Although similar to the TV buzz, which is sometimes called intercarrier buzz, the mechanical type is more haphazard and random, and is easily identified. In either case, buzz is a defect and requires correction.

Before attempting to correct this defect, it is necessary to ascertain which type of buzz it is. The easiest to eliminate are the mechanical type buzzers, such as the buzzing from the loudspeaker or anything loose in or near the cabinet. A dangling wire, a tube shield not seated, or even a piece of paper or cardboard near the set can easily be identified, located, and remedied by removal, tightening, etc. Loudspeaker buzz, although actually mechanical, is probably due to a defect in the speaker assembly—a tear in the speaker cone, a rubbing of an out-of-round voice coil (usually caused by warping due to heat), or even a simple loosened mounting. Any of these types can be recognized by comparison (keeping in step) with the audio, as well as total disappearance when the audio momentarily pauses. While a loss component is easily corrected, and a tear in the speaker cone can be judiciously repaired by application of quick drying cement to the tear (not patching with tape), a rubbing voice coil is almost never satisfactorily repairable. But it is quite within the ability of the average do-it-yourselfer to replace the whole speaker by following a few simple directions.

- Disconnect speaker from set. This will usually be a pair of wires connected by push-on clips, automotive style. Sometimes it may be a two-pin plug. Others will be soldered into the circuit. Dismount speaker from cabinet to gain access to cone (speaker front).
- Carefully depress cone near its center so that it moves into assembly, at the same time listening and feeling for any scraping or binding. There should be absolutely none. If this is the case the speaker is not at fault.

- Obtain a duplicate speaker. Any parts supply house will identify the speaker and furnish a replacement. Of course, the physical dimensions (4 by 6, 5 by 7, 4 by 10, 6 by 9, etc.) should be the same, otherwise there will be a fitting problem in the cabinet. Make sure that the impedance rating is the same, that is, 4 ohms, 8 ohms, 16 ohms, etc.
- Unsolder the old cable, if used, from the defective speaker and solder on to the lugs or terminals of the new speaker, taking care not to scorch or burn the cone in the vicinity.

If there is any question about the characteristics of the old speaker, a reference to the data package for the TV set (the parts list or the schematic diagram) will sufficiently identify the speaker to enable the replacement parts house to find a replacement.

The preceding examples were either mechanical types of buzz issuing from the offending part, or generated by the speaker in step with the sound. The following types are nonmechanical; but emanating from the speaker as does every other sound—music, speech, or noise transmitted as part of the program. What is most characteristic about these types of buzz is their pitch; they are all of 60-hertz pitch, although their timber is different. It is not essential for the purpose of this explanation to consider the technical reasons for this; it suffices to say that both hum (usually called ac hum) and buzz peculiar to the TV circuit operation discussed here are of the same basic frequency, which also is the frequency or repetition rate of the vertical sync. And although it was not intended that sync pulses reach the audio portion of the TV set, and ultimately the speaker, such is sometimes the case, as detailed below. When this occurs, the audio system responds as to any other sound and reproduces it in audible form.

As to the difference in sound between these, the musician would perhaps explain this by calling the ac hum a pure (single) tone, and the buzz a complex tone, one having an admixture of many overtones. As in case of hum, there are some characteristics of buzz which help locate its origin within the TV set. These are discussed in the following paragraphs.

Warm-up buzz During the first seconds (up to a minute or two) of the warm-up period, a buzz does not necessarily indicate any malfunction; a stabilization of some of the automatic gain circuitry usually reduces this buzz to an inaudible level. No action is required in this case.

Constant buzz If the buzz remains for any length of time or is a constant annoyance, simple adjustments may remove it. The checks for the specific cause should be made. Often it is necessary to set the fine-tuning control, not for the best sound consistent with a good picture, but for the optimum sound with good picture and minimum buzz. Once this control is so adjusted, it should hold for most stations without individual adjustments. If this does not reduce the buzz to inaudibility, overloading may be responsible.

Assuming the tubes to be in good condition, that is to say, not very old, and the contrast control not set unreasonably high, check the AGC pot. This rear chassis adjustment regulates the overall amplification of the picture and sound IF stages. Too high a setting will produce a buzz and sometimes a glaring picture. As indicated earlier, the AGC pot is adjusted by choosing a medium-good station. Adjust the contrast control about midway or until a good range of light and dark areas is obtained. Now advance the AGC pot until the buzz is audible; then reduce this setting just past the point where the buzz stops. Check the range of black-and-white areas to see that neither extreme has been lost, that is, the darker areas are not prematurely becoming a solid black while the very light areas are not becoming a glaring white. Adjustment of both the AGC and *contrast* may have to be made to obtain a happy medium—a full range of illumination from black through all shades of dim through full brightness with a minimum of audible buzz.

Sync buzz This fairly infrequent phenomenon of sync buzz is due to interaction between the sync tubes, vertical oscillator and amplifier and the video amplifier. Assuming that the TV set operated normally in the past, such a buzz may stem from two sources. One is the rearrangement (unintentional) of the wiring within the chassis so that the vertical sync and video amplifier circuits are too close to each other. This could happen only after a repair; therefore, it may be necessary to correct the *lead dress* to eliminate the buzz.

Practically every manufacturer has appropriate Caution notes to the servicer to replace any soldered lead in exactly the same position (as far as possible) as originally. We shall therefore assume for our discussion that any buzz pickup due to improper lead dress came about after a shop repair. The remedy is obvious. Unsoldered resoldered leads should be of the same length and in the same location as before.

Another source of sync buzz is a malfunction in the video amplifier tube. Substituting a new video amplifier tube will prove or disprove this suspicion. Incidentally, a tube tester check of this tube may not be conclusive because a rather small leakage or amount of gas within the tube may be causing the trouble; and the defect won't show up until the tube warms up.

Intercarrier buzz This is a buzz inherent and characteristic of almost all TV sets except the ancient ones known as the split-sound types. In a great many TV sets, there is a rear chassis adjustment (usually a slotted-shaft type) marked simply *buzz*. The control is rotated very gradually until the buzz is at a minimum. This should be done with a station tuned in while listening to the sound (music or speech) accompanying the program.

In some TV sets there is a variable coil (quadrature or discriminator) which is adjusted to eliminate buzz or tunable hum in the sound. A manufacturer's schematic and parts layout will help you identify the correct coil. (Do not attempt to adjust the sound coils unless you know exactly which one it is.) The sound coils are often located next to the sound output tube, transistor, or IC output (Fig. 7-7). Usually, the tallest shield coil is the interstage, and the short one is the discriminator or quadrature coil.

After locating the correct coil, check for the type of adjustment tool needed. Some coil slugs have screwdriver slots while others require a hexagon shaped alignment tool. Choose a strong TV signal and tune in the best picture, color, and sound possible. Remember, a touch-up of this coil is all that is needed. A slight turn clockwise or counterclockwise may be sufficient. Simply adjust the coil until all or most of the buzz has disappeared. Re-check the picture and sound reception from several different stations.

Multiple pictures

I refer here to the existence of two or more duplicate pictures one on top of the other. Usually each picture is not very stable and is only a fraction of the full screen height. The

7-7 The sound coil could be located next to the sound output transistor or IC circuit. Tune in a station. Adjust the fine-tuning control for best picture and sound. Then locate the sound coil and turn it slightly either way for best sound.

fault here is unmistakable: the vertical oscillator circuit is operating out of frequency. There are three possible causes, two reasonably likely, and the third very rare indeed.

Vertical hold misadjusted

Whether this control is an operating control on the front panel or a semiaccessible one behind a little door, it usually has sufficient adjustment latitude to cause oscillator operation at a fraction of its normal frequency or well above it. In other words, it is possible to produce such a malfunction by rotating the control to its extremes. The first step to correct a case of multiple pictures is to adjust the vertical hold control first in one direction, then in the other. The picture will roll up or down, and after a while should jump from three to two and finally to one picture across the screen. Careful adjustment is required after a single picture is obtained to leave the control at the optimum position. This is checked by switching stations and observing whether or not the picture tends to roll (Fig. 7-8).

Horizontal multiple pictures

When the horizontal control knob will not stop the picture rolling or moving sideways, suspect a defective horizontal oscillator circuit (Fig. 7-9). Look for a defective horizontal oscillator tube, transistor, IC, or poor filtering of the horizontal circuits. Within the transistor circuits the horizontal sync can be scoped, indicating poor or missing horizontal sync. Dried-up decoupling capacitors filtering the horizontal oscillator circuits may cause poor horizontal lock in.

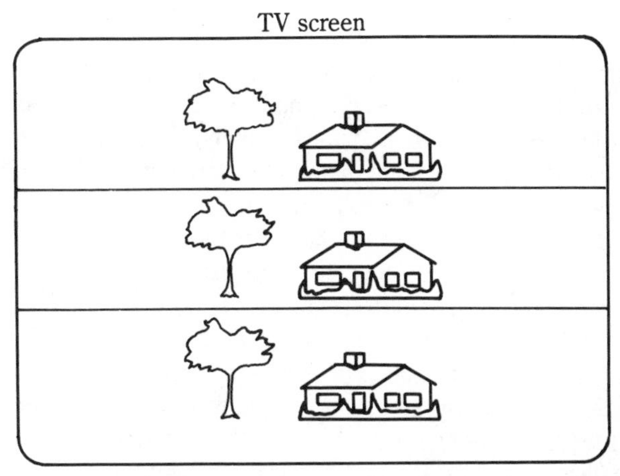

7-8 When the picture rolls upward or downward, readjust the vertical hold control until it locks in.

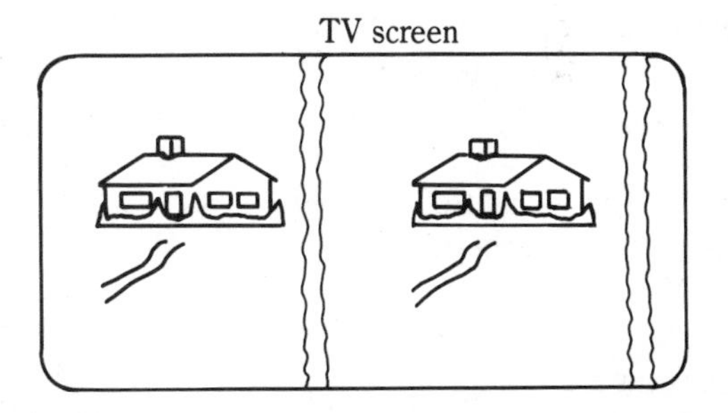

7-9 A defective horizontal sync or lock circuit lets the picture flop either right or left and cannot be straightened with the horizontal hold control.

Horizontal drifting may occur when the chassis is normal for a few seconds and then the picture goes into horizontal lines. The horizontal hold control may straighten up the picture for a few minutes until the picture starts tearing in another direction. Check each bypass and filter capacitor in the horizontal hold circuits. Sometimes spraying with coolant may locate the defective component.

Defective component

A defective capacitor or resistor (changed value) in the vertical oscillator circuit will cause something other than one picture to appear on the screen. In rare cases, the vertical hold control can compensate for this change, but this is most unreliable. If the control has to be moved to an extreme position, a component is at fault and the cure at best will be very temporary.

Defective tube

In very few cases, a defective tube may cause this difficulty. If the picture can be stopped by setting the vertical hold control to its extreme position, you may try replacing the vertical oscillator tube; however, just a slight improvement, that is, requiring almost as extreme a position of the control for a single picture display, still points to a defective component.

Other causes

What has been said about multiple pictures or scenes in the vertical direction, applies to some degree also to the horizontal direction; however, in case of horizontal off-frequency operation, either due to misadjustment or (more likely) to a defect, the whole image may be much less stable, going through any number of variations of Fig. 2-31, during which no recognizable picture may be displayed.

What is common to both types is the fact that there is a tendency for the scene to lock or stabilize at a sub-multiple of the correct repetition rate; thus, the vertical repetition rate is 60 hertz, and the transmitted sync pulses normally ensure locking at exactly this rate, provided the circuit is initially set as closely as practicable to 60 hertz. In other words, the transmitted sync information can and will correct minor deviations from exact frequency. Should this deviation be excessive, however, sync correction is no longer effective and the picture rolls or tears. Apparent locking or sync at a frequency which produces two or three pictures across the screen is what might be called a technical quirk (the explanation is beyond our scope here) occurring at exactly half or a third or sometimes even a fourth of the correct frequency. Should a malfunction cause the vertical sweep to operate at approximately 30 hertz, the sync will more or less hold, but two pictures will now be seen. Similarly, three images would appear if the frequency is around 20 hertz. Sync is less effective at each lower sub-multiple.

In case of the horizontal direction, the basic frequency is much higher (15,750 hertz). The number of the heavy stripes in Fig. 2-31 corresponds to the frequency the sweep circuit is operating.

Multiple pictures in the horizontal section might result from improper adjustment of the horizontal hold control. For example, after a loss of horizontal sync, the H-hold control is turned the wrong way. Instead of the picture straightening up, or locking in horizontally, all you see is lines and multiple pictures in a horizontal plane.

This condition can occur in small black-and-white or color portable TVs where a single coil adjustment is used as the horizontal hold control. These controls have no stops in either direction and can be turned either way until the adjustment slug drops out (Fig. 7-10). If the coil slug or core has fallen out of the coil at the back, restart the core from the back side. Often, the plastic shaft end has broken off, letting the core fall out. Replace the plastic shaft after threading the metallic core into the coil. Use the hot end of a soldering iron to make a flat end on the protruding plastic shaft so that it can not pull out.

Turn the shaft until the core is in about the middle of the coil. Connect the antenna and tune in a local TV station. Rotate the horizontal hold control while observing the TV screen to determine if the lines are spreading apart. Slowly adjust the control until the picture locks in. Next quickly switch the channel selector off channel and back on again to observe if the picture loses horizontal sync. Readjust the horizontal hold control until all stations remain locked in with the set hot or just turned on.

The preceding explanation of multiple picture causes is intended to aid the do-it-yourselfer to analyze and locate the defect in such cases. Keep in mind that, by design, the sweep frequencies are pretty close to the exact values, and the adjustments are fine adjustments made to compensate for *minor* deviations due to normal wear and tear. It should be obvious to the troubleshooter that a severe or extreme condition cannot be *adjusted* back to normal.

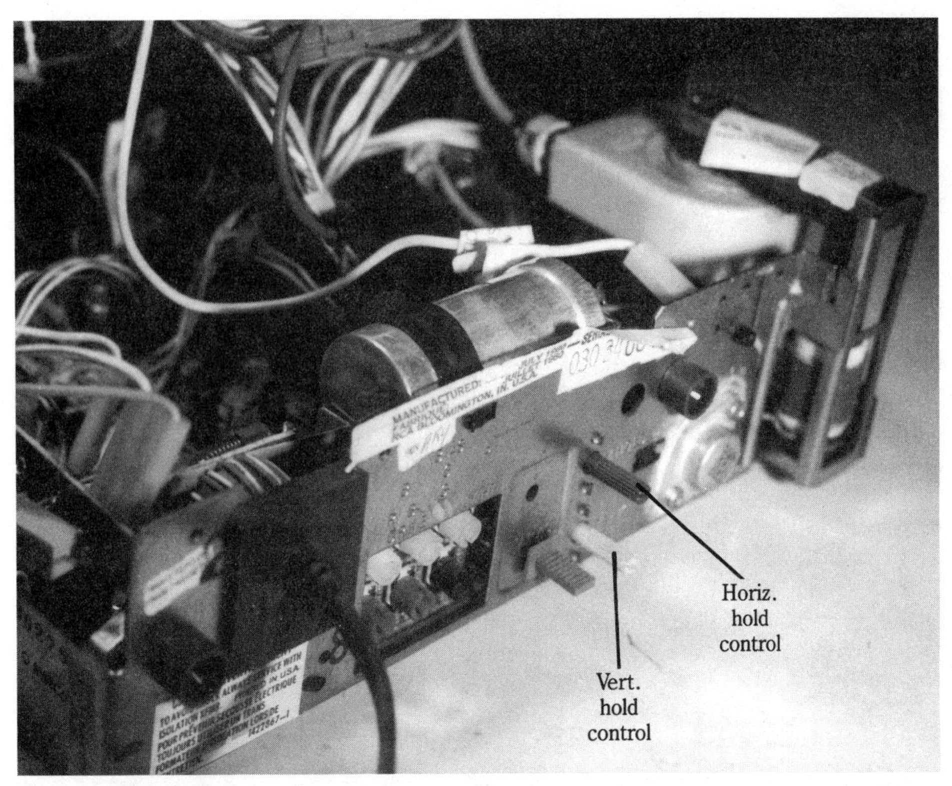

7-10 Locate the horizontal hold control on the back of the TV chassis. Rotate the control or coil in either direction. Observe if the horizontal lines open up. You are going in the right direction when the lines spread wider apart.

Various suggestions given here, particularly the adjustments for purposes of correction, are intended and will be effective only in case of wear and tear deterioration or gross misadjustment through carelessness. In case of a component failure, whether a tube, transistor or passive part such as a resistor or capacitor, no amount of adjustment will effect a permanent cure.

Even in those cases where setting the control to or near one of its extreme positions of rotation brings back synchronization, the cure will only be temporary, since, as explained previously, the setting of the control is only approximate, to enable the sync pulse to take over. With the control at its extreme, no correction latitude remains, except in the opposite direction of rotation. In other words, the control will no longer serve its normal function of being adjustable *up* or *down* from its center position. With but a slight further aging, no adjustment will be possible. Locating the defect and its correction is obviously the answer.

Vertical and horizontal lock-in

In today's solid-state TV chassis, you may find a horizontal frequency control inside the chassis and no vertical or horizontal hold control to be adjusted by the operator. This is usually accomplished with processor countdown and crystal-controlled circuits. The circuits are so well regulated and controlled that no hand adjustments are needed. Of course, when the TV pictures roll, bounce, or slip sideways, internal sync circuits must be repaired.

Smoking chassis

The smoking chassis may be found within the tube or solid-state chassis. Shorted or overloaded power transformers may put out a pungent smell. The arcing yoke winding may burn intermittently (Fig. 7-11). In the earlier solid-state chassis with integrated high-voltage transformer, the enclosed molded diodes may arc over. The early tube flyback transformers had a tendency to arc and burn with no drive voltage on the horizontal output tube. Besides a burning smell, the leaky tripler unit may arc over between plastic shell and chassis.

Although, the overloaded resistor and capacitor are small in size, they can short and burn holes in PC boards. Overloaded power transistors mounted directly upon the PC board have burned small holes. Often, the board can be repaired by cutting out the charred section and rerouting the wiring with external hookup wire. After being struck by lightning, the chassis may begin to smoke in several sections. Usually, the chassis can be repaired by replacing the smoking component.

7-11 The components indicated have a tendency to smoke, burn, or arc over in the TV chassis.

Line noise and filters

In connection with the problem of electrical noise or interference such as streaks, dashes, etc., some words about noise filters are in order. There are many such devices or gadgets advertised to cure any and all TV ills. While they are not all completely useless, many of their claims are exaggerated. Before buying and installing one of these, you should know what you can reasonably expect.

There are three sources of noise associated with a TV set, whether visual (in the picture) or audible. Two of these are external, while the third is internal. To dispose of the last one first, noise in a TV set can come from a defective component; an old resistor can sometimes develop what the TV servicer calls an intermittent, one with poor internal continuity. What is even more likely, due to the high operating temperature involved, is a tube may become a noise source. If it's a tube, it is often possible to locate such a culprit by gently tapping the tube with the rubber eraser of an ordinary pencil; if the tube is noisy the tapping may produce flashing streaks in the picture or bell-like sounds from the speaker. Incidentally, a tube tester is of little value in performing reliable tests for noise. In case of a defective component, a professional will have to be called upon for analysis and repair. The same holds true for operating controls which become noisy with age. This includes contrast, volume, tone, fine tuning and channel selector.

Line noise includes all types of electrical noise generated by appliances which operate on ac. While not all appliances are potential noise makers, some of them are. Their electrical noise travels along the ac wiring in the house and enters the TV set. By contrast, noise from utility poles equipment and devices outside the house seldom reach the TV set this way. But this is nothing new, and radio and TV manufacturers have, almost without exception, provided simple noise filters inside the receiver for this purpose. Nevertheless, it is possible that an additional filter may be of some help if it is the proper kind and is properly installed.

It is very doubtful that the simple little gadget that looks like an ac plug on one end and an ac receptacle on the other can be much use. The fact that the built-in filter does not seem to help suggests one of two possibilities. Either the noise does not enter via ac line, or if it does, that the elimination of the noise must be accomplished at the source, that is, at the offending appliance, be it the washing machine, oil burner, electric drill, etc.

The third source of noise, and a very likely one, is the antenna. Of course, an old corroded antenna with poor lead-in connections is a potential source of noise. But what I have in mind here is noise generated elsewhere and picked up by the antenna along with station signals. Other than relocating the antenna and transmission line, nothing can be done here, except enlisting the aid of the utility company if the power lines, pole transformer, etc., prove to be the cause. In fact, for any but the simplest type of local noise the average TV owner is hardly in a position to move his or her antenna any appreciable distance, especially horizontally. A change in the antenna height may be helpful if it can be first found experimentally that a different height (higher or lower) proves noise free.

With regard to the transmission line, the problem is easier to solve. If it is apparent that the transmission line runs along a noisy structure or device or power line, either relocation or shielding should produce some improvement. Relocation is simpler and should be tried first. If this is not feasible or it does not help, substitution of a shielded (metal clad) transmission line is almost 100 percent certain to be effective.

There are two simple types of shielded transmission lines suitable for practially all TV sets. One is unbalanced 75-ohm line, sometimes called a coaxial or coax line. This contains an insulated center conductor and a metallic (braided) outer conductor, sometimes rubber covered. When installing this lead-in the outer conductors must be thoroughly grounded by a solid connection to the TV chassis. In addition the TV set must be equipped (some are) for a 75-ohm transmission line. Otherwise, a little accessory known as a matching transformer must be connected between the coaxial line and the TV set.

A second type of shielded transmission line is the 300-ohm twin-lead shielded line. This is the usual two-wire flat line with the addition of an external shield. Assuming that the TV set requires a 300-ohm antenna (most sets do), the connections are the same as with the unshielded transmission line, except that the external shield must be tightly connected to the TV chassis at one end and to the TV mast at the other. There's usually a slight signal loss because of this outer braid or shielding, but this is not significant, provided the antenna proper is fairly good.

In extremely noisy areas use a shielded coaxial lead-in and place the antenna as high and as far away as possible. If a noisy power line is radiating noise across the reception path of the TV antenna, place the antenna upon a higher tower. The height of the noisy power line and its proximity to the TV antenna, should be considered when deciding if the additional cost of relocating the antenna is warranted. In some areas, it may be impossible to eliminate all noise from the low VHF TV channels. Always use shielded cable to keep the power-line noise from being picked up by the lead-in wire.

A final note on noise prevention or elimination. The notion that an indoor antenna is immune to outdoor noise is completely false. No compact, simple, abbreviated antenna can improve reception over a good outdoor antenna. In fact, the *window improvements* can seldom approach the performance of even the simplest outdoor antenna properly installed. Except for locations with extremely high signal strength, these window gadgets are fairly useless. And as to devices which employ radar principles and convert one's housewiring into one giant all-direction antenna, the best that can be said for them is that they will work where any scrap of wire will work; they will not be much use where a normal antenna is required.

Spike block protectors

The spike block protector may protect the present-day TV set, stereo, word processors, computers, test equipment, and telephone systems from transient spikes from static electricity, motors, lightning, etc. Some spike block protectors suppress radio frequency interference along with surge voltage protection. The spike block protector cleans up the ac power for computers and helps prevent data loss, false printouts, monitor distortion, and computer component damage.

Some units are very small, with a three-prong plug that plugs into the ac power receptacle and then the TV set is plugged into it. The large units may have four or six outlets with indicator lights (Fig. 7-12). The clamping response time may be within 5 nanoseconds and suppresses up to 220 joule spikes at current levels up to 13,000 amps. Often, the unit will handle up to 15 amps of load power.

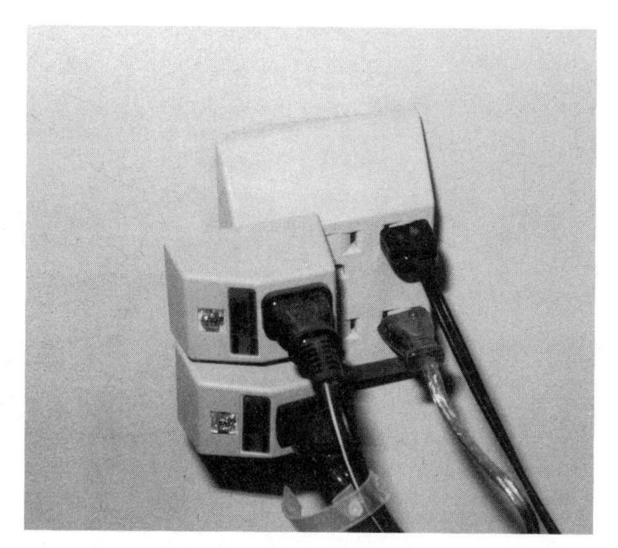

7-12 The spike surge and transient protector with several outlets between TV and power outlet.

When your new TV set keeps breaking down every few months, suspect transient spikes are destroying IC or critical processors in critical TV circuits. In today's busy world, many power motors and high-voltage power lines contain transient spikes that may destroy the small control processor. If you live in an industrial neighborhood, place a spike block between TV set and ac outlet.

Power-line surge protectors

In some of the lastest TV chassis and microwave ovens, a surge protector prevents extensive damage to the ac components. A high-voltage line surge or storm line damage may place more than 120 volts upon the TV chassis. The surge protector may prevent damage from a lightning storm. A TV set may be protected by a surge protector when lightning strikes a tree or power pole two blocks away.

The surge protector may be included in a spike block, separate unit or soldered across the ac line in the TV chassis (Fig. 7-13). These surge protectors look like ceramic disk capacitors except when a higher voltage surge is found on the power line, they will short out and destroy the ac fuse. Actually, the protector takes the surge of voltage, so to speak. These single surge protectors can be added to the chassis of any unit that is powered by the ac power line.

FM traps

Unwanted interference from strong local FM stations can be eliminated with one or more FM traps. The broadcast signal from a strong local FM station will sometimes modulate (ride on top) the TV station signal and cause FM oscillating lines or curved patterns across the picture. The FM signal may be picked up by the TV antenna as a harmonic of the TV broadcast frequency.

To eliminate it, place an FM trap across the antenna terminals at the rear of the TV receiver. If the FM signal is still seen in the picture or heard in the sound, a second FM

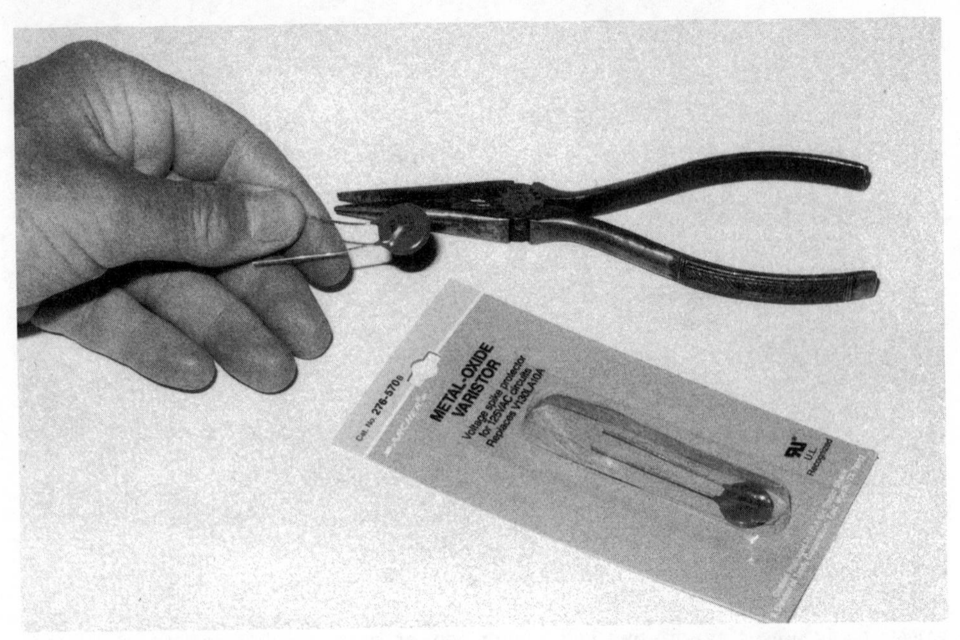

7-13 Solder a single disc-type surge protector in the ac circuit of the TV chassis for higher voltage surge protection.

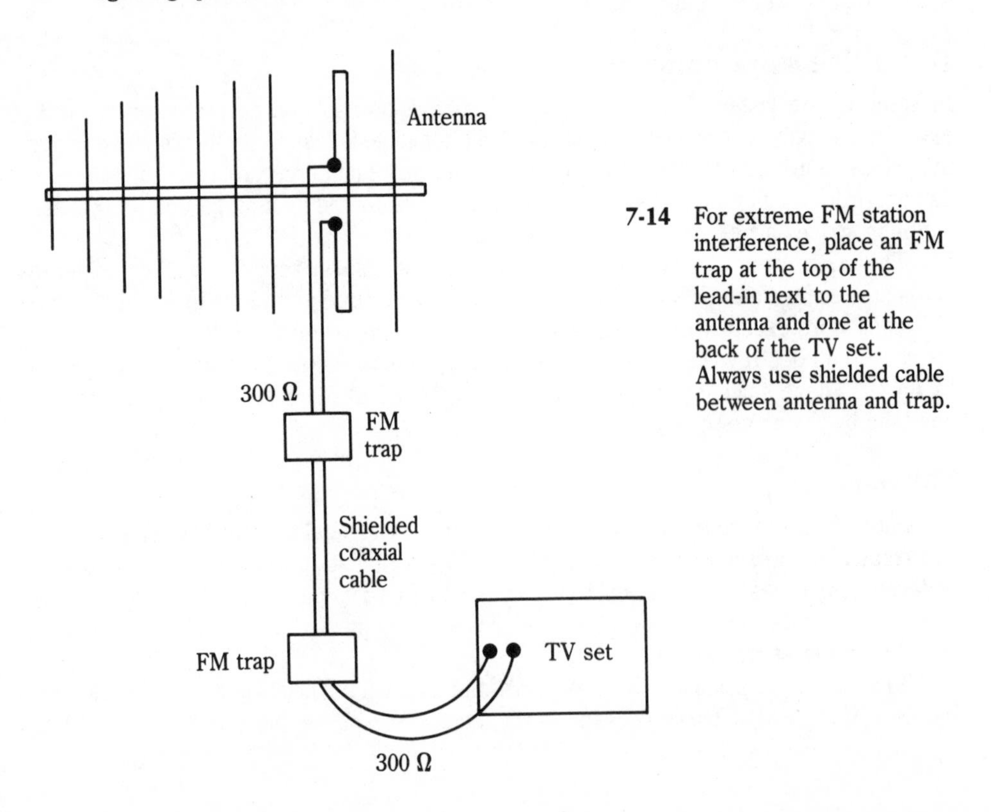

7-14 For extreme FM station interference, place an FM trap at the top of the lead-in next to the antenna and one at the back of the TV set. Always use shielded cable between antenna and trap.

trap must be placed at the TV antenna terminals. This means the FM trap must go high up the antenna mast. Change all the flat lead-in wire between the two FM traps to a coaxial shielded cable to help eliminate FM or unwanted noise from being received by the lead-in acting as a TV antenna (Fig. 7-14). When a booster antenna system is used, make sure the top side of the booster has an adjustable FM trap to help eliminate unwanted FM signals.

8

Troubleshooting catastrophic failures

THE VARIOUS ADJUSTMENTS AND CORRECTIONS COVERED SO FAR HAVE BEEN primarily of the preventive-corrective maintenance type. They are intended to keep a TV set in normal operating condition, as well as to correct deviations due to aging, wear, deterioration, etc. Our concern now involves catastrophic failure where the TV set ceases to operate either partially or totally.

Completely dead receiver

As obvious as this may seem, it is sometimes overlooked. If there is absolutely no evidence that the power is on (pilot light on, tubes slowing, etc.) a few simple tests will provide the answer. They should be made before going further.

- Check for a loose plug, or a disconnected plug at the wall outlet.
- Check for a defective wall outlet by plugging in lamp or other appliance.
- Check for a defective fuse or open circuit breaker. In the earlier sets, a power line fuse can be found on the rear apron of the TV chassis (the metal strip at the bottom of the set). More recent receivers are likely to have a resettable circuit breaker that may be reset or closed by a simple push. This requires a few seconds for cooling off after the set goes off. Should the circuit breaker pop open again immediately after resetting (or, for that matter, should the replacement fuse blow similarly), there is a defect, such as a severe overload or short circuit, and further troubleshooting is required.
- Check the on-off switch on the TV set. It may be defective.
- Check for loose or not completely inserted interlock plug of the ac cord. If the pilot light and picture seem to operate intermittently, suspect poor ac plug connections (Fig. 8-1). In older models, the interlock plug is disengaged when the

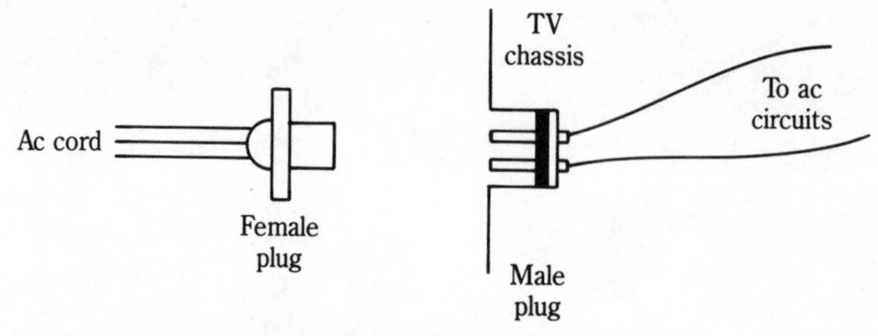

8-1 Check for a loose ac interlock plug at the back of the TV set. Intermittent operation of pilot light and TV could be caused by a dirty or burned ac interlock plug and male socket.

back cover is removed for customer protection. The ac contacts may become worn or corroded. Lightning may cause a no-picture or intermittent operation. In the later solid-state chassis, the ac cord is soldered right into the ac circuit. Replace the female chassis plug if excessively burned or has poor contacts.

If the answer to the first two checks is negative for transformer type sets, the fault lies either in a defective on-off switch or a defective power transformer. In either case, the repair is not for a beginner. The switch is almost always part of a control assembly (volume, tone, etc.) and requires professional attention. A transformer replacement is even more out of the reach of an amateur. Do not attempt either. Fortunately, both of these possibilities are rather improbable.

In a transformerless (series tube hookup) TV set the next step is to proceed according to the detailed instructions given in chapter 5. Since any one tube (or tube group, see Figs. 5-3 and 5-4) or fusible resistor will stop all tubes from lighting, the procedure suggested will almost certainly lead to the open circuit. The remedy is then obvious.

Isolation transformer

The isolation transformer is used by electronic technicians to prevent blowing fuses and chassis damage when connecting test equipment to the ac TV chassis. The TV chassis that works directly from the power line, without power transformer, may have a hot chassis and should be protected with the isolation transformer (Fig. 8-2). Sometimes when applying the ac-operated test instrument, something has to give. The common ac ground of test instrument and common hot ground of TV chassis may blow more than fuses in the ac TV chassis.

The ac TV chassis is plugged into the isolation transformer and the transformer is plugged into the power receptacle. Besides isolation of the power line, the transformer may have adjustable taps to raise or lower the power line. Raising and lowering the power line in the ac TV chassis may be required to service chassis or high-voltage shut-down. Slowly raising the power line may prevent damaging another horizontal output transistor in the solid-state chassis. An intermittent problem may be located by raising or lowering power-line voltage.

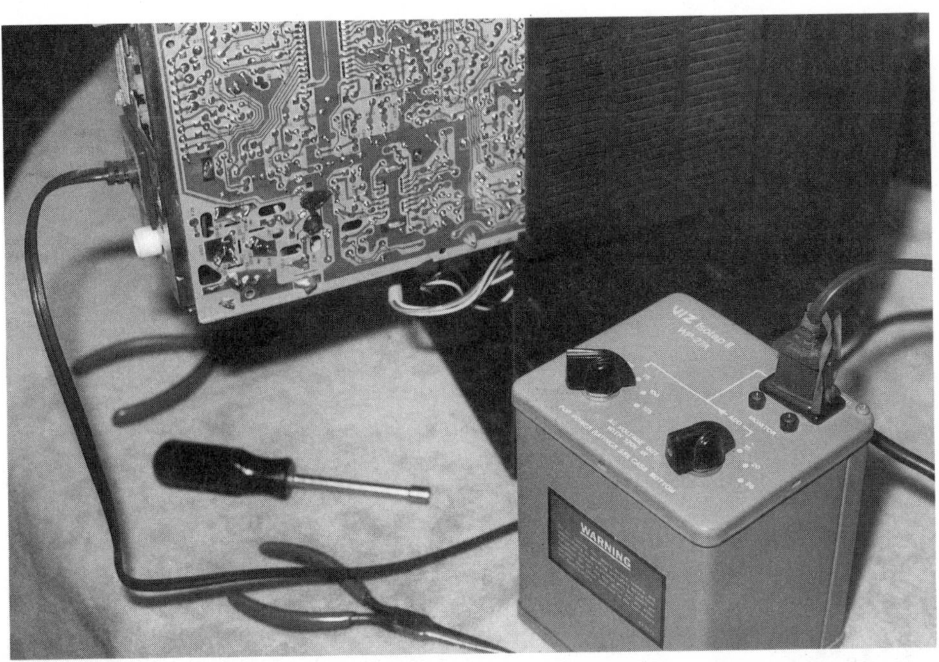

8-2 With today's ac-dc TV chassis, always connect a variable isolation transformer to the TV set before attaching any test equipment.

Tubes light only

There are two inferences to be drawn if only the tubes light. One, the fault lies in a part of the set common to both video and audio. Second, since the screen is dark, which is a fault common to both the signal circuits and the deflection circuits; you can have high voltage with a defective video system.

A power supply failure can be very simple or it can be beyond the capability of the beginner. If the power supply uses vacuum tube rectifiers, simply replace the rectifier tube. Suspect a leaky component in the power supply circuit if the plate of the rectifier tube becomes red. Turn the set off immediately or quickly remove the rectifier tube. The overloaded condition of the rectifier tube can be caused by a shorted or leaky tube in the horizontal output and deflection circuits. Silicon and selenium rectifier power supply circuits and their problems are discussed in chapters 10 and 11.

Raster only

This symptom eliminates the horizontal and vertical deflection circuits and points strictly to sections common to the video and audio portion of the set. Of course, this failure rarely exists in transformerless sets, as any tube failure usually disables everything—video, audio, and raster. Referring to the block diagram in Fig. 6-3 all tubes in the tuner

(V1 and V2), IF amplifiers (V3A, V3B, and V4) and the video amplifier (V5A) are suspect. Sound IF tubes (V5C, V7A and V7B) are not likely to be involved, even if the sound is missing because the simultaneous absence of the picture points to one or more of the tubes from the video amplifier (inclusive) back toward the antennas. By substituting tubes as suggested earlier, you should quickly locate the problem if it is a tube (Fig. 8-3).

8-3 A white raster (without picture) but with normal sound indicates problems in the video circuits.

No sound but picture OK

Failure of the sound may involve any of the stages and tubes beginning with the sound IF, just following the video detector, up to and including the speaker. This includes V5C, V7A, and V7B in Fig. 6-3, V5A, V6, and V7 in Fig. 6-4, etc., and applies equally to parallel and series filament tube sets, since the presence of a picture precludes any tube burnouts. Sound failure can also be due to a failure in the sound detector, which in some sets consists of a pair of semiconductor diodes; however, diode failure is most infrequent and may be anticipated as a wear-out type of failure, taking much longer than tubes. In those rare cases when the sound is weak even if all the tubes identified above test good, the detector diodes may as a last resort be suspected.

There is an important caution that must be made in connection with the sound section of a TV receiver and it pertains to the handling of the speaker. Since most speakers on TV receivers have quick disconnect clips or pin-and-jack connections (for convenience in removing the chassis from the cabinet) it is quite possible to deliberately or accidentally pull the speaker leads off, leaving the speaker electrically disconnected from the receiver. This is potentially damaging.

In tube-type sets, disconnecting the speaker with the set on may damage the output tube. With transistor sound sections, disconnecting the speaker with the set on spells certain death to one or two expensive transistors; therefore, *never* disconnect the speaker leads while the TV is on. In the troubleshooting procedure just discussed, first switch off the set, substitute for the suspected tube, then switch the receiver on again. While damage is not 100 percent certain in some tube sets, it is not worth risking such damage. With transistors, damage is certain.

Now and then there will be a TV set (or for that matter any solid-state radio or record player) in which the power transistors are said to be protected from such damage. My advice is still valid—play it safe and do not disconnect the speaker while the piece of equipment is on.

Although extremely rare now, in some TV sets, known as the split-sound type (as distinguished from the current intercarrier type), there may be a sound failure in stages not identified on the block diagrams in Figs. 6-2 through 6-5. As an illustration, Fig. 8-4 is a block diagram of an old split-sound TV set. Notice that the sound and picture information channels split apart (hence the name) immediately after the tuner. In such a set, if any are still around, the absence of sound must be due to a fault in V9, V10, V11, V12, V13, or V14. None of the tubes in the upper string (the video IF), including V3, V4, V5, V6, V7 or V8 can be involved. Incidentally, the two (or three) tubes in the tuner, V1 and V2 (and V3), cannot be responsible for the absence of sound unless both sound and picture are missing.

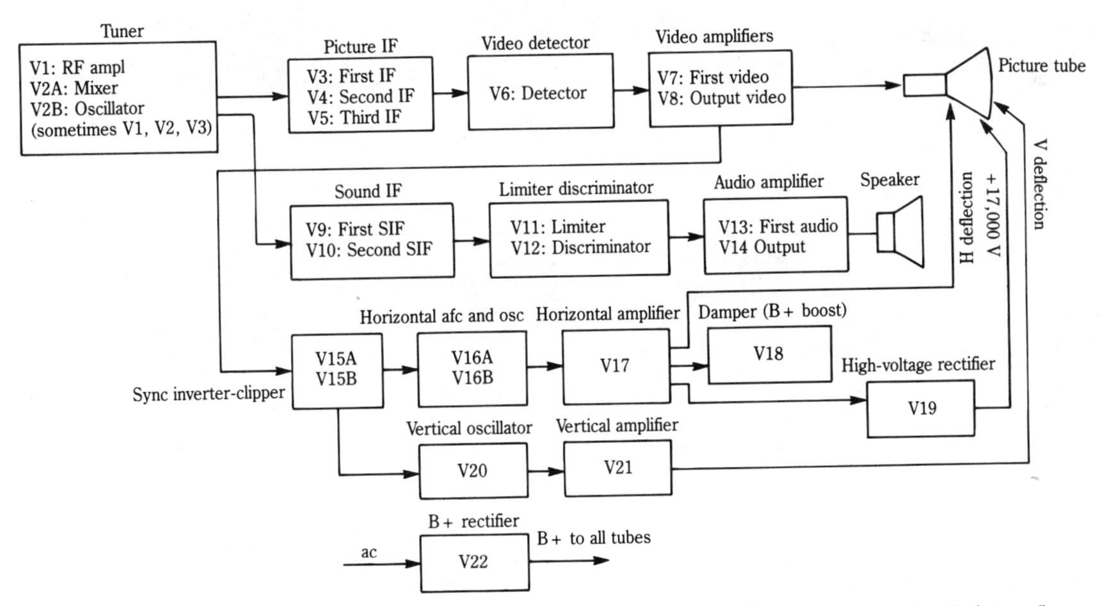

8-4 Block diagram of an older split-sound TV receiver. Notice the separate IF stages of sound and picture (hence the name).

In connection with this type of relic, it should be pointed out that a failure of picture only, but with a raster present, must be charged to the upper leg of the split, that is, the video IF channel comprising V3, V4, V5, the video detector, V6, and the video amplifiers, V7 and V8.

No sound, but picture OK (solid-state chassis)

Often, the no-sound symptom with normal picture indicates problems within the sound stages. In present TV chassis, the horizontal stages must perform before sound will come up to normal operation. The solid-state sound problems usually occur in the output stages. The sound stages may consist of only one large IC component (Fig. 8-5).

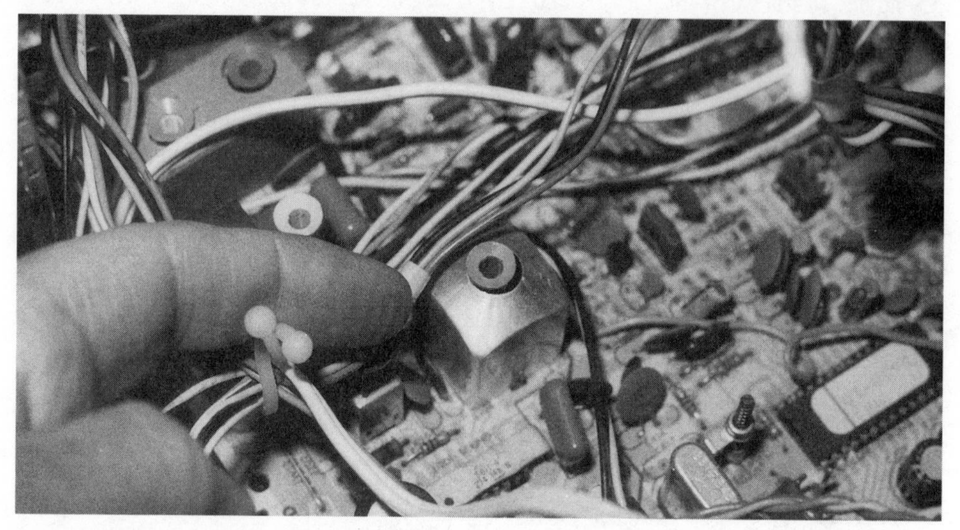

8-5 The finger points to the sound discriminator coil, which makes a distorted growling type of sound when out of adjustment. The complete sound IC is located nearby.

The early solid-state sound circuits consisted of transistors and then integrated components were added. In some chassis, one IC chip may contain the audio IF amplifiers, discriminator, preamp audio, and audio output circuits. Today the audio IF, FM detection and audio amp may be included in one large IC (Fig. 8-6). The sound output is amplified by either transistors or another IC audio output component.

The audio may be signal-traced with another audio amplifier or signal traced from signal point A. When no audio is found at point A, suspect a defective IC or improper supply voltages. If sound is found at the input of the audio output transistor and not at the speaker, check for a defective transistor or speaker. Use a signal tracer and check the audio signal right up to the speaker terminals.

Snow

This picture defect, snow, may vary from good sound, snowy picture to fairly good sound, very faint picture, or no picture at all. In addition, the snow may appear at ran-

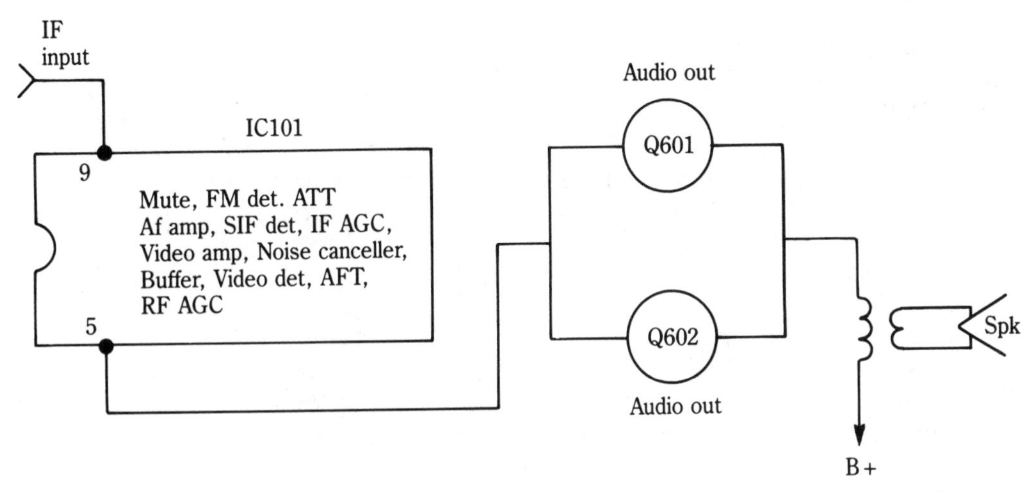

8-6 Block diagram of a solid-state sound and AF amp inside IC101, which has other circuits. The sound output at pin 5 drives two audio output transistors.

dom, suggesting (improperly so) that the fault lies outside the set. Actually, this is not so. Snow is a name given to a picture which seems to be broken up into small flecks or pieces which seem to shimmer or waver. They do not fall or move down the screen, although in a general way they are reminiscent of snow-filled air.

Regardless of which of these snow problems a particular TV set has, it is distinctly different (in fact, opposite) from the weak picture cases described earlier. In the vast majority of cases the presence of snow is evidence that the TV set is working hard to produce a picture from a poor signal. This means that either no signal is received by the TV set, or the received signal cannot get through due to a break in the functional continuity in the set. The procedure for isolating the fault and applying the appropriate remedy is simple.

Examine the antenna lead-in system. An old or otherwise poor antenna system will not necessarily cause a sudden appearance of snow. But a break in one or both conductors of the lead-in at the point of connection to the antenna, at the point or points of fastening on the way down, at the entrance through a window or wall or finally at the connection to the antenna terminals on the rear of the set—a break at any one of these vulnerable points is a likely cause for snow on an otherwise normal TV set. To check for a suspected broken lead-in, a substitute antenna, rabbit ears, an indoor wire antenna, or even a length of wire connected in place of a suspected antenna will be sufficient to verify whether or not the antenna is at fault. Of course, allowance must be made for the fact that the substitute antenna is, at best, a mediocre performer. But if the picture improves and the snow effect is diminished, the antenna lead-in is probably at fault, and that means a further check for a break in the wires.

A snowy picture may be caused from improper orientation of the antenna or from missing antenna elements. Check and see if the wind has turned the antenna away from the desired broadcast signal. Double check the antenna's direction in respect to the control box indicator, if a rotor is being used to orient the antenna in fringe areas. If snow is absent from the picture on some channels but is visible in the picture of a lower numbered channel, suspect that several of the back elements of the antenna are missing. The signal could be cut in half if just the back element rod is missing.

Check the RF amplifier tube (by tester or substitution) in the tuner (V1 in all block diagrams). This tube, more than any other, is likely to be the culprit. Even when it has deteriorated very severely this tube will still allow some signals to get through; hence, it may not be suspected. But the selfcompensating functions of a modern TV set will try to make up for the severe reduction in signal, amplifying a lot of noise which appears as snow. If no improvement is attained, similarly check the first IF tube (V3 in all block diagrams). To a lesser degree than V1, this tube is also potentially responsible for snow and should be checked accordingly.

Check the oscillator tube in the tuner and the remaining IF tubes if the RF and first IF tubes seem OK. It's best to substitute a new tube for the oscillator tube. While the probabilities are low that one of these tubes is responsible, they nevertheless should be checked with a good tube tester or by substitution.

While the tuner is a most likely suspect for this defect it is assumed that tuner is operating normally since, as discussed earlier, tuner abnormalities evidence themselves in other ways and their repairs will automatically preclude them as a source of snow. Nonetheless, and especially in cases of minor deterioration in tuner operation, which many TV owners are willing to tolerate for a while, imperfect contact in one or more of the tuner switch contacts will also cause a snow effect on the screen. Usually, in the early stages of tuner wear, it is but necessary to jiggle the tuner channel selector to obtain a better contact and eliminate the snow. Incidentally, in such a case this is an obvious hint that the tuner needs maintenance—cleaning at least.

Snowy picture

Do not overlook lightning damage if the picture is snowy. Sometimes lightning may strike a large tree near the power line and cause only minimal damage to the TV set one block away. Inspect the antenna terminals for black or burned spots. Sometimes the one antenna wire may be blown clear off. The center wire inside the 75-ohm shielded cable may be fused to the connector or outside shield.

Check the balun coils located upon the tuner. Notice if the small coupling disc capacitors are burned. These two capacitors isolate the antenna lead-in from the antenna input to prevent damage within the TV chassis (Fig. 8-7). Sometimes the coupling capacitors are burned or blown away with damaged balun coils. You may find only one or two ends of the coil burned off. Simply soldering back the burned leads may cure the snowy picture. Continuity tests with the ohmmeter can easily locate the open balun.

Sound OK but no picture

A totally blank screen with normal sound points to trouble in the video portion of the TV set, including the picture information and horizontal sweep circuits (if there's no raster) but not the vertical or sync circuits. The horizontal stages are involved because they are responsible for the presence or absence of any light (raster or picture) on the screen.

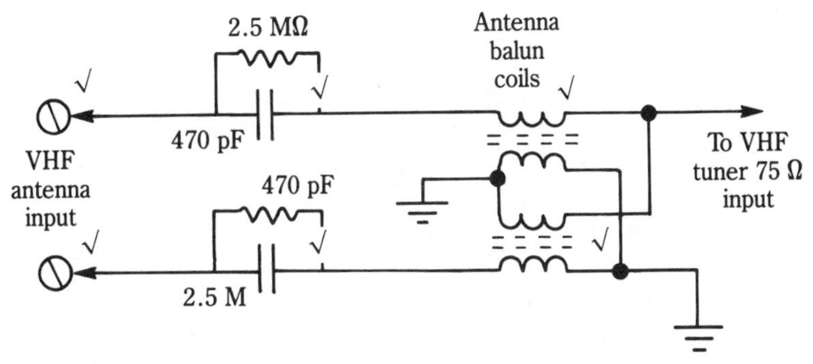

8-7 Check for burned input capacitors or antenna balun coils with lightning damage, which produces a snowy picture.

The troubleshooting procedure is based on some logical assumptions. First, the picture tube is not burned out (Fig. 8-8). In a transformer-type set, a burned-out picture tube is identified like any other burned out tube-absence of any light near the base of the tube. Of course, it is assumed that the picture tube socket is properly seated on the tube base. Once in a long while it may work loose or simply lose proper contact. Second, the tube did not become defective since the last time it was on; it usually doesn't happen that way, certainly not suddenly. If the tubes are wired in series and all tubes light, the procedure is no different than for a parallel tube set. If no tubes light (remember, this is

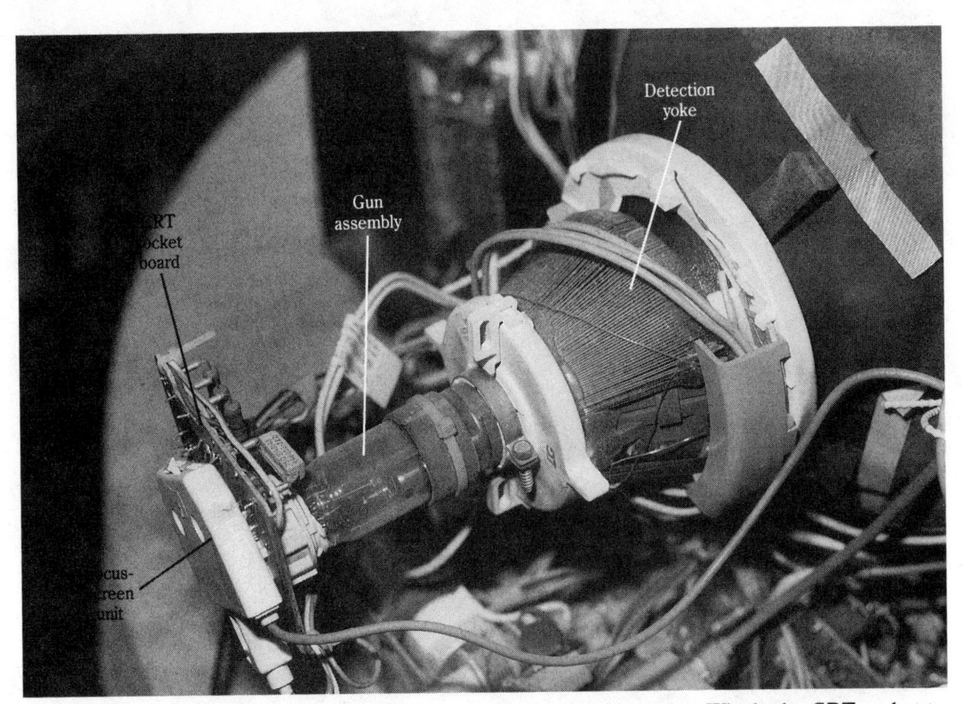

8-8 Notice if the filaments light up at the rear neck of the picture tube. Wiggle the CRT socket to see if the tube will light, indicating a defective tube socket or dirty element pins.

an all-or-none situation) the procedure outlined previously should be followed until the bad tube is found. If this does not solve the problem, that is, tubes will light but still no picture, the troubleshooting is the same as for the parallel filament sets.

The heater or filament may not light up with a defective socket, poor wire connections, or PC board connections. Sometimes slightly moving the CRT socket and watching for the heater to light up may turn up a defective socket (Fig. 8-8). Some of these sockets are enclosed while others have spark gaps on the picture tube elements.

In the early TV chassis, the thin wire from the picture tube heaters would have a poor contact and cause intermittent tube filaments. The TV set may operate for hours and the screen go black. Simply soldering the pin socket on CRT solved the intermittent heater problem.

The present TV socket may have bent or burned socket pins and caused the heater to not light up. Check where the filament pins solder directly to the tube socket. Sometimes soldering up these pins solved the intermittent picture. Check where the chassis filament wires connect to the picture tube socket. Also, check filament wire connection upon PC board or chassis for poorly soldered connections. Take a continuity ohmmeter test across the filament pins of socket for low resistance measurement. No or high reading indicates open or poorly soldered connections (Fig. 8-9).

8-9 Check the CRT board for poorly soldered connections or dirty tube pins. Sometimes the focus pin gets corroded, producing poor focusing.

The step-by-step sequence for locating the cause of a dark screen is easily followed. Replace the horizontal output tube (V11 in Fig. 6-2, V10 in Fig. 6-3, V11 in Fig. 6-4, V10A in Fig. 6-5), etc. Since this tube is directly responsible for generating the high voltage (15,000 to 25,000 volts) necessary for producing light on the screen, failure of this tube will result in a totally dark screen. Precautions outlined previously regarding the handling of this tube, especially if it happens to be inside the high-voltage cage, must be followed at all times. If a new horizontal output tube does not solve the problems, the next step is in order. This is also assuming that the horizontal output stage is not working, although the tube itself is okay.

Carefully remove and replace the damper tube (V13 in Fig. 6-2, V12 in Fig. 6-3, V13 in Fig. 6-4, etc.) This tube, if faulty, breaks the path of the B+ boost (250 to 650 volts), again disabling the horizontal circuit responsible for producing light on the screen. In most cases this would have to be a catastrophic failure; a weak or even poor testing damper tube will not remove all light from the screen.

If the damper tube proves to be good, proceed to carefully remove the high-voltage (15,000 to 25,000 volts) rectifier (V12 in Fig. 6-2, V11 in Fig. 6-3, or V12 in Fig. 6-4). This tube is always inside the protective cage and it always has a top cap connection which may be reluctant to come off. An additional difficulty, even if a minor one, is the fact that this tube lights very dimly, sometimes almost invisibly (depending on its physical position) and hardly feels warm to the touch, hence should not be assumed to be bad just by appearance alone. *Do not try to feel this tube while the set is powered.*

Incidentally, this is the only tube in a transformerless (series filament tubes) TV set that is not part of the series string; it can be burned out even if all the other tubes in the set are lighting. The foregoing procedure, should restore the picture or at least the raster to the screen; however there is one other possibility in some makes of TV sets.

As stated earlier, some TV sets make use of an internal fuse, usually inside the high-voltage cage, for the protection of the horizontal output portion of the TV set only. This is not the main fuse which protects the whole TV set. When this internal fuse blows, the whole horizontal output circuit, including the high-voltage system, is interrupted, resulting in no light on the screen. Replacement of the fuse is relatively simple, even in those cases when the fuse is soldered into position by pigtails (Fig. 8-10A).

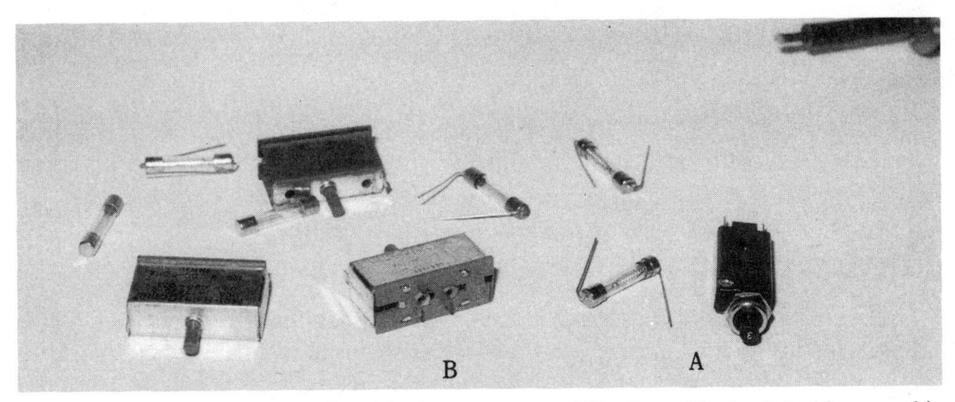

8-10 The different type fuses found in the ac power and B+ line with circuit breakers used in older TV chassis.

Visual inspection must be very carefully made as the metal conductor inside the glass fuse is very thin; therefore, an open is not too obvious. In replacing the fuse, if it is of the snap-in type the procedure is quite obvious. The fuse should be of the same physical length in order to fit into the fuse holder clips. Its electrical value is very critical. If the fuse is of the soldered-in type, a slightly different replacement procedure is called for. All radio repair supply houses sell a replacement fuse assembly which looks somewhat like Fig. 8-10B; as can be seen, this is a double fuseholder or two fuseholders fastened back-to-back. One side is slipped over the old presumably burned out fuse, while a new snap-in fuse is placed on the opposite side.

Fusible or isolation resistors

In the early black-and-white and TV chassis, fusible resistors were used in series with the ac power line instead of with the fuses. The fusible or low-ohm isolation resistors provide time lag and get red hot before opening in overloaded tube and transistor chassis. The fusible resistor may be off 2.8, 4.8, or 7.5 ohms. These fusible resistors were found in the early high-filament or heater-type tube series circuits.

A small resistor may be found in the ac-powered TV chassis. When a leaky or shorted component was found in the low voltage or HV circuits, the low-wattage resistor went open to protect the other TV components (Fig. 8-11). The low-ohm resistor may be found in series with other fuses. Sometimes when lightning or a component shorts within the power supply or horizontal output circuit, the fuse is found blown and the low-ohm resistor is also open. Always check for a shorted low-voltage diode, regulator, or horizontal output transistor when a low-voltage resistor is found open.

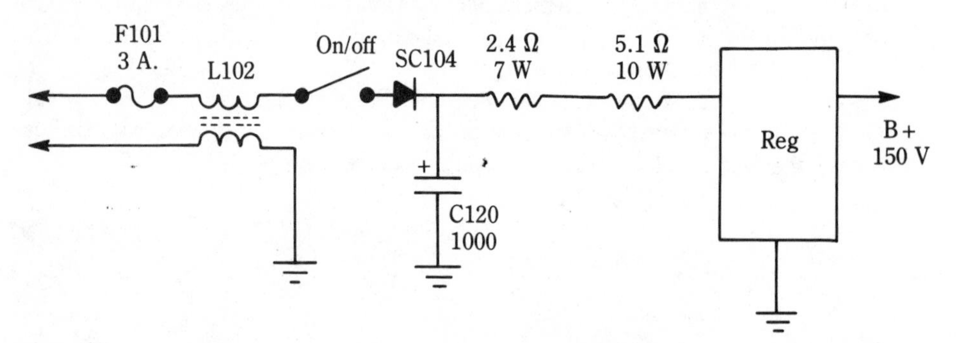

8-11 When the ac fuse and 5.1Ω or 2.4Ω resistors open, suspect defective regulator or horizontal output transistors.

Dim picture not covering full screen

When operation is just a little below normal, that is, when the picture could be a little brighter and have a little more contrast and when the picture does not quite reach the edges, the fault is probably a poor rectifier tube (or tubes). Referring to the block diagrams again (Figs. 6-2 through 6-5) we find the rectifier tubes identified as 5U4G, 5U4GB, 6AX4, 5AX4.

In some very large TV sets, particularly of earlier vintage, two rectifiers (a pair of the same or two different types) are used. Since in every case the rectifier supplies the power to all other tubes in the set, a reduction in output here will reduce the performance of most tubes in the set to some degree. Furthermore, rectifier tubes operate at high temperatures and full load at all times and are more prone to deteriorate than any other tube in the set. Replacement of weak rectifier tubes usually corrects insufficient brightness, poor contrast, and insufficient height and width.

Besides the rectifier tube, replace the horizontal output, damper and high-voltage rectifier tube causing insufficient width. The sides of the picture can be pulled in due to a gassy horizontal output and damper tube. While the set is operating watch, what effect tapping the side of the tube with a pencil has on the picture. A defective tube may cause the width to change or sparks in the tube. These tubes will sometimes test okay in a tube tester but they will not operate in the circuit. Always, substitute the low voltage rectifier, horizontal output and damper tube to correct this problem.

Low ac line voltage

Low or high ac line voltage may cause problems within the TV set. Low line voltage may cause the sides to pull in of the raster and make components work harder. Excessively high line voltage may shorten the life of TV components or cause high-voltage or chassis shutdown. The correct line voltage should be around 120 volts ac. In farm or country power lines when the TV set is the last one on the power-line transformer, low voltage may occur (Fig. 8-12).

Suspect high power-line voltage when horizontal output or low-voltage power supply components repeatedly break down. Check the power line voltage with the DMM. If the

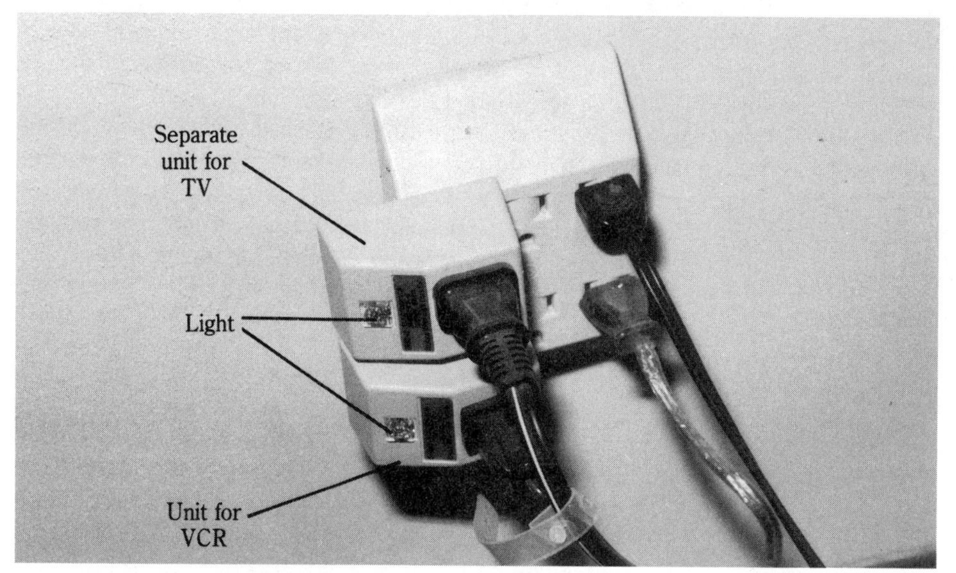

8-12 Measure the outlet voltage where the TV is connected for correct ac line voltages (117 to 120 Vac).

power line voltage is over 125 volts, notify the power company. Most power-line sources will place a standard voltage tester to be monitored over a 3- or 5-day period. Suspect high power-line voltage when the chassis shuts down, kicks out circuit breakers, or blows the fuse for no apparent reason.

Picture dim and brightness/contrast ineffective

There is another case of improper brightness or contrast control where the picture size is normal. Unlike the previous example where the brightness and contrast controls are functioning but at their maximum settings there is still not sufficient picture brightness, this case identifies a condition in which the brightness and contrast controls are not functioning properly. Sometimes increasing these settings, or advancing the controls, will produce a photographic negative effect—blacks are white and vice versa. More often the gradual variation from dark to light to bright, as is the case with the average TV screen, will not be obtainable, except perhaps at very low (dim) settings of the brightness and contrast controls. At all other positions, as when attempting to get a normal picture, the light and dark areas appear flat, muddy and very dull. Sometimes such a picture suggests looking through a very dirty window or through a gray filter at a normal scene. Any of these symptoms, alone or in combination, suggest a defective picture tube (of course, it is assumed that the problem is not the one just described previously). The electron emission has deteriorated to such a low fraction of the normal amount that internal adjustments or small tube replacements will no longer help. The only permanent remedy is the replacement of the picture tube, and this is a job for the professional servicer.

Before deciding to have a new picture tube installed, determine if the cost of the repair is warranted. A frequently used tube-type TV set that is 8 years old is considered to have reached its life expectancy. Replacing the picture tube is not prudent because you will soon have to replace the filter capacitors or the flyback transformer, which can be rather expensive. The life of a solid-state TV receiver is from 10 to 12 years.

If you're shopping for a replacement picture tube (plus installation) some money can be saved by buying a rebuilt tube. In fact, if no specific instructions are given to the contrary, the replacement tube may well be a rebuilt one. This is quite satisfactory, both ethically and technically, provided it is a tube rebuilt by one of the standard tube manufacturers. The 100 percent brand new tube today is still available, but its use is the exception rather than the rule, and it offers absolutely no advantage over the standard rebuilt tube.

When it has been established that a dim picture is due to a weak picture tube, the life of such a tube can be extended, often for a number of months, by a simple technique commonly called rejuvenation. What it amounts to is a rejuvenation of the electron-producing element of the tube. It can be restored, for a while at least, by the application of a higher-than-normal voltage, raising the temperature of the electron-emitting surface on the tube element.

Most professional servicers use a one-shot remedy by applying an overvoltage for a short while, thus reactivating the electron emitter. After this, the tube reverts to its normal operating voltage. You may also install a booster device which is attached to the tube and left there for the remaining life of the tube. It is simple, less expensive, effective, and most important it may be less likely to shorten the tube life. Since a rejuvenation overvoltage is applied for only a relatively short time, it is necessarily more drastic and may have some delayed action effect on the remaining life of the tube. By contrast, the boost, which continuously operates with the tube, applies a much more modest overvoltage and is, therefore, less detrimental to the remaining tube life.

Figure 8-13 is a sketch of a common picture tube booster. At one end is a socket (or section of one, as found on most TV sets) which fits over the picture tube base, exactly like the original one. At the other end is a plug which contains a small transformer. Connection to the TV set is simple.

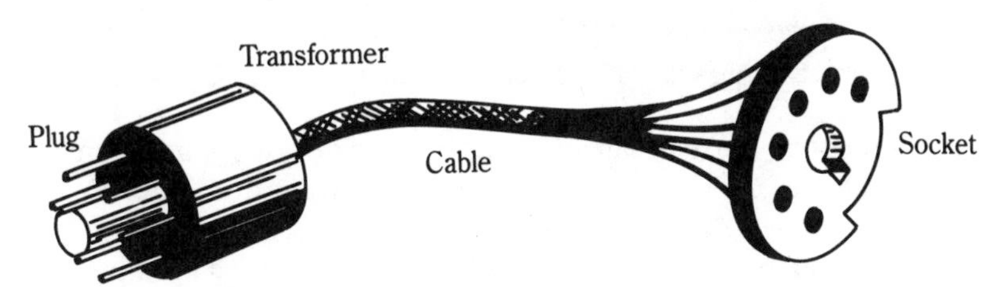

8-13 Sketch of a typical picture tube booster. The socket goes over the picture tube base.

Switch off the set and carefully pull the socket from the picture tube base. Mate and connect this socket with the base plug on the booster. Connect the socket on the other end of the booster cable to the base of the picture tube. The booster may be left hanging as is.

One important reminder: there are two different types of boosters for sale in radio supply stores and they are not interchangeable, even if they look alike. One is exclusively for parallel (transformer-type) tube hookups and the other is exclusively for series tube hookups (transformerless). Make absolutely sure which type of TV set yours is and purchase the correct type. Some types are for both filament arrangements and by virtue of a moveable wire connection.

Heavy black-and-white bars

This symptom, heavy bars, looks somewhat like Fig. 8-14A. There may be either one dark and one light horizontal bar, each covering approximately 50 percent of the screen height, or, as in Fig. 8-14B, there may be three bars, one wide and two narrow. Often these bars will slowly drift up or down the screen. To locate the cause of this malfunction, more than one step is usually required.

Understanding the problem will make the solution so much easier. The vast majority of such bars are caused by what is commonly called ac hum, although the word hum usually refers to the audible manifestation of the unwanted presence of ac in a circuit. One of the most likely sources of such ac hum (or at least those within the ability of the beginner) is a tube.

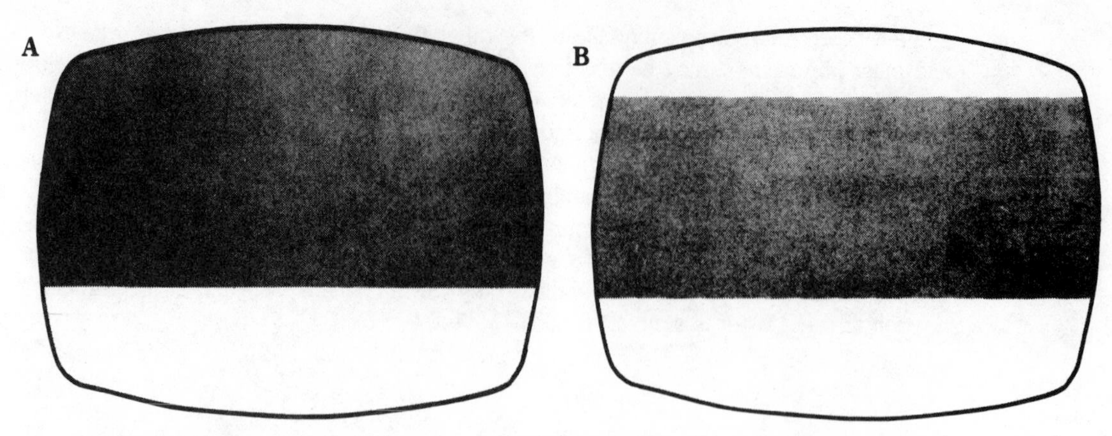

8-14 Ac hum can cause the raster to vary from black to white (A) or from white to black to white (B).

Hum bars may be caused from a poor filter capacitor. Heavy lines can be produced by a nearby high power line. If the bars remain in the picture with the TV antenna disconnected from the set, the problem is in the TV chassis. Interference or hum bars picked up by the antenna will not be seen after removing the lead-in wire.

One or two vertical hum bars seen in the raster can be caused by capacitors. Capacitors have a tendency to dry out and lose their capacitance, resulting in poor filtering of the low voltage power supply (Fig. 8-15). A defective electrolytic capacitor can be located by shunting a good capacitor of the same value across the suspected one. Look for a white or black substance near the defective capacitor terminals. Check for correct polarity before turning on the set.

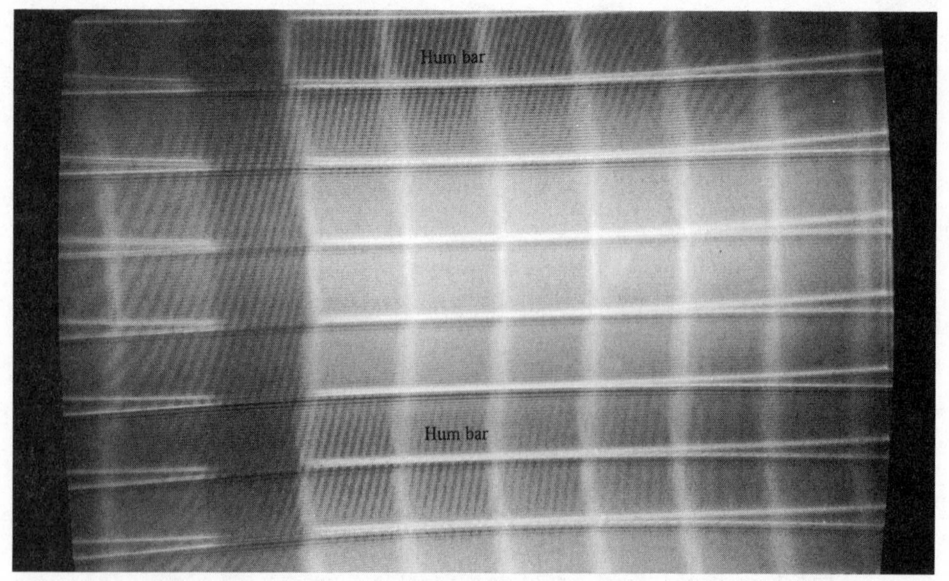

8-15 Defective filter capacitors can cause hum bars in the raster. These capacitors are located in the low-voltage power supply.

A defective tube in this case is one which has developed what is called heater-to-cathode leakage. The technical significance of this statement is of no particular importance at this point, except insofar as corrective actions is concerned. Specifically, when one or more tubes are suspected of causing the horizontal bars, you should test them very carefully on a tube tester having a leakage indicator, usually in the form of a small neon bulb. Total emission tests, as indicated by the needle swinging into the replace, weak or good portion of the scale, are of no value here, and may, in fact, be misleading, since tubes with heater-to-cathode shorts or near shorts often read high.

Not only must the neon glow be looked for very carefully, but enough time (at least a few minutes) must be allowed for the suspected tube to reach a high enough temperature to produce the leakage or short. The same precaution applies to the tube substitution procedure. Merely switching off the TV set often allows sufficient time for the culprit tube to cool somewhat and operate normally when the set is switched on again.

Therefore it is imperative for the beginner to follow two simple steps in order to eliminate this ambiguity of indications. First, observe how soon the bars appear after the set is turned on initially. Second, when switching tubes allow at least that much time to see whether or not the bars reappear before concluding that the job is done. It should be noted here, that contrary to most other malfunctions in a TV set, which usually can be traced to one particular and exclusive tube or stage, the ac hum responsible for the bars under discussion here can originate almost anywhere in the TV set, although under different circumstances and with different symptoms. Because of this, the troubleshooting procedure must be able to locate the defect without a random hit-or-miss, trial-and-error, let's-see-what-happens type of procedure.

To locate a defective tube, it is most helpful to isolate the fault to a particular portion of the TV set, keeping in mind the fact that a number of tubes can be removed from the transformer-powered TV set without affecting the raster. For example, referring to a typical block diagram as in Fig. 6-3, all the tubes, individually or in a group from V1 through V7, can be removed without removing the raster, although the video picture itself may disappear. A systematic isolation procedure for this TV set is very simple.

Remove the first IF (V3 in most sets); both picture and sound will be disabled but not the raster. If the bars disappear, the defect is either in the tube just removed or in the tuner itself. To determine which is at fault, replace V3 and remove V1. If this removes the defect, V1 is at fault. If the bars persist, the fault lies either in the converter tube, V2, or in the previously removed V3. If neither of these tubes affect the bars, the trouble lies further on, in the IF stages, V4 and V5, the video stage, V6, or in the picture tube itself, as we shall determine presently (Fig. 8-16).

Remove the last IF tube (V5 in Fig. 6-2). This also disables the picture and sound, but since the video amplifier (V6) is still in its socket, it is most likely suspect if the bars remain. If the bars disappear, the fault is somewhere in the IF circuits, such as V4 or V5. To determine which, replace F5 and remove V3. If the bars disappear, V3 is at fault; if the bars persist, V4 is at fault. If removing V5 makes no change the IF is not at fault. If the bars persist, the video amplifier is almost certain to be responsible.

Another possibility, although a much less likely one, is the sync tube(s). If the "high probability" video amplifier does not seem to be defective, the sync tubes should be checked. In fact, after eliminating the video amplifier as the cause, the sync tubes become the prime suspect.

8-16 The dark section of the picture indicates a leaky video tube or transistor.

Remove V6A, the video amplifier. This will disable both the picture and sound, as was discussed earlier. If removing the video amplifier also removes the bars the logical step is the replacement of this tube. If this cures the malady, this tube was at fault. If the bars persist, remove and replace the sync tube, V8A. If the bars are still present, the fault lies beyond this point either in the picture tube or in the power supply to the picture tube.

Very often, the fault will be in the power supply portion of the picture tube circuits (that is, filter capacitor) but most likely the picture tube itself has an internal short or a near short. Occasionally, even if rarely, it is possible to burn out the short with a tube rejuvenator. If a professional servicer offers to do this, he or she will probably stipulate that it must be done at your risk. This is not unreasonable and entails no loss to the owner, since the tube is not serviceable and its trade-in value will not diminish even if it is burned out in the attempt.

Picture snaking

This defect in the TV picture is mentioned here because it is usually caused by the same type of tube defect as the ac bars just discussed, namely, leakage between elements in one of the tubes. All vertical lines—objects, picture edges, or people—seem to weave in shapes of S curves. Almost any tube in the TV set may be at fault, but it's most likely to be tubes in the horizontal, sync, and video circuits (V10A, V10B, V11, V8A and V6A in Fig. 6-2), and least likely to be those in the tuner (V1 and V2) or in the IF amplifiers (V3, V4 and V5). The procedure for locating the bad tube is the same as for the just-discussed hum bars.

Sound bars

In connection with this phenomenon, mention should be made again of this type of defect stemming from two other sources (overloading, AGC adjustments, etc.). The first is internal and is under the control of the user, the second is atmospheric, and virtually nothing can be done about it. Advancing either the contrast or the AGC adjustment beyond their normal settings as described in chapter 6 will give rise to various degrees of instability and even snaking. The remedy is obvious.

In case of the external cause, little can be done. Normal TV reception is based on line of sight signal travel from transmitter to receiver; that is, the FCC allocated station locations and power outputs is such that two stations of the same channel number are separated geographically so as not to interfere with each other. At infrequent times, however, and often incident to some radical weather change, signals travel not only directly (line of sight), but also via reflection from an electrically charged layer above the atmosphere, covering much greater distances than intended. Such a signal will play havoc with the local station of the same channel (since there are minute differences between the two stations despite the fact that the same FCC standards apply to both), affecting sync, AGC, and other vital functions. The effect may differ from set to set and from station to station (the higher frequency channels are usually not affected), but while it lasts, little can or should be done about it. Any attempt at correction through some adjustment will simply misadjust the set that when conditions return to normal the set is actually out of correct adjustment.

One final possibility need be mentioned: overloading due to the proximity of a powerful local station. Although relatively rare, this has happened in the past. In fact, an offending radio station actually distributed little devices—filters, of a sort—to residents in the vicinity of the powerful transmitter, in order to eliminate or at least minimize the problem. There are simple little signal reducers, which connect to the antenna terminals of the TV set. Of course, the simplest remedy, even if not 100 percent effective, is to reduce the contrast control when the particular station is used.

Sound bars are somewhat similar to the heavy black-and-white hum bars except they are much thinner (Fig. 8-17), more numerous (there may be a dozen or more), they appear somewhat wavy, and, instead of slowly drifting up or down, they seem to waver

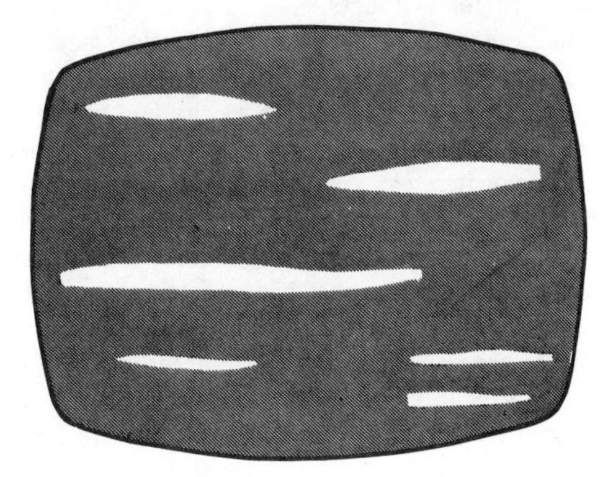

8-17 The effect of sound bars is similar to hum bars, except there are many more of them.

with the sound from the speaker, as if animated. Sound bars appear when some of the sound energy reaches the picture tube and is reproduced as light. The cause of this malfunction is seldom a defective tube, rather it is due to an improper adjustment either of the fine-tuning control or, more seriously, one of the tuned circuits inside the TV set; however, only the fine tuning is within the capability of the beginner to correct. Try to adjust the fine-tuning control for a clean picture even if it means some reduction in volume.

If this is not effective, it is an indication that either one or more of the IF circuits is improperly adjusted, or, in some TV sets, the sound trap (called 4.5 MHz trap) circuit is improperly adjusted. This should not occur if the TV set ever functioned properly, as it is extremely unlikely although possible that these circuits would drift out of adjustment; however, after a repair, or especially alignment in the repair shop, some of these circuits may have been improperly adjusted.

Herringbone weave

A herringbone pattern on the screen is seldom due to internal causes, but these cannot be automatically discounted. One likely cause of the phenomenon is interference from another TV set. Although the FCC has established some very clear standards for TV receiver radiation, many sets will make their presence known in other neighboring sets, especially in apartment houses and in master TV antenna installations. This can easily be identified by its random, intermittent nature and absence from most channels. It appears only when the offending receiver and the victim sets are tuned to the same station. There is little that can be done about it, other than attempt to minimize it by touching up the fine-tuning control or switching to a different channel for the duration.

Where there is an adjustable fine-tuning control, a herringbone pattern may be due to improper setting of this control. In older sets, adjustment of this control is required each time the channel is changed. The proper setting is the point which gives the best picture, not the loudest sound (these two were not always coincidental). In more modern receivers where the fine tuning is not critical and where usually one setting serves for all channels, there is considerable leeway in setting this control, but it can still be set (improperly) so as to produce a herringbone weave on one or more stations. An optimum setting (good for all stations, not best for any one) should be made by trial and error to avoid herringbone effects.

Incidentally, since many TV sets may receive the same channel on two adjacent selector positions (that is channel 2 on dial positions 2 and 3, channel 7 on positions 7 and 8, etc.) it is important to have the fine tuning optimized for the best results on the correct position. It is perfectly normal for some sets to have a herringbone pattern on the picture when, say, channel 2 is examined on the channel 3 position of the selector. In such a case, the fine tuning is most probably incorrect.

A herringbone pattern may, in some few cases, stem from internal causes. If, after exhausting the various possibilities listed above, the problem still persists, a substitution of one or more tubes in the video detector and video amplifier section of the receiver should be tried. This includes V6A in Fig. 6-2, V8A in Fig. 6-3, V5A in Fig. 6-4, V5A in Fig. 6-5, etc. In those cases where the video detector is a semiconductor diode, you may have to call in a professional servicer to do the job because careful soldering is required. This also applies to those cases where tube substitution is of no help, suggesting that a defective component other than tubes may be responsible; however, semiconductor video detectors are seldom at fault.

Barkhausen interference

Barkhausen interference appears as one or more ragged vertical lines, usually in the left portion of the picture. Sometimes Barkhausen interference may appear as a small group of adjacent lines somewhat as in Fig. 8-18. Such interference is most prominent between scenes (station breaks, etc.) or when the received station is one of the more distant (weaker) ones. This interference is invariably caused by a peculiarity (not really a defect) of the horizontal amplifier (V11 in Fig. 6-2, V10 in Fig. 6-3, etc.) tube. No tube tester will discern anything abnormal in such a tube, and nothing short of substitution will prove whether or not the problem is within the tube. Of course, it is just barely possible that the second tube may behave in the same manner, but this is highly unlikely. There are one or two other possibilities for which the corrective steps are simple and straightforward.

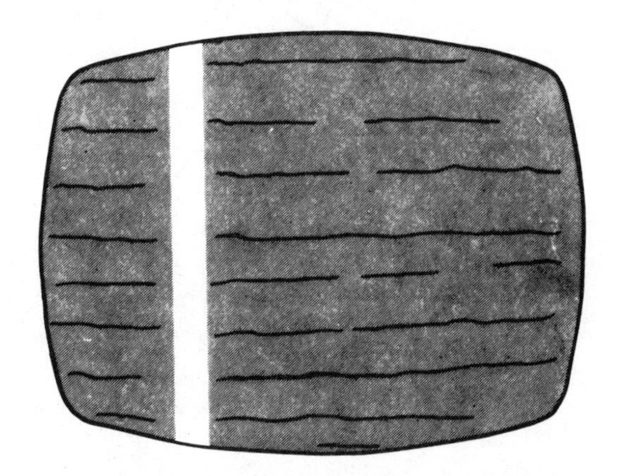

8-18 Ghost lines or Barkhausen interference appears vertically toward the left side of the screen.

Check the tube caps on the horizontal amplifier and the HV rectifier (V11 and V12 in Fig. 6-2, etc.). They should be well seated and otherwise not intermittently connected. Check the door or cover of the cage containing the high-voltage components. It should not be loose or open. Check for the rare possibility that the TV lead-in is somehow bunched, possibly because of the surplus length, near this cage or near the horizontal amplifier tube or general vicinity. Substitute a new tube for the horizontal amplifier.

There is one other and less preferable method for remedying this effect without replacing the horizontal amplifier tube. It involves the purchase of a little device, sometimes called a Barkhausen suppressor, and consists of a small magnet with an attachment (spring) band. This gadget is slipped over the horizontal amplifier tube's glass envelope and its position adjusted by rotating and sliding it up or down, while observing the Barkhausen lines on the screen. At the top cap of the horizontal amplifier tubes, there is a potential of thousands of volts. It is not necessary to come in contact with it since the magnet is slipped over the tube with the cap off and the receiver off. But care is the watchword here, nevertheless.

Picture ghosts

This malfunction is another one seldom caused internally, but it is quite easy to correct. A picture ghost is a duplicate picture, usually weaker than the original and usually displaced somewhat to the right. Figure 8-19 is one example of this type of ghost. We are employing the word usually because there is also an unusual version of the same thing. A brief explanation will make it much easier to correct the fault.

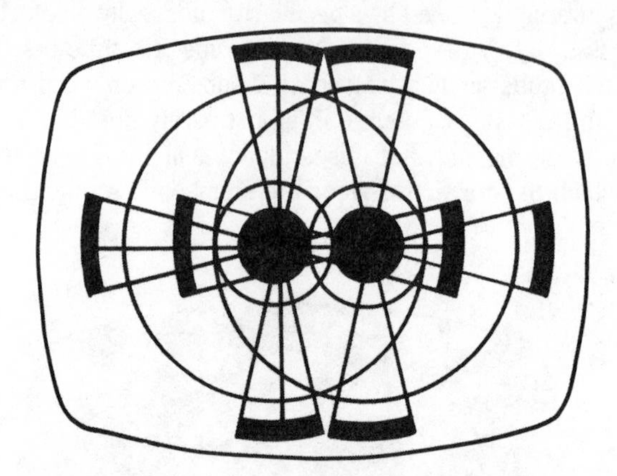

8-19 A ghost image, caused by multipath reception of two signals (one delayed slightly), appears to the left of the desired image.

It is common knowledge that a TV signal travels from the transmitting antenna to the receiving antenna. It is also a fact, although not so commonly known, that the TV image is reproduced on the home screen from left to right; that is, the picture is generated much in the same sequence as the writing (or reading) of a line on a page. It follows that portions of the TV picture on the right appear later than the portions on the left. Of course, the difference in time is extremely small (millionths of a second) and the eye sees the whole picture simultaneously; however, should two signals from the same TV station arrive one behind the other because one comes in a roundabout way the second signal will arrive a little later. This is known as multipath reception; that is, one signal travels the shortest path between the transmitter and receiver antennas, while the other follows a somewhat zig-zag path by striking some object on the way, then bouncing off, etc. This latter path is obviously longer, and takes more time. Figure 8-20 is a simple illustration of this occurrence.

If the TV receiver is far enough away from the transmitting center (such as a central metropolitan location) so that all stations lie in approximately the same line of path, the surest way to avoid or eliminate the type of ghost just described is to reorient the antenna. It probably has shifted so that it no longer points directly at the transmitters. It should be remembered that TV transmitters radiate in all directions (omnidirectional), while all receiving antennas are relatively sharply unidirectional; therefore, it is possible that the receiving antenna now is aimed at a tall building or a hill which acts as a reflector for transmitted signals. In that case the antenna is picking up a direct and a reflected signal.

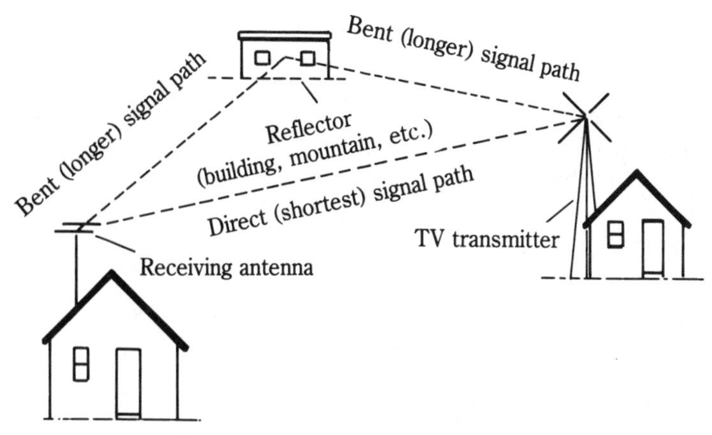

8-20 Multiple signal paths between transmitter and receiver cause ghost images.

If the transmitting location is not in sight, a map and compass should be used for correct pointing, although an aim-and-try (rotate antenna, observe picture, repeat if necessary) procedure is quite satisfactory. It is also fairly safe (although not absolutely certain) to assume that the correct direction is that in which most TV antennas in the immediate vicinity are pointing.

In cases where different TV stations are not in line but are in definitely different compass bearings, an antenna rotor is essential. This will permit the receiving antenna to be pointed directly at the transmitter without any compromise.

As more and more TV receiving antennas are made to receive a combination of VHF (channels 2 through 13) and UHF (channels 14 through 83) signals, an additional adjustment may be required for elimination of ghosts and for maximum signal, especially on the UHF stations. This adjustment consists of the raising or lowering of the antenna, while simultaneously rotating it, for minimum ghosts and maximum signal. Contrary to most cases of VHF antenna installation, the highest possible location is not necessarily the best for UHF. In fact, lowering a VHF-UHF antenna system as little as a few inches may make the difference between a poor UHF signal and a very strong signal. Since the VHF is not so critical it is safe to adjust the height for best UHF performance only. The direction of the antenna is fairly critical for all stations so this should not be neglected.

Tubes may cause picture ghosts, but such faults are confined almost exclusively to the front end of the TV set, specifically in the antenna stage (V1 in all block diagrams or V1, V1A in VHF-UHF TV sets). In those TV sets where the UHF tuner uses no tubes, the problem is virtually eliminated automatically, as the transistorized antenna circuit is seldom susceptible to this problem.

Arcing in the picture or raster

Small white and dark lines going horizontally across the picture may be caused by outside interference or arcing within the TV chassis. Disconnect the TV antenna lead-in wire to determine if the noise is being picked up by the antenna system. Automobile ignition, neon signs, power lines, and motors are a few of the manmade noises picked up by the antenna.

Arcing within the TV set may be caused by the high voltage anode connection arcing over. Disconnect the power cord and discharge the picture tube by connecting the picture tube's high voltage anode connection to chassis ground. Use a wire jumper attached to a long screwdriver that has an insulated handle. Slip the anode lead and clip out. Notice if arcing marks are found under the rubber cover and check the clip to see if it is rusty. A poor ground connection between the outside area of the picture tube and chassis ground may cause small arcing that can be seen in the raster. Check and install a new ground if required. Replace the horizontal output or damper tube if they are suspected of internal arcing.

Loss of fine detail

This defect may be either component deterioration or due to misadjustment. Any portion of the picture where fine variations are present, such as a person's eyes or hair, a fine pattern on paper or clothing, and even the fine scan lines (raster) will be degraded by this defect. There are a few probable causes.

IF alignment incorrect

If the TV used to have good picture detail, a loss due to IF circuits misalignment will seldom occur, except and unless those circuits have been improperly adjusted. The remedy is obvious—complete IF alignment by a professional.

Fine-tuning control

Especially in very old TV sets this can materially degrade picture detail since the ultimate effect is similar to that of a misalignment. The remedy here, too, is obvious, except that the setting of the fine-tuning control also affects the sound and, as discussed previously, ghosts. An optimum, possibly compromise setting is recommended.

Automatic fine tuning

Automatic fine tuning (AFT) may be found in the latest TV chassis. When the AFT button is engaged, the sound, picture, and color will automatically track, even if someone plays around with the fine tuning control. Always, keep the AFT button pushed in for normal reception. Each channel may be manually fine tuned while the AFT function is off. With the AFT function off rotate the fine tuning control until the picture and sound is normal. Now, push in the AFT button and notice any change. If there is no change, go to the next channel and repeat the procedure until there is no change with the AFT button in or out.

Defective tube

Either the video amplifier or the video detector may be responsible. Checking the suspected tube is of little value here and substitution is recommended. As in the case of ghost problems, should a semiconductor diode be involved, or should tube replacement fail to solve the problem, the services of a professional will be required.

Improper focusing

Although this misadjustment should be fairly obvious, the beginner may not be aware of it at all times, since, in mild cases, defocusing can be tolerated, especially in action scenes (which applies to most TV programs). Picture focus deterioration with long use is not unusual and it should be checked whenever picture detail seems deficient.

As detailed earlier, focusing is best done by closely observing the TV screen, using a mirror if necessary, while adjusting the focus control for the sharpest horizontal lines on the screen. In those older receivers, changes in the ion trap magnet position should be made with circumspection as it is easy to degrade picture brightness, picture size and coverage of the screen edges by gross changes in the ion magnet position. Very slight and gradual rotation and sliding will easily show whether the focus is affected.

Smearing

This defect, smear, is best observed on the larger areas of uniform illumination, such as a wall, the border between a person's white shirt and dark suit, etc. Practically every cause listed previously for loss of fine detail applies here, with two exceptions. First, focus misadjustment is not nearly as evident on large areas of uniform illumination. Second, there is a good likelihood that a defective component in the video amplifier circuit is responsible for the loss of good quality in the heavy areas.

One other potential cause for loss of heavy detail is misadjustment of the AGC. Since the AGC function sets the amplification level of the TV set, it is, in a manner of speaking, a coarse presetting for the contrast control, while the latter is, comparatively speaking, a fine control. As stated earlier, the AGC should be set so that the contrast control is set about midway while receiving an average signal. Any setting of the AGC requiring the use of either the maximum or minimum extreme setting of the contrast control for an average signal is also likely to cause a type of distortion which ultimately evidences itself in picture smear.

A smeary-looking picture may be caused with open peaking coils or coupling capacitors in the video tube or transistor circuits. The picture is not clean but the whole raster appears smeared. Check for open peaking coil or delay line within the video circuits (Fig. 8-21).

Often, the voltage is quite lower than normal on the plate of video tube or collector terminal of video transistor. These coils are very low in resistances and can be located with the DMM. If the picture smears intermittently, suspect a broken or poorly soldered connection of the coil winding. The defective delay line in the video circuits may cause video smearing. Check for defective video IC when all other components seem normal.

Line pairing

This is a picture quality defect causing severe image deterioration both in the fine detail and in the large solid areas, but particularly in the fine detail area. It is due to an accidental overlapping of adjacent raster lines. In appearance, the number of horizontal lines on the screen is effectively cut in half, while the thickness of each new line and the spacing between them is doubled.

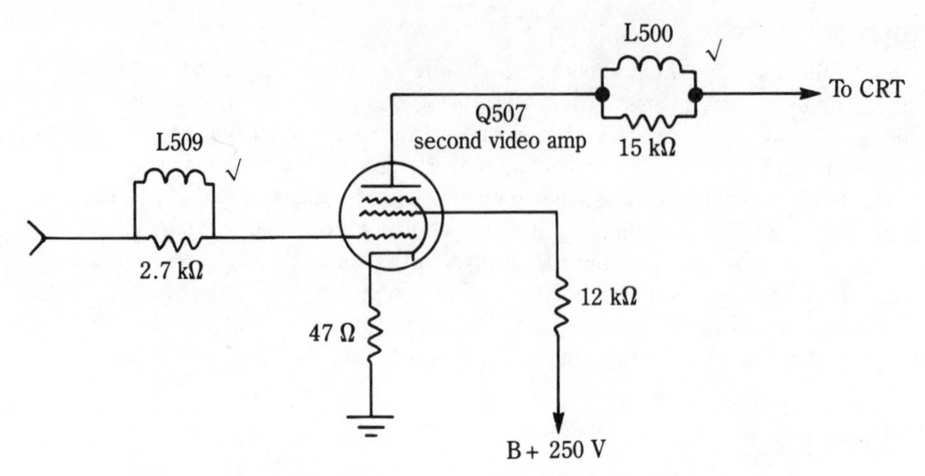

8-21 Open peaking oils in the video circuits can cause video smear.

In chapter 1, it was established that the visible rectangle (raster) on the screen consists of approximately 480 interlaced lines; that is, lines "painted" by the electron beam, first the odd-numbered set (lines 1, 3, 5, 7, etc.) then the even-numbered lines (2, 4, 6, 8, etc.). During normal operation, the sync or precise timing pulses from the transmitter keep the spacing between the lines exactly right, so that, for example, line 2 falls exactly midway between lines 1 and 3, line 4 fits exactly between lines 3 and 5 and so on. In this manner any picture information on one of these lines is clear and distinct from any information on an adjacent line. Should, however, due to a malfunction, two adjacent lines overlap, two different sets of picture detail will overlap and neither will be clear. The term *pairing* of lines refers to just such an overlap. A failure in the sync circuits, due to either internal (TV set) or external (freak reception or transmitter malfunction) causes will often produce this phenomenon. To verify whether or not the TV set is responsible, a simple procedure is involved.

Switch to a different channel, preferably three or four away. If the malfunction persists, the fault probably lies within the set. Since it is extremely unlikely that two or more TV stations are having technical difficulties at the same moment, this source can be discounted. Similarly, freak reception, with possible interference from a more distant station on the same channel will produce just this effect, but is most unlikely to occur on a substantially different frequency. Thus, if the pairing occurs on channel 2, switch to channel 5 or 7 and observe. If freak reception is responsible, the other channels will most likely not be affected. If pairing persists, the TV set is at fault.

Replace the horizontal control tube (V10A in Fig. 6-2), the horizontal phase detector tube (V9A in Fig. 6-3), or the horizontal AFC tube (V7C in Fig. 6-5). Each of these tubes, although somewhat different in name and operation, performs a basic function in a unified effort of controlling the timing of the individual raster lines. A defect in any of them may produce pairing and similar sync malfunctions. In testing these tubes, careful checking for leakage after a few minutes of being in the tester is important. Of course, substitution of a new known-to-be-good tube is always conclusive.

Remove the sync tube (V8A in Fig. 6-3, V8B in Fig 6-4, etc.). This tube can be equally responsible for pairing, except that a serious malfunction here would also show up in a loss or poor performance of vertical sync. The picture would tend to roll or slide up or down one or more frames at random times; however, the malfunction may be not be very obvious here.

Check the vertical hold control. Although this control setting is not very critical in the presence of a strong signal, a fine touchup may sometimes cure mild case of pairing. Adjust this control for best interlace; picture rolling will automatically be corrected.

Check the AGC pot setting. The AGC adjustment cannot be overemphasized. It may seem paradoxical to the beginner and even to some TV servicer that too strong a picture single may go hand in hand with too weak a sync signal. Since the AGC adjustment serves to establish the strength of the picture signal or contrast, it is quite clear that in an effort to obtain a good strong picture where one did not exist (quite often for other reasons), the AGC setting can be advanced to the point where the sync pulses are reduced below a safe minimum. Adjustment of the control should be made until good sync performance (no more pairing) is restored, even if this produces a somewhat weaker picture, which can be corrected by the contrast control.

Missing adjustments

In a generalized description of causes and effects of TV malfunctions and their remedies, provision must necessarily be made for examples of controls and adjustments found in a great variety of makes and models of sets. Frequently, therefore, the TV owner may find that his or her set has no such control or adjustment. One good example of this is the ion trap magnet. While it is found fairly generally in older sets, such is not the case in more recent models. The same may be said for such an adjustment as horizontal lock or focus on black-and-white receivers. There are two explanations to this. In case of a ''missing'' control or adjustment, it is fair to assume that it is not required; it should also be added that in such cases the problem will not arise in the first place, at least not from this cause. In the other case, an exactly identical control or adjustment may not exist, although there seems to the nonprofessional TV do-it-yourselfer to be one with a similar name. Various horizontal adjustments are in this category. I have therefore provided in the various block diagrams a variety in which the ultimate functions are the same although the identifications are somewhat different, either due to small design differences or simply because of different terminology used by individual manufacturers.

The sync blocks in the various diagrams are a good example of this practice. Furthermore, in suggesting possible causes for one particular defect, I refer to these seemingly different stages for examination, indicating that they are functionally the same. Finally, in reading the instruction booklet or examining the various descriptions in the service information for a particular TV set, the owner will be able to identify the exact function of the different component groups or blocks even though the name may be slightly different from some typical TV set described here.

Shorts or fire

This type of failure, short or fire, in a TV set, although it may be relatively hazardous, is usually easy to diagnose, especially when the set stops working completely. There are a number of possible failures of this type, each requiring a different procedure and remedy.

Fuse blows

If the main fuse (the one which is accessible without removal of the back cover) fails, it may be due to either a momentary overload, a weakness of the particular fuse or an overload due to a defect in the TV set itself. It is permissible to replace such a blown fuse with another one of exactly the same value. A lower value will blow without provocation, while a higher value may not protect the TV set and may even cause a fire. Carefully watch the behavior of the set immediately after replacement and for a while afterward. If no further difficulties arise (TV set works normally, no smells of burning appear and no crackling or frying sounds are heard), the incident may be forgotten. The assumption is that the fuse or a sudden surge in the line voltage was responsible, and the TV set is not at fault (Fig. 8-22).

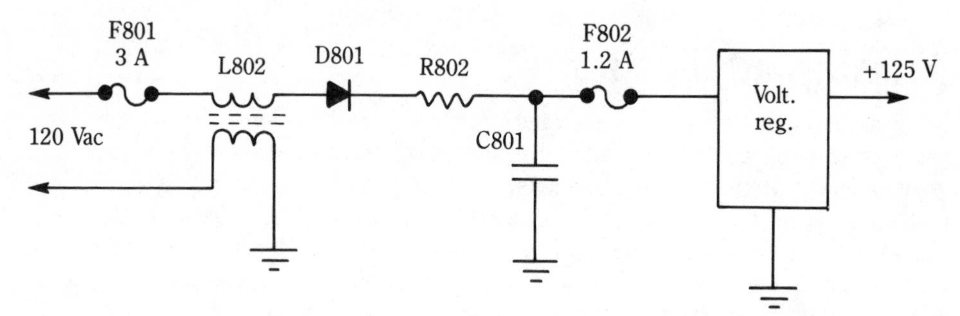

8-22 Suspect overloaded components in the regulator or horizontal output transistor when F802, F801 and R802 are open.

Replacement fuse blows

If the second fuse blows, whether immediately or soon thereafter, it must be assumed that the set is at fault. No further fuse replacement should be attempted.

Circuit breakers

What has been said about the main fuse of the TV set applies equally to the circuit breaker—a device that has replaced the fuse in many TV sets. In simple terms, the circuit breaker is a type of lifetime fuse. While the ordinary fuse blows during an overload and has to be replaced with a new one, the circuit breaker opens and disconnects the TV set from the ac line. To restore operation, it is merely necessary to reset the circuit breaker by pushing a button on the outside of the device.

There are two common types of circuit breakers in use today. One type can't be reset as long as the overload persists, so that there is little danger of unknowingly causing the TV set to operate under damaging or even dangerous conditions. If a breaker can't be reset, as evidenced by the fact that the pushbutton does not respond to normal pressure, it is an indication that the overload is still there.

Thermistors

Many TV chassis have a thermistor located in the ac low-voltage power supply. A thermistor's internal resistance changes with its temperature making it useful to control ac voltage. In older TV sets, a 120 ohm (cold) thermistor was placed ahead of the degaussing circuits (Fig. 8-23). On later TV chassis, thermistors may be found with only a few ohms of resistance. A few seconds after current flow begins, the resistance of the thermistor will decrease to only a few ohms, letting the ac through to the power supply. When a TV chassis with a good fuse or one with the circuit breaker reset is dead, suspect a lead melted off the body of the thermistor. These units are easily replaced by unsoldering two connections.

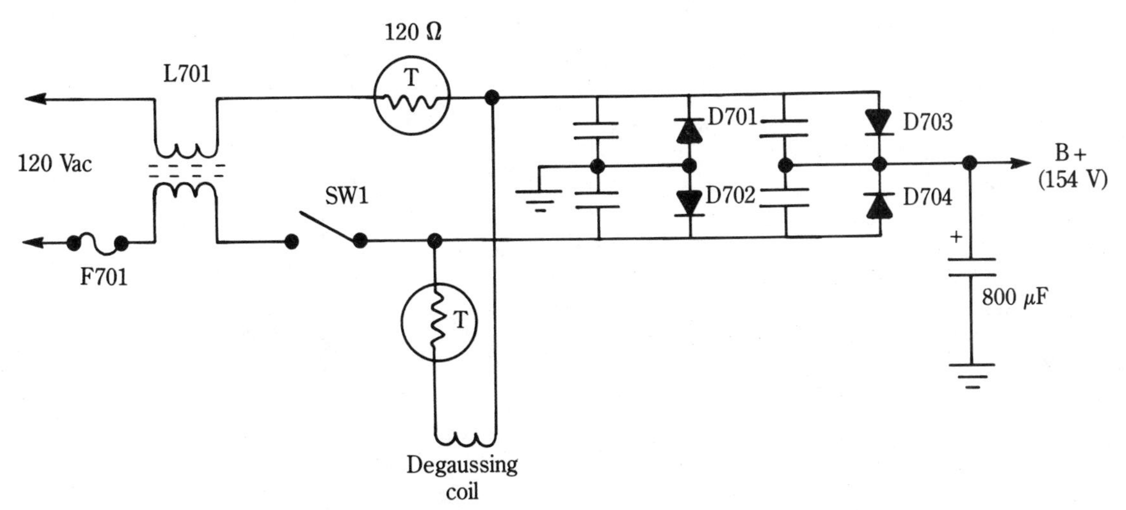

8-23 The thermistor is in the ac portion of the low-voltage power supply circuit. A 120Ω thermistor is placed ahead of the degaussing circuit.

Burning smell but fuse OK

It will sometimes happen that a fuse will not blow when it should. Fuses are not very accurate devices; their behavior is not 100 percent predictable. In such a case, damage to the TV set is possible, depending primarily on how long the overload exists before the fuse goes (the correct and safe procedure is to immediately switch the set off) and whether it's a transformer or transformerless type set.

In transformerless sets, a *local* overload may be sufficient to start a resistor to overheat or actually burn out. A burning resistor emits some intolerable odors. While the remedy is beyond the scope of the beginner, the damage is seldom of major proportions. It is usually caused by the total failure of another component (a capacitor, for example).

In transformer-type sets, in addition to the same possibility as just outlined for a transformerless set, repeated blowing of the correct size fuse is very likely to be due to a defect (short circuit) in the power transformer. It too may overheat and smell for a very short time (seconds, not minutes) without being permanently damaged. This, too, is a repair job for the professional, and his or her advice on whether or not replacement is required must be taken.

As in the case of a burnout in a transformerless set, a failure here seldom takes other parts with it. In other words, a burned out transformer usually does not cause damage to other parts; however, as a matter of cause and effect, a transformer failure, in addition to being due to an inherent defect, may also be caused by an earlier failure of another part. A defective rectifier tube (V14 in Figs. 6-2 and 6-3), semiconductor diodes (Figs. 6-4 and 6-5), or a short-circuited filter capacitor in the main power supply may be responsible, in which case, they too, will have to be replaced when the TV set is repaired.

In addition to a defective power transformer or shorted capacitors, there are many other potential causes of the fuse blowing, even if less common than these two; however, since in all cases the services of a professional are required, it is only necessary here to classify all fuse failures into two categories. One is a random failure not due to any permanent defect. This one is identified by the fact that the correct replacement fuse does not blow. The second is a causative failure due to a permanent internal defect or damage, which should be repaired by a professional.

In connection with random fuse blowing, it should be pointed out that it may sometimes be caused by injudicious placement of the TV set. Since most TV sets draw hundreds of watts, it is absolutely essential to allow the heat to dissipate. Pushing a TV cabinet tight against a wall or some other piece of furniture is sure to cause overheating. Regardless of whether or not the fuse blows, the heat buildup shortens the life of the tubes and other components, not to mention the poor performance caused by thermal instability.

9

Troubleshooting the color section

AS HAS BEEN POINTED OUT, COLOR AND BLACK-AND-WHITE RECEIVERS ARE essentially the same up to the video amplifier. In other words, all circuits and functions from the tuner through the video detector plus sync, sweep, and sound are common to both monochrome and color TV reception; therefore, the troubleshooting procedures discussed so far apply equally to color and black-and-white receivers. In fact, color TV troubleshooting, except for problems in sections that deal solely with the color signals, is easier when the color functions are disabled.

Referring to the color TV block diagrams (Figs. 9-1 through 9-4) you'll see that in each case an adjustment is provided for setting the color level of the set. If the color control is turned to one extreme (counterclockwise) it shuts off the color completely, leaving a normal black-and-white picture on the screen. Turning the control in the opposite direction (clockwise) gradually increases the color signal up to full color (and beyond, into distortion).

The color control is called different names in various receivers. In Fig. 9-1 it is called simply color. In Fig. 9-2 it is called chroma gain, while in Figs. 9-3 and 9-4 it is called color control and color level, respectively. In each case, this adjustment should be turned to the extreme no-color end of rotation before starting any of the troubleshooting procedures outlined previously.

No color but black-and-white normal

The first and most obvious reason for the absence of a color picture is the setting of the chroma or color input control. Since this affects all three primary colors, all color will disappear if the color control is accidentally turned off. A simple check of the control setting will determine whether or not it is the cause of no color. If it is, the control is simply reset (assuming no other adjustments have been changed) by eye while observing the screen.

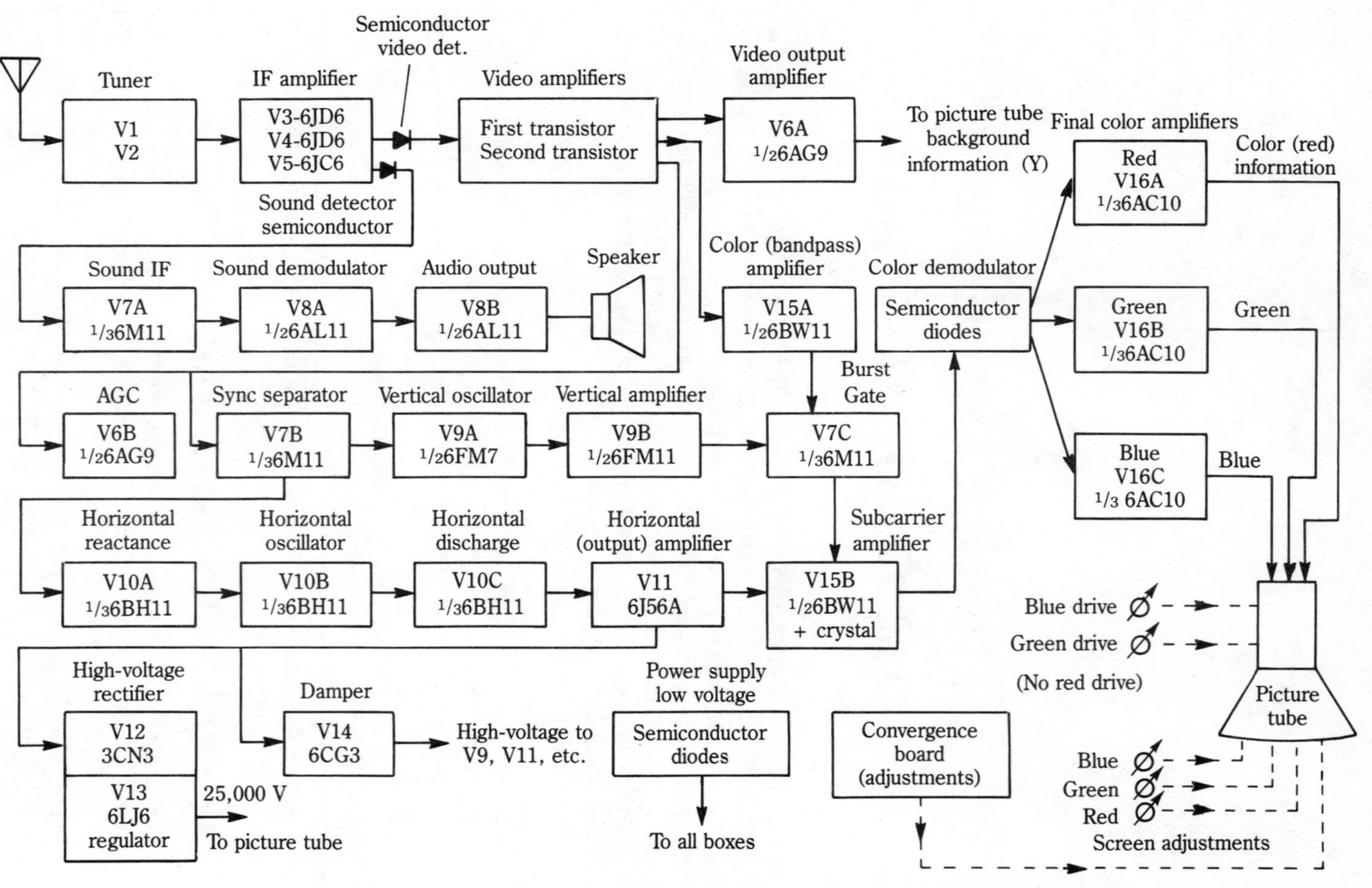

9-1 Block diagram of a typical tube color set. Semiconductor diodes serve as color demodulators.

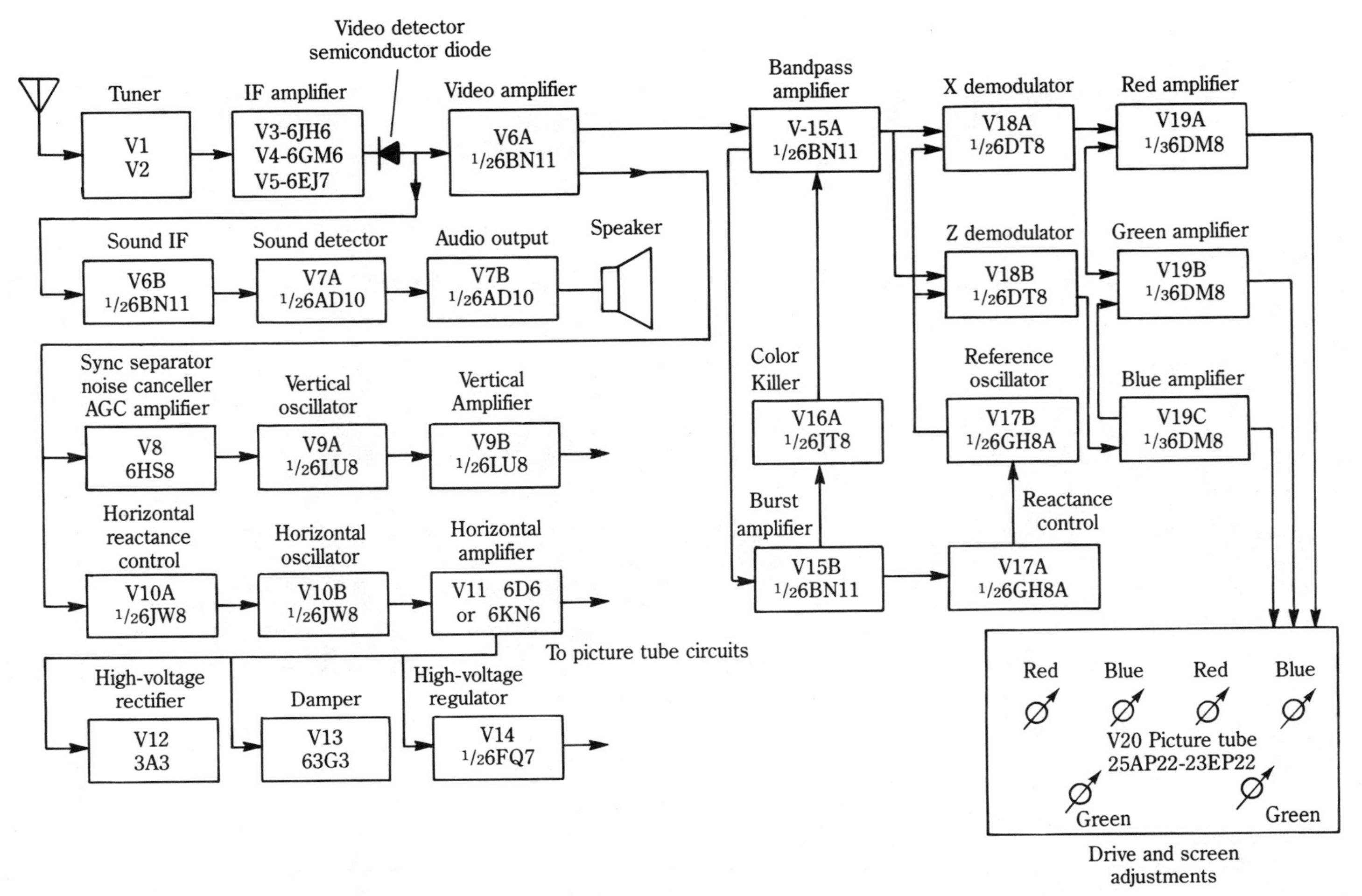

9-2 Block diagram of a color receiver using dual-purpose tubes as a color demodulator. Note that the drive and screen adjustments are part of the tube settings (they are not necessarily physically on the picture tube assembly).

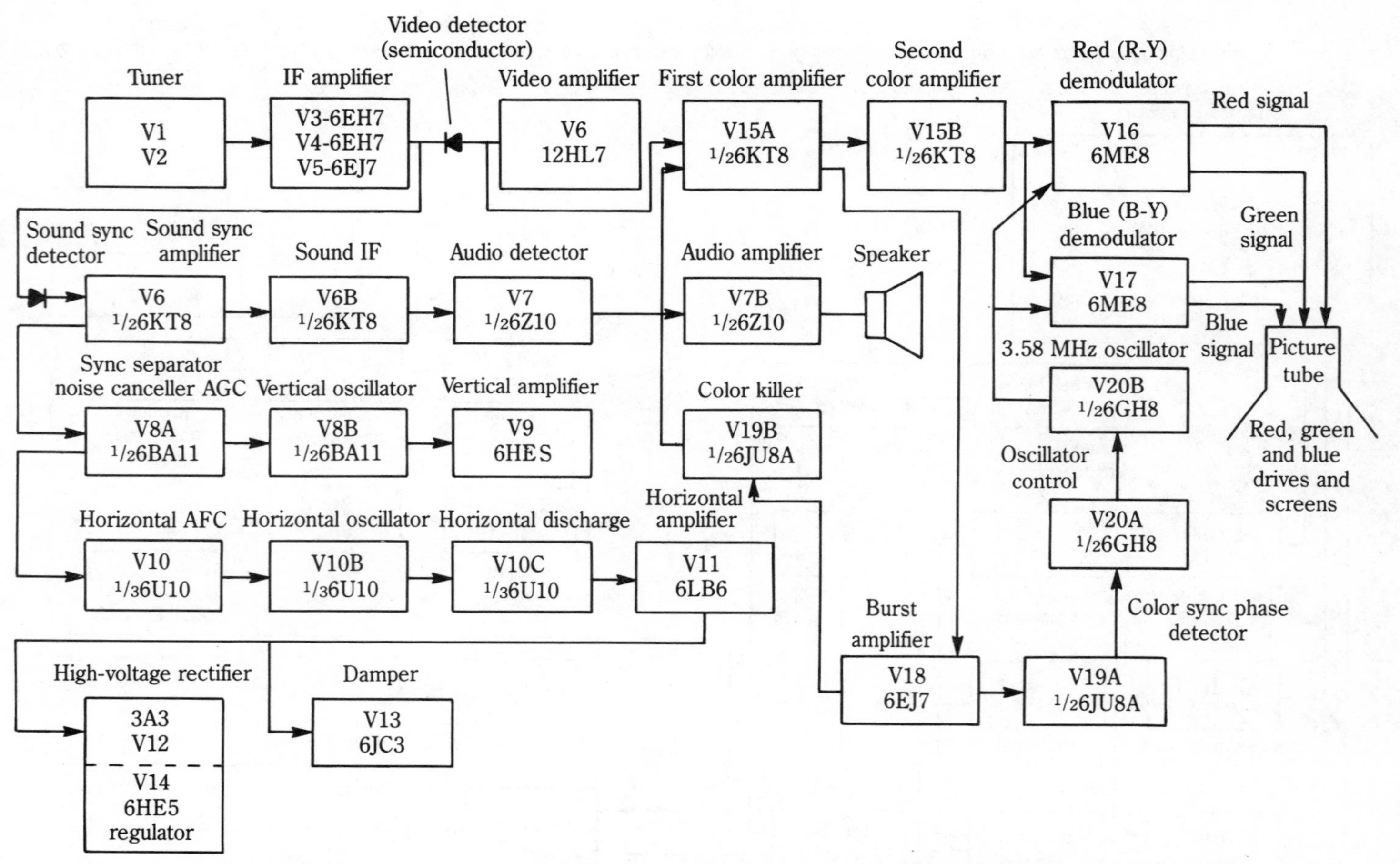

9-3 In the color set represented by this block diagram, the color demodulators drive the picture tube directly.

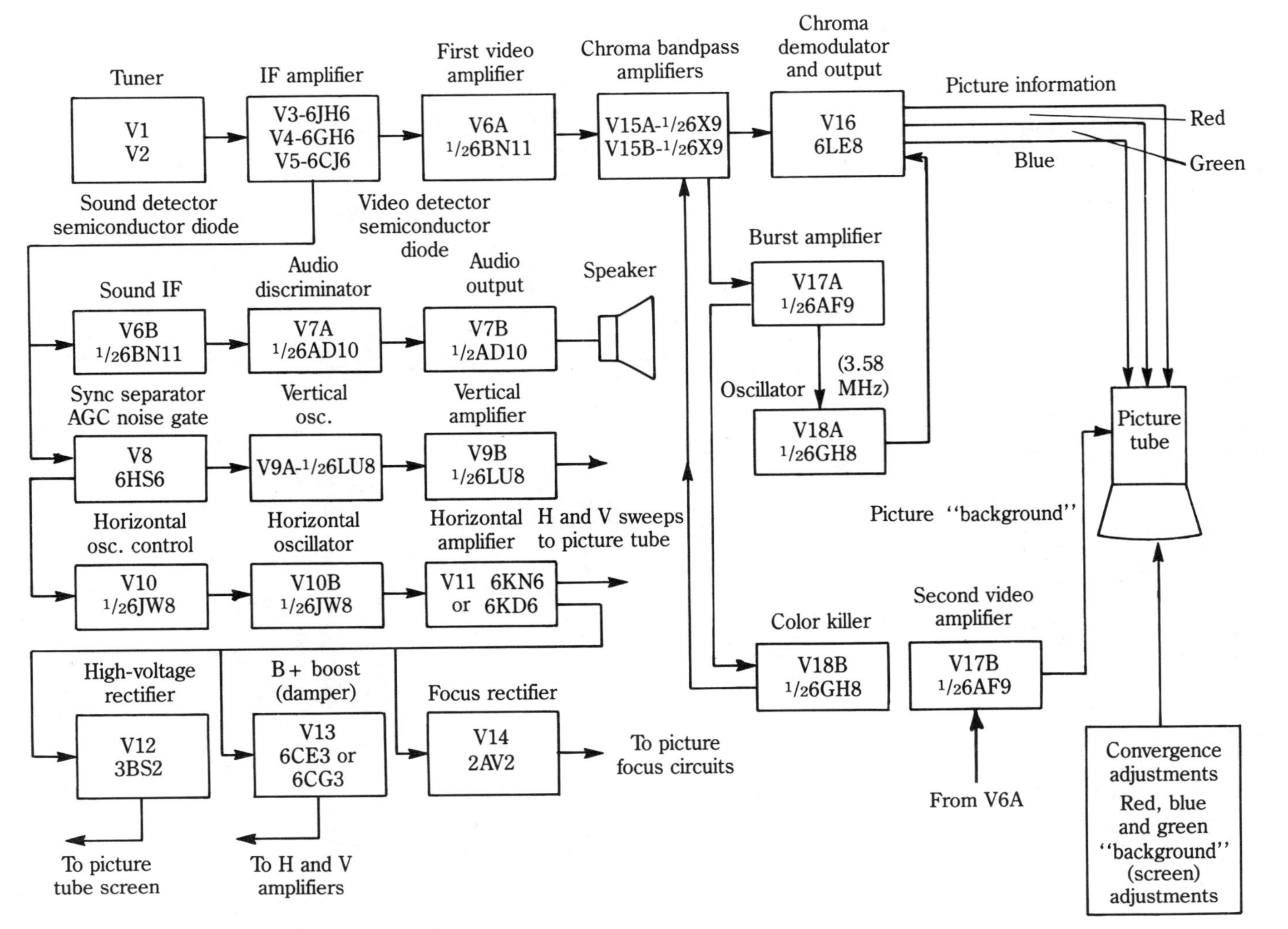

9-4 In this block diagram, one multifunction tube demodulates the color signals and drives the picture tube.

Fine tuning and color level

In addition to the obvious matter of setting color level from no color to any desired intensity by the chroma or color control, there is another factor affecting the presence and intensity of color—the fine-tuning control. As discussed earlier (chapter 4), reception of color depends also on the burst signal from the station and the 3.58 MHz color oscillator (block J, Fig. 4-1) in the receiver. Since the setting of the fine-tuning control naturally affects the received signal, it is possible, especially in a weak station, to have the fine-tuning control off its optimum position. While a strong enough picture is still received, the color will be absent or very faint. Adjustment of the fine-tuning control will bring the color component information up to an acceptable level. Since in all but a few TV receivers, fine tuning is preset for each individual station, such an adjustment will not affect any other station.

Chroma control ineffective

If after advancing the chroma control there is still no color on the screen the failure probably lies beyond this point (between the control and picture tube). Furthermore, since none of the colors are present the defect must be in a circuit common to all colors, including the bandpass amplifiers (V15A in Figs. 9-1 and 9-2, V15A-V15B in Figs. 9-3 and 9-4), the 3.58 MHz oscillator (V15B in Fig. 9-1, V17B in Fig. 9-2, V20B in Fig. 9-3 and V18A in Fig. 9-4), the chroma demodulators and the final amplifiers. To find out which stage is defective, begin by substituting tubes.

Bandpass amplifiers In some cases, the tube location diagram inside the TV cabinet may indicate two separate stages using a common dual-purpose tube. This is illustrated in Figs. 9-1 through 9-4, and is reflecting the fact that many tubes in modern TV sets are multiple-function or multiple-section types. Two, three, or even more tubes in one envelope is not uncommon; so, when you remove a tube of this type, another function may be disabled, too. For example, in our typical block diagram in Fig. 9-1, half of V15 is the bandpass amplifier, while the other half is the subcarrier amplifier; however, since the whole tube must be discarded if one half is defective, it does no harm to disable two functions at once. When a replacement tube is substituted, both functions will return to normal if the original tube was a fault. If the trouble is still there, then this particular (dual) tube is not the cause.

3.58 MHz oscillator As was mentioned in the general discussion of the color receiver process in chapter 4, this function is essential for the recovery of color. If this oscillator fails, all color will be absent. Replacement with a good tube will prove or disprove this suspicion. As in the case of the chroma amplifier, a twin tube may be involved here, but this does not affect this trial-and-error check on the subcarrier oscillator.

3.58 MHz crystal A brief explanation should be helpful. As mentioned earlier in the general description of the TV system, certain very precise timing (or sync) signals are sent from the transmitter for cueing or timing the various signal sequences on the receiver. One of those signals exclusive to the color TV set is the subcarrier *burst* signal. This is a precise timing pulse that keeps a locally generated signal at its exact timing, and it is commonly called the 3.58 MHz reference. Without this reference signal, no color reproduction can take place. The heart of this oscillator is an electromechanical element called a crystal. In some TV sets, it is a plug-in device and resembles Fig. 4-2.

To remove the 3.58 MHz crystal, grasp it and pull upward, after first noting carefully the appearance of the picture tube screen. Often, a very poor crystal may still allow some color to appear on the screen. If the removal of the crystal makes no difference, then the fault does not lie here; however, if the slight coloration previously seen does disappear, the crystal is probably defective (or intermittent) and should be replaced. Classically, when an inoperative 3.58 MHz crystal is encountered the screen will appear to have only blue and green—no red. Also the tint or hue control will have no effect on the color present on the screen.

This is a standard item found at most radio supply houses catering to the TV servicers; therefore, a replacement crystal of the same type is readily obtainable. Where the service data lists a crystal by a particular manufacturer's part number, it may be more practical to obtain a replacement from the service department of that manufacturer.

In most recent TV sets, the 3.58 MHz crystal is a pigtail type; that is, it is not like the plug-in type just described but is intended to be soldered in the circuit similar to a resistor or capacitor. In such a case, the crystal is "worked around" in the troubleshooting procedure until all the other alternatives are exhausted, and if none solves the problem of the absence of color, it is assumed that the crystal is at fault and a professional servicer is called. (See chapter 10 on do-it-yourselfer procedures on soldering printed circuit components as an alternative to calling a TV servicer.)

Color killer tube There is just a possibility, although not very probable, that a malfunction in the color killer tube would cause the color absence. As explained in chapter 4, the function of the color killer stage is to shut off the chroma amplifiers in the absence of a color signal. At first glance this may not sound very logical; why shut off the color amplifiers when there is no color to amplify? Actually, however, this is logical as can be easily shown.

On all color block diagrams, Figs. 9-1 through 9-4, each of the three guns of the picture tube receives a signal from a color amplifier. The signals are electric currents or voltages, not actual colors. If any of the color amplifiers feed an electrical signal into one of these guns, that gun will produce a color on the screen, the particular color depending on which gun is being fed. To ensure, therefore, that no color is shown when a black-and-white picture is transmitted, the color killer keeps the color amplifiers shut off until a color burst signal (present only when a color picture is transmitted) commands the color killer to let the amplifiers perform normally. If, due to a malfunction, the color killer tube keeps the chroma amplifiers permanently shut off, no color will come through. Replacement with a good tube will prove or disprove this suspicion.

Color killer setting In all cases where there is an adjustable control marked *color killer adjust* or just plain *killer*, the setting of this control should be checked before suspecting the color killer tube. It is just possible that this control is set so that the bandpass amplifier could never pass a signal. Incidentally, the color killer control is normally set so that all traces of color just barely disappear on a blank channel; then advance it ever so slightly beyond this point.

Component failure If all the preceding procedures fail to restore the color to an otherwise normal TV set, the fault must lie in one of the internal components in the path of the color signal, such as a coupling capacitor or a failure of a supply voltage to one of the tubes. In the first case, the tubes function normally, but the signal path is interrupted. In the second case, one or more tubes are inoperative due to the failure of the correct operating voltage. Whichever the case, the services of a professional servicer are required.

Simplified solid-state TV block diagram

The simplified solid-state TV block diagram consists of a varactor, band decoder, and multiband tuner in the latest TV chassis. The tuner may be controlled with a microcomputer, system control, push-button, or remote control systems. The IF amp from the tuner is amplified and passes through a SAW filter to the video amp stage. The AFT, RFAGC, IFAGC, sound IF, video amp, sync separator, luminance, and chroma processing are found in one large IC component (Fig. 9-5).The chroma amp and demodulator circuits connect to the CRT bias or color output transistor. The output from the three color output circuits feed to the cathodes of the picture tube.

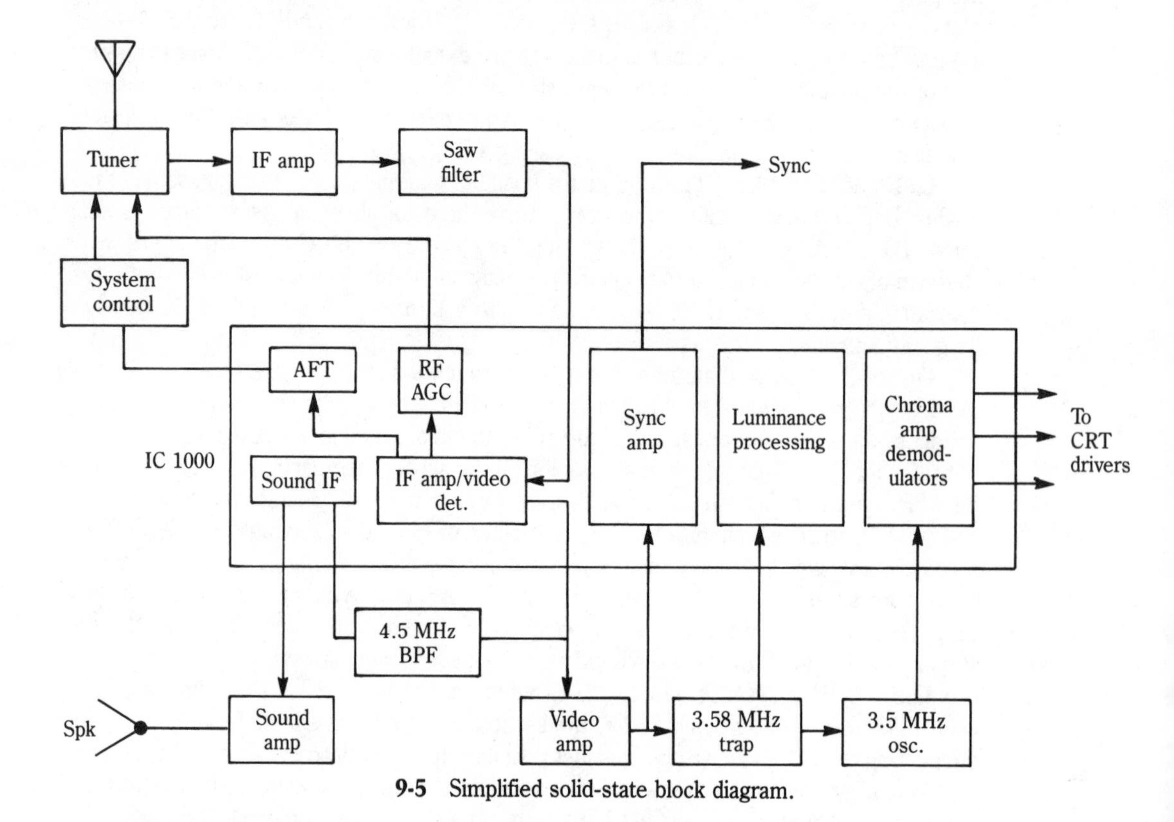

9-5 Simplified solid-state block diagram.

Solid-state color circuits

In the early solid-state chassis, all color stages were made up of transistors. Then, the novel integrated circuit (IC) was added to the color circuits. At first, one small separate IC component contained most of the color stages (Fig. 9-6). The first and second

9-6 The PC board side of a color IC component.

chroma amp, burst amp, ACC detector, R-Y demodulator, B-Y demodulator, killer amp, 3.58 MHz color oscillator, and color matrix were found in one IC, along with the luminance components (Fig. 9-7). The red, green, and blue color output transistors were connected between color IC and picture tube.

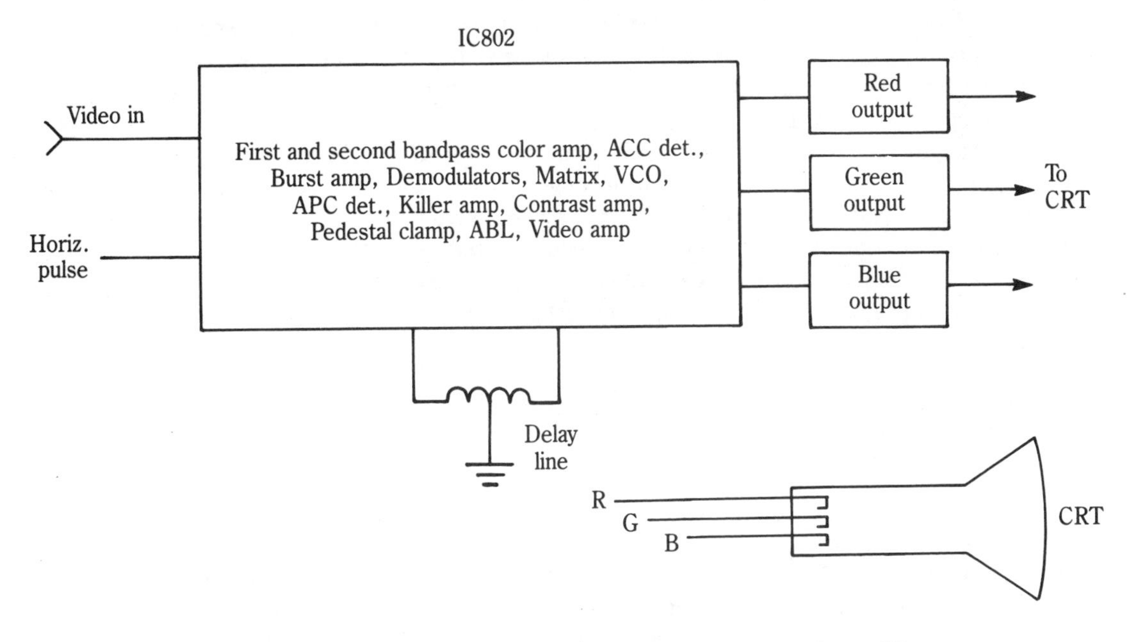

9-7 Often, many different color, luminance, and IF stages are contained in one IC component.

First and second chroma amps The chrominance signal is amplified by the first and second chroma amp and sent to the burst amplifier. Usually, the first chroma amp gain is controlled by the automatic chroma control (ACC). The color control proceeds external bias to vary the voltage and gain of the second chroma amp.

Burst amp The color signal from the second color amp is applied to the burst amplifier. The tint control is connected to the burst amplifier and applies external bias and voltage to the burst amp changing the current ratio and color tint.

Automatic chroma control (ACC) The ACC circuit stabilizes the chroma output against fluctuation of the chroma/burst signal level. The peak detection of the burst signal controls the gain of the first chroma amp with a voltage. When the burst output is large, this voltage reduces the gain of the first chroma amplifier, thus reducing the burst (chroma output) and maintaining a fixed amplitude.

Color killer The color killer circuits prevent colored snow or noise from appearing in the raster when a black-and-white picture is being televised. This killer circuit switches the second color amp on and off according to the transmitted burst signal. Color killer adjustment control was made in the early TV chassis; now the color killer circuit operates automatically within the luminance/chroma IC.

R-Y and B-Y demodulator The chroma signal is applied to the B-Y and R-Y demodulator circuits. Often, the R-Y demodulator and G-Y demodulators operate through a lag circuit and feed to mix in the matrix circuits. In some circuits, the matrix color feeds directly to red, green, and blue output transistors, while in others a red, green, and blue amp or bias stages are found between matrix outputs and red, green, and blue color output amplifiers.

Automatic phase control (APC) The automatic phase control circuit regulates the color oscillator frequency and phase to ensure correct color sync and to provide signals to the various control circuits. The APC circuit detects the phase difference between burst signal and the 3.58 MHz oscillator, providing detection voltage. The detection voltage is controlled so that the voltage controlled oscillator (VCO) is in phase and frequency with the burst signal.

RGB output circuits The color signal from the large IC may be fed directly to the red, green, and blue output transistors. In some later color chassis, the red, green, and blue amplifiers plus the RGB output circuits are also found inside a large IC component. When the red, green, and blue output circuits are used separately, they are found upon the back of the CRT socket board. Actually, the RGB output circuits combine the color signal at the base terminals, and the luminance signal at the emitter terminals of each output transistor.

Color IC processing

Today, one large IC component may include the AGC, IF amp, video detector, AFT, sound IF, audio signal processing, sync separator, vertical and horizontal deflection, X-ray protection, and luma/chroma processing circuits (Fig. 9-8). The video signal is applied to the bandpass filter allowing only color to enter the first and second color amplifiers. The gain of the first color amp is controlled by the automatic color control (ACC) circuits. The gain of the second color amp is determined by the color control dc

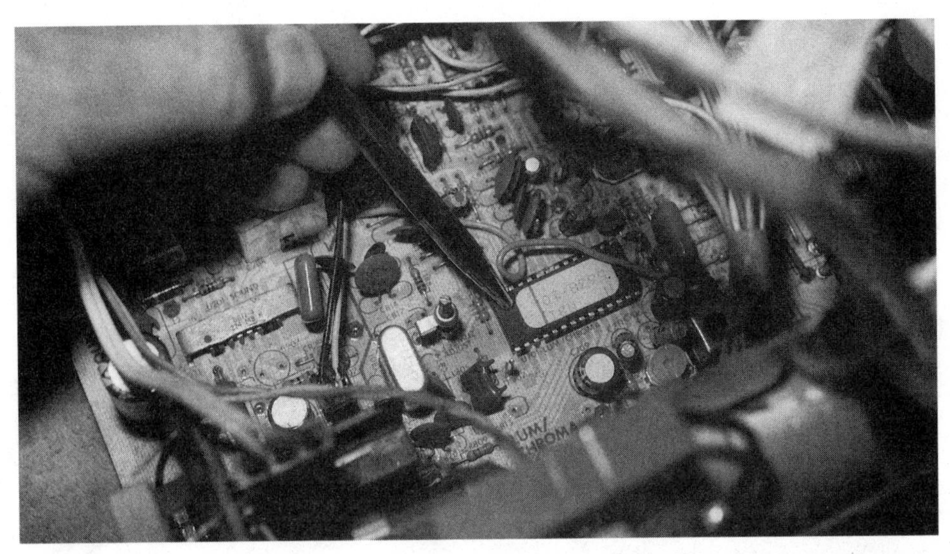

9-8 A large IC is found in one of today's color chassis that includes color processing.

voltage developed by AIU microprocessor. The second chroma amp is turned off and on when a black-and-white signal is received by the color killer circuits. The output of the second color amp is connected to the color demodulator circuits.

The VCO color oscillator circuits are used in demodulating the chroma signal to the chroma demodulators (Fig. 9-9). The frequency is controlled by a 3.58 MHz crystal oscillator. The APC circuit compares the phase of the color oscillator to the color burst

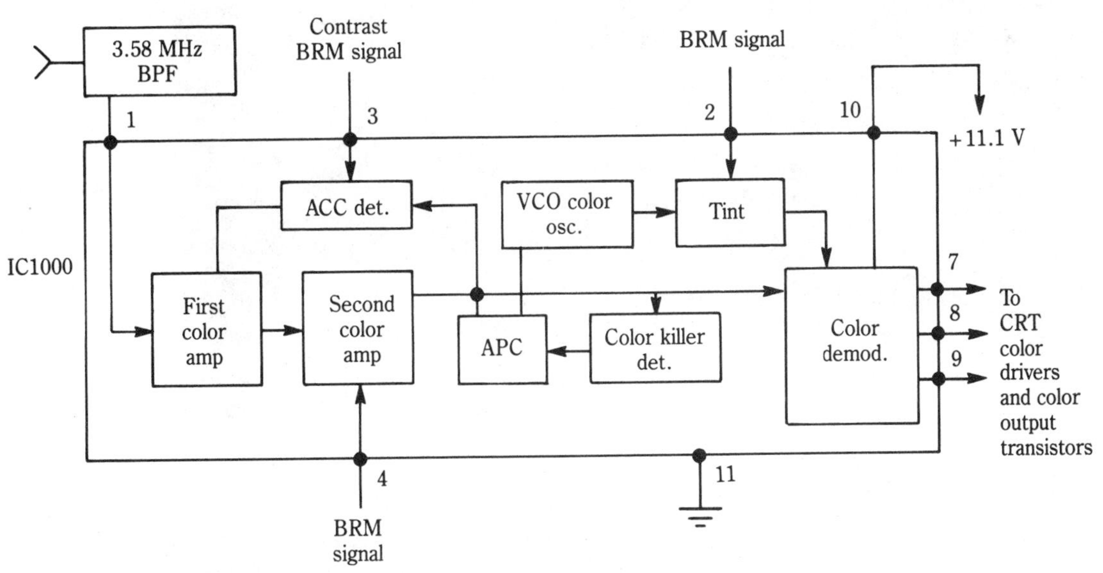

9-9 Block diagram of a late-model color-processing IC circuit.

signal, generating a phase error signal. This error phase signal keeps the color oscillator locked to the burst signal.

The chroma demodulator circuit uses the dc tint control voltage and color signal from second color amp to develop the R-Y, G-Y and B-Y signals. The color difference signals from the chroma demodulator circuit are fed to the CRT color driver and amplifiers where the color and luminance signals are mixed and fed to the cathode elements of the picture tube.

Color modules

Color modules were found in many color chassis when modules were used throughout the TV set. Some RCA chassis had two separate color modules. In the Zenith chassis, the color circuits were also included with the AGC, IF, and sync circuits (Fig. 9-10). When the picture has no color, intermittent color, or mixed colors, simply replace the color module. Today, most of the color circuits are soldered directly into one large PC board, although one large IC may include all of the color, luminance, IF, AGC, and sync circuits.

9-10 You might find one or two color modules within the TV set. Here the color section is included with IF, video, luminance, AGC and sync circuits in one large, shielded module.

Comb filter

The comb filter circuit separates the luminance (brightness) and chroma (color) video information so when displayed upon the screen, the cross-color lines are eliminated (Fig. 9-11). The video signal is supplied to the comb filter luminance processing channel and

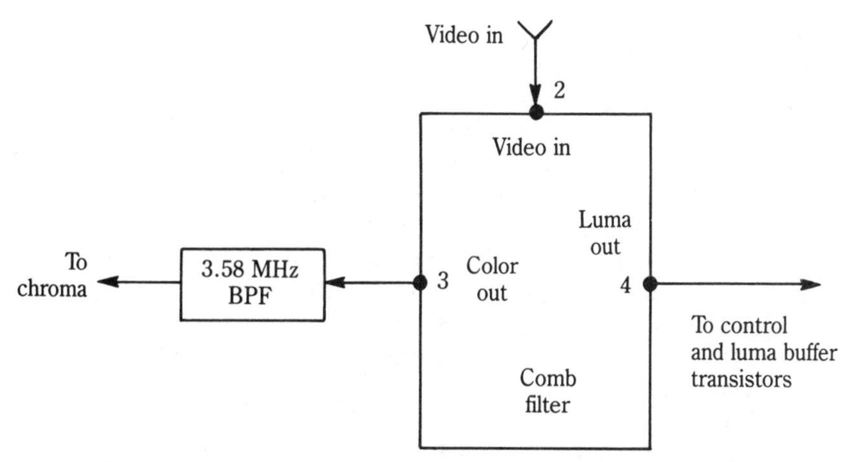

9-11 Block diagram of the comb filter in the luma/chroma circuits.

amplified by a gain stage. This composite video signal is first inverted and then amplified. The combed color output signal of the comb filter IC goes through a chroma peaking circuit to the luma/chroma processing IC and the other path through a vertical filter network.

Other comb filter circuits may consist of a delay line driver, delay line, luminance buffer, luminance inverter, luminance equalizer, and peaking buffer (Fig. 9-12). The output of the peaking buffer transistor connects to the peaking circuit of the chroma/luminance IC. The luminance signal from luminance equalizer transistor feeds to the contrast amp. The luminance inverter provides color signal to the chroma processing of IC 650.

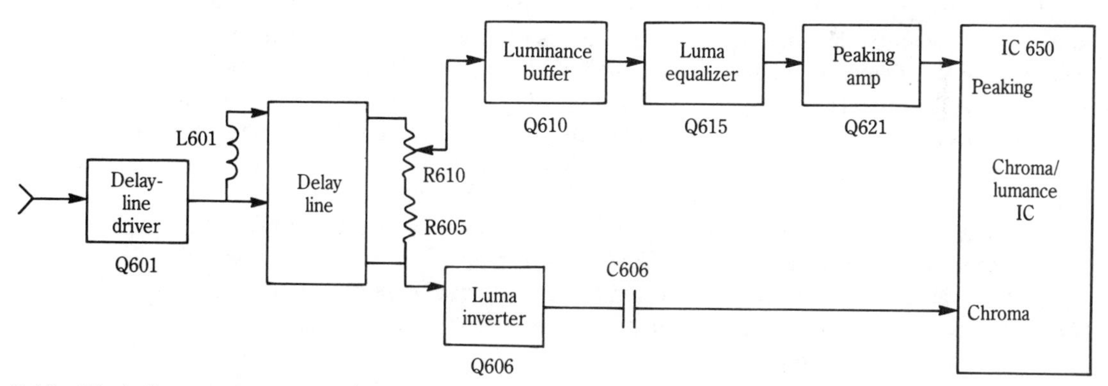

9-12 Block diagram of a comb filter circuit consists of transistors, delay lines, and luminance stages in several TV chassis.

Weak colors

If all the colors are more or less uniformly weak, that is, no one color is much worse than the other two, the fault again lies in a common color circuit. The troubleshooting sequence here is logical.

Bandpass amplifier A weak tube in either of the bandpass stages, if there are two, will most likely decrease the color contrast.

Color control setting If the color control has not intentionally or accidentally been backed off, in other words, if this control is in its usual position or setting, try advancing it clockwise. Advancing the color control will probably restore the color to a good level; however, if it does, it is an indication of weak tubes, especially if advancing the control to its maximum position still does not produce a normal color picture. Incidentally, it should not be necessary to have this control beyond its approximate mid-position if the bandpass amplifiers are in normal working condition.

Color killer setting An improper setting of the color killer control can produce the same symptoms as weak bandpass amplifiers. Unless one is sure that the setting of this control is correct from previous observation under normal set operation, the possibility exists that the setting of this control is responsible for reduced color performance. A quick check on the proper setting is to switch to a station transmitting no color at the moment and adjust this control as outlined previously for the normal setting to obtain a clean black-and-white picture.

Solid-state color killer adjustments In the early color chassis, the color killer control was found on the rear chassis or within the area of color components. Tune in a black-and-white picture and adjust color killer control until the colored snow or noise is eliminated. No color killer controls are in present-day color TV chassis. The color killer circuit operates automatically within the chroma circuits.

Selective color troubleshooting Up to now we have discussed what might be called total or catastrophic color failures. These involve either all of the color portions of the receiver or a vital link in the chain which, if broken, stops any color from getting through. In the following pages, we shall consider local color defects, when a partical break in a secondary chain is responsible. For this purpose, it is necessary for you to be able to clearly identify the function and tube or tubes involved in each case, although you will not have to understand the technical version of how and why. Such identification is a bit confusing because different terms are used by different manufacturers for the functions and because of differences in functions between different TV sets; however, a study of the block diagrams of color receivers, beginning with Fig. 4-1 and continuing with Figs. 9-1 through 9-4, will enable you to identify the basic functions in each diagram, regardless of the individual differences. Since Fig. 4-1 is a *typical*, not necessarily *actual*, receiver its various blocks show *functions* as distinguished from actual examples. By contrast, all the other diagrams just listed represent *actual* receivers. Furthermore, there is considerable variation between them to afford a good sampling of different methods of achieving the same end result.

Color functions

After the color detection take place the three primary colors exist individually, up to and including the picture tube; thus, when it is stated that one color is absent, it is understood that the various shadings or combinations of this color are also missing. A missing red signal denotes a picture made up the remaining two colors only. This implies that the absence of blue would show no shades of blue and also an abnormal green, since the green signal is largely dependent on the existence of a normal red and blue. This may not follow from the logical understanding of colors in nature because it is based, instead, on the technical structure of the color system in the TV set as we shall presently show.

An examination of the various block diagrams in Figs. 9-1 through 9-4 will disclose some important facts. There is a group of function blocks called demodulators and an adjacent group called amplifiers. Further observation will disclose that the demodulators are identified as the X demodulator and the Z demodulator, as in Fig. 9-2, and consist of two tubes or two sections of a twin tube. In the latter application they are marked to indicate that they are one-half of a two-in-one tube (for example, V18A is one-half of a 6DT8). In other cases, as in Fig. 9-4, they are part of a three-in-one tube, such as a 6LE8. In one case (Fig. 9-1), no tubes are used; instead four semiconductor diodes (D1, D2, D3, and D4) are employed. In Fig. 9-3 the terms R-Y and B-Y are used instead of X and Z. In the next group of function blocks, in practically all cases, they are marked R-Y amplifier, B-Y amplifier and G-Y amplifier.

In some cases (Figs. 9-3 and 9-4), there are no such amplifiers at all. In one example (Fig. 9-4) one multipurpose tube (V16, 6LE8) performs all the functions of the demodulators and final color amplifiers. In the second group of functions, those of final color amplifiers, there are also a number of different arrangements.

Returning to the names and functions of the various stages, the demodulators serve to extract or separate the color signals from the combined electrical signals which served to carry the color information through the TV set. After the demodulators, the signals are the ultimate color information, which in some TV sets requires only additional amplification before application to the three color guns of the picture tube. As to the color identification, both the demodulators and the following amplifiers carry the designations R-Y, B-Y and G-Y.

Notice that the hyphens between the letters are intended as minus signs and should read G minus Y, R minus Y, etc. These names describe a method of color processing technically called *difference color system*, that is, difference color demodulators and difference color amplifiers. In the beginning, at least, it is not necessary to know the technical reasons for this. Instead, it is perfectly satisfactory to assume that R-Y stands for red, B-Y for blue, and G-Y for green. What needs to be made clear here is that in all TV sets there will be no G-Y demodulator because this is achieved as part of the R-Y and B-Y functions. Furthermore, in some TV sets the demodulators are marked X-demodulator and Z-demodulator. The first of these is equivalent to the R-Y demodulator and the second to the B-Y demodulator.

Individual color adjustments

In addition to the previously mentioned overall color gain control marked *chroma control* or *color*, there are two other sets of adjustments in the color-only portion of the receiver which must be taken into account when partial color deficiency is evident. One set of controls is called *drive*, the other set is called *screen*. There is usually one in each set for each color, namely, red drive, blue drive, and green drive and red screen, blue screen, and green screen. Sometimes the red drive is omitted.

Drive and screen controls

Drive and screen adjustments are important to the overall representation of a color picture. Unless these *setup* controls are properly adjusted it will be impossible to achieve a good color representation of the color telecast. Although these two sets of controls seem to have a similar effect on the color quality, their purpose is in two different extremes. Drive controls govern the highlights while the screen controls set the operating point for each gun. That is, misadjusted screen control will have a more dramatic effect on the picture than a misadjusted drive control.

Improper adjustment of the drive controls may produce, depending on whether the adjustment is too little or too much, either too faint a picture in the dimmer portions of the scene or too strong on the brightest portions of the scene. By comparison, the screen adjustment can make the whole picture look either too low or weak (insufficient) or too high (oversaturated). The normal adjustment for these twin-type settings is, first, to get the complete range of variation from dimmest to brightest, and second, to get this complete range up to a level which is an acceptable picture.

Gray-scale adjustment

The previous method for adjusting these two sets of related controls is based on purely subjective criteria—you adjust the controls until, to the best of your knowledge and observation, the color picture is most acceptable. A somewhat more impartial method, and one that the professional servicer usually follows, is what is known as the gray-scale adjustment. This simply means that the three guns of the picture tube are balanced to produce as nearly a perfect black-and-white picture as the design of the TV set allows.

As explained earlier in connection with the characteristics of light, white light is a balanced combination of all colors. In a color TV system this is modified to read: a normal black-and-white picture is the result of the correct combination of the outputs of all three color guns. It is the purpose of the drive and screen adjustments to produce such a correct combination. The term gray-scale adjustment is borrowed from photographic terminology and denotes a continuous range from pitch black through all shades of gray to full white. A normal black-and white photograph should show all these levels, provided, of course, that they existed in the original scene. In a three-gun color picture tube proper setup of the gray scale is a prerequisite for a balanced color picture. But even in a gray-scale adjustment carried out by a professional servicer there is a certain amount of *magic* involved, except in those cases of more recent sets where a detailed adjustment procedure, applicable to a particular set only, is provided by the manufacturer. In such cases, there is often an auxiliary device marked *normal-service switch* or similar terminology, indicating that the switch is left in the normal position for use of the set (some normal functions are disabled in the service position), and temporarily switched to service for adjustment purposes. But even in these cases final touchup is still done by the servicer by magic. Based on all these factors, the best procedure for the beginner still is to use careful observation and judgment, proceeding as outlined in the steps to follow.

1. Turn the receiver on and let it warm up for about 30 minutes.

2. Turn the color or chroma control to minimum. If the set is equipped with a control that varies the gray-scale adjustments, set it to midrange. Some brands of receivers call this control picture quality or tint. Do not confuse this control with what is called the tint control on some makes. In one case we are discussing the hue control and in the other it is the raster coloration control. If it varies the coloration of the raster with the color control at minimum, it is the control in question. Set it to midrange so that when this procedure is completed you will have the desired and intended range of the control.
3. Adjust the front-panel controls for normal viewing with the exception of the foregoing controls. Tune in a medium-strength station and fine tune the tuner.
4. Vary the brightness control from minimum to maximum, observing the entire screen for discolorations or tinting. If there is a normal black-and-white picture from low brightness to high brightness, it is not necessary to go any further. If this is not the case, proceed to step 5.
5. If the TV set is equipped with a normal-service switch as described earlier, set it to the service position. If the receiver isn't equipped with this switch the procedure will be slightly more difficult, but it can be accomplished with a little extra care. With the switch set to the service position a horizontal white line from side to side will appear.
6. Set all of the screen controls to minimum, noting their original position.
7. One at a time, adjust the red, blue and green screen controls so that each color just begins to appear on the screen. Adjust all three so that each gun is turned on at approximately the same level. This is best accomplished in a darkened room because red will not appear so bright as the other colors and can be more readily seen in this manner. In receivers not incorporating a normal-service switch, you must eyeball the screen at a low-brightness setting to determine when each color just appears on the screen. The screen controls are set so they have the same threshold level for each gun.
8. If the receiver is equipped with the normal-service switch, set it to the normal position. If it is the type without the switch set the brightness control to a normal viewing level.
9. One at a time, bring each of the drive controls down from a maximum position to a point that results in a true black-and-white picture. Most often, the red drive is not adjustable and the end result will show the blue at maximum and the green slightly reduced.
10. Check the picture for gray-scale tracking at all brightness levels. When you are satisfied the job is as good as can be obtained, check the color presentation by turning up the color control.

Color interdependence

Although separate final amplifiers may be used for each of the three primary colors, there is a certain amount of interdependence between them. So much so that a failure in one color will probably affect them all because of the technical nature of the color image structure. As detailed earlier, most TV sets have but two color demodulators, an R-Y (or X) and a B-Y (Z). The G-Y does not exist separately because it is in a sense a composite of the other two as far as extraction of the signal is concerned.

We do not, of course, mean that a simple mixture of red and blue will produce green. What is meant is that the color transmission standards are such that proper blending of the R-Y and B-Y signals will also result in a G-Y output for feeding the G-Y final amplifiers. Furthermore, just as R-Y and B-Y (red and blue) components combine to produce the G-Y (green) component, a defect or malfunction in one of the first two will produce a corresponding deficiency or abnormality in the third one. For example, loss of the red signal will, primarily, remove virtually all the red from the picture and secondarily, distort the remaining color and particularly the green, into some other hue or combinations of hues. A detailed troubleshooting sequence for the absence of a particular color follows.

Red, green, and blue color output circuits

The color output signal is fed from the luminance/chroma IC component to the base terminal of each color output transistor. The color signal is coupled to the base terminal, while the luminance signal is tied to the emitter terminal (Fig. 9-13). Some TV chassis have color buffer or bias transistor stages between color output transistors and picture tube elements. Often, the CRT board contains the three color output transistors and required components.

The color output transistors have a higher dc voltage applied to the collector terminals (150 to 250 V) than most transistors. When one color is missing, either the color output transistor, components in the path to CRT cathodes, or the picture tube gun itself is defective. If the picture goes entirely one color (red), suspect a leaky red color output

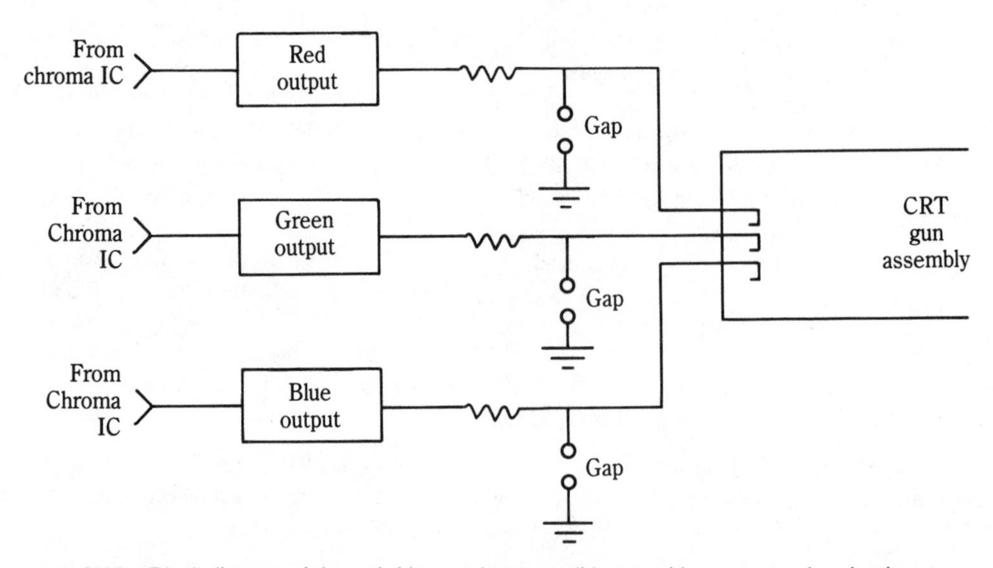

9-13 Block diagram of the red, blue, and green solid-state video output color circuits.

transistor or red gun assembly in the picture tube. Don't forget to check for dust inside a spark gap or defective spark gap when one color flashes on and off. Check voltage measurements and scope waveforms to locate the defective component in the color output circuits. Usually, TO-220 power transistors are used in the color output circuits, and collector voltage can be easily measured on the metal body of each transistor.

Red missing

There are three areas that affect the red color in the picture—the demodulators, the individual color amplifiers, and the picture tube. Based on some possibility of side effects on the other primary colors as well as on the in-between shades, the test sequence is simple.

Blue and green appear fairly normal Since you probably do not have access to a color-bar generator (except in the case of a built-in color generator), the judgment of color must be done subjectively. Look for known blue picture elements (such as sky) or known greens (such as grass) to judge. If such areas of known color seem to show normal color, the trouble is probably not in the demodulators.

Replace the R-Y final amplifier This is V16 in Fig. 9-1, V19 in Fig. 9-2, etc. If normal color returns, you've found the cause of the trouble. If there's no substantial improvement, replace the original tube.

Locate the red screen adjustment This control affects only the red gun of the picture tube. Before manipulating this control, mark the starting position (knurled or slotted shaft) with a pencil or a piece of masking tape. Advance this control gradually, usually clockwise, to increase the red color. If no noticeable effect is observed, restore the shaft to the starting position. If gradual improvement results from turning the control advance the adjustment further until red is normal.

A note of caution is in order here. First, if the red screen control had accidentally been backed off (this is a possibility, but not a probability), advancing it to normal color production is perfectly proper; however, if the control has to be turned all the way to barely obtain a fairly good red, a defect elsewhere in the system is responsible and compensating for this fault with the screen control is not advisable for the good of the picture tube. While it is quite possible that the red gun in the picture tube had deteriorated, this is not too common since in normal use all three guns deteriorate more or less equally.

Replace the R-Y demodulator This is not much of a possibility in view of the fact that the green seems normal; it would not be normal if the R-Y (or the X) demodulator were defective. If this replacement does not cure the red deficiency and advancing the red screen is only partly effective, a defect in one of the components (a capacitor or resistor) is most likely the cause. Incidentally, the red screen adjustment is helpful if one of the color guns is defective and a new picture tube will ultimately be required. Under these circumstances no harm is done in advancing the control and obtaining as much use of the tube as possible before a replacement must be purchased.

Touch up the red drive control This is important because the drive and screen controls for any one color are interdependent and may seem to act alike, especially in the hands of the inexperienced. As in the case of the screen adjustment outlined previously, it is good practice to observe the initial setting of the drive adjustment so as to be able to return it to its original setting if no change is apparent. The original setting of the drive control is important since, unless there is evidence to the contrary, the control was set for the proper operating conditions for the particular gun (red in this case); however, you're safe in making a slight adjustment if there is clear evidence that such an adjustment, combined with a corresponding adjustment of the screen control, improves the red color sufficiently to make it acceptable. If such adjustment fails to show a substantial improvement in the red color, a defective component is probably responsible.

Blue and green missing

As in the case of red, the blue channel has its own demodulator (B-Y or Z demodulator) and its final amplifier (B-Y amplifier). If the screen shows a color picture containing mixtures of red but not blue or green, the indications are that a defect common to both blue and green is responsible, since blue is one component of the green signal. The first and most likely suspect is the B-Y demodulator (or the Z demodulator in some TV sets). The first step is simple.

Replace the B-Y demodulator If the blue color appears, the green will also be present. If replacement of the demodulator tube does not help there is one other tube suspect; however, since both blue and green are missing, both the B-Y and the G-Y amplifiers would have to be defective. This would be rather unlikely if the B-Y and G-Y amplifiers were two separate and distinct tubes. A close look at the typical block diagrams in Figs. 9-1 through 9-4 will show that in almost every case the two amplifiers are two sections of a multisection tube. A failure in both sections, therefore, is no longer unusual. Consequently, the next step is to check these tubes.

Replace the B-Y and G-Y amplifiers Replace the two-in-one or three-in-one tube comprising the B-Y and G-Y amplifiers. This is V16 in Fig. 9-1, V19 in Fig. 9-2, etc. If the fault lies in the color amplifier this step should be effective. If it does not restore the blue and green colors try the next step.

Reduce the chroma gain Reduce the chroma or color control to minimum so that no color is present. If the black-and-white picture is normal the picture tube is okay and the fault lies in a defective component, an open capacitor or resistor. Of course, there is just the possibility that both the blue and green guns of the picture tube are defective, but you'll need a professional servicer for positive diagnosis and repair.

Blue missing

This malfunction is evident by a predominantly red and green picture with an almost total absence of blue. Since the green is present, the blue signal must exist in the output of the B-Y (or Z) demodulator. The only possibilities for a loss of the blue color are indicated in the following paragraphs.

Defective B-Y amplifier Replace the B-Y tube whether it's a separate tube or one section of a multifunction tube. A failure here will be cured by the substitution of a good tube. If no satisfactory improvement is obtained here try the next step.

Blue screen adjustment Adjust the blue screen control on the back of the set. As discussed earlier, a partial deterioration of any one color gun in the picture tube can be (temporarily, at least) compensated by advancing the screen control. If, however, only a slight improvement (or none) is realized from a major adjustment of the screen control the fault lies either in a defective component in the blue signal path or, less likely but still possible, the blue section of the picture tube is gone. In either case, outside help will be required.

On the other hand, should the blue screen adjustment produce a substantial improvement a further correction may be attempted by a small adjustment of the blue drive control. Since this points to a deterioration of the blue gun, these two adjustments can restore the tube to a reasonably normal operating condition for as long as the tube will hold out. Again, this precaution is worth repeating: no decision to replace a picture tube should be made without the concurrence of the professional TV servicer.

Color presentation unstable

This could be either tearing of the color picture, similar to the diagonal zig-zag tearing of a black-and-white picture when the horizontal frequency adjustments are off (see Fig. 2-12), or in less severe cases, a partial breakup with the color wavering and general picture instability, often only intermittently.

It is assumed that the do-it-yourselfer is following the general admonition given earlier with regard to any color troubleshooting—before attempting any work in the color portion of the receiver, the performance in monochrome is checked. If the symptoms are the same as in color operation, the fault lies not in the color-only portion of the set, but in the common section, and repair should be made there. Only if the monochrome performance is normal should work on the color section be attempted.

As a general rule, this type of behavior suggests improper operation of one of the color sync stages. This includes, depending on the particular TV set, the burst amplifier (V15B in Fig. 9-2, V18 in Fig. 9-3, etc.), the burst gate (V7C in Fig. 9-1), the 3.58 MHz oscillator (V17B in Fig. 9-2, V20B in Fig. 9-3, etc.) and the 3.58 MHz crystal itself. In some TV sets, the sync and oscillator functions may be called color sync, burst gate, subcarrier amplifier, and chroma oscillator control. In all cases, the functions are the same even if the names differ. These functions are the color sequencing and positioning precisely in step with the pictures being transmitted. Any malfunction here will throw the received picture out of kilter. The procedure in troubleshooting these malfunctions is the next step.

Replace color sync tubes

Since the number of sync-related functions and tubes used in each case differ widely, it is necessary to determine the sync lineup in each particular set. Using typical block diagrams as examples the sync stages in Fig. 9-1 consist of burst gate V7C, and subcarrier amplifier V15B in the order given. The corresponding sequence in Fig. 9-3 is gated burst amplifier V18, phase detector V19A, reactance control V20A and 3.58 MHz oscillator V20B. In Fig. 9-2 the sync lineup is burst amplifier V15B, reactance control V17A and reference oscillator V17B. In Fig. 9-4 it is burst amplifier V17A and 3.58 MHz oscillator V18A. These typical diagrams are based on actual TV sets and cover a large portion of all color sets in use today. Should one exceptional case have a somewhat different arrangement or designation you can identify each function by comparing the names and functions with those given here. The word *chroma* may be used for the word *color* and *control* is often used for *sync* or *burst*.

Although the replacement or substitution of tubes in the order given is the most direct way of finding a malfunctioning stage it does not mean that random substitution will not ultimately produce the same results. But for logical procedures the sequence should be followed, as it proceeds, in all cases from the color IF (bandpass) amplifier to the 3.58 MHz oscillator. A final general reminder: in all the sets represented by the block diagrams in this chapter the 3.58 MHz stage contains a crystal, and it is just as vulnerable a link in the sync chain as all the other tube stages enumerated here.

Replace first tube in sync lineup

Very often the first tube in the first sync lineup will be one section of a multifunction tube, but this does not change the procedure. If only a minor improvement in stabilization results you have not found the trouble. In Fig. 9-1 the first tube is burst gate V7C, one-third of a 6M11 tube. In Fig. 9-2 it is burst amplifier V15B, one-half of a 5BN11. In Fig. 9-3 it is burst amplifier V18, a 6EJ7. In Fig. 9-4 burst amplifier V17A is one-half of a 6AF9.

Replace second color sync tube

In Fig. 9-1 this is sub-carrier amplifier V15B, one-half of a 6BW11. In Fig. 9-2 it is V17A, one-half of a 6GH8A, the reactance control. In Fig. 9-3 the second tube is V19A, the sync phase detector; in Fig. 9-4, it is 3.58 MHz oscillator V18A. If no better than a slight improvement is apparent, the fault lies elsewhere and the procedure continues.

Try a new crystal

Where the second tube involved the 3.58 MHz oscillator, as in Fig. 9-4, replace the 3.58 MHz crystal if it is a plug-in type. It is also advisable in this case to observe whether the color breakup is always present from the moment the TV set is switched on or if it appears later after the set warms up. If the latter is true, the crystal is a most likely suspect, as these units sometimes stop functioning (oscillating) after a period of normal operation but may return to normal if the set is switched off, then immediately switched on again.

Another possibility is poor contact at the crystal connecting pins. As described earlier in many TV sets the crystal connects to the TV set by means of two thin wire pins, similar to those on miniature tubes. Applying a gentle pressure on the crystal case may momentarily restore normal operation if poor contact is the cause. In other TV sets, the crystal is soldered in, like any other part, and replacement is not so easy (see chapter 10). As a final note it remains to be said that a crystal failure or a 3.58 MHz oscillator failure usually, even if not always, causes a total loss of color information, not just breakup.

Solid-state color oscillator problems

The defective color 3.58 MHz crystal may cause no or intermittent color. Most color crystals are soldered into the PC board of the present-day TV chassis. Before replacing the 3.58 MHz crystal, take voltage measurements upon the oscillator IC pins (Fig. 9-14). Do not try to make any adjustment of crystal frequency without special test equipment.

9-14 When color is missing from the picture, observe voltage and scope waveforms to detect the defective color IC.

Often, pushing around on the color IC or capacitors within the oscillator circuit may cause the color to become intermittent. Suspect poorly soldered IC connections when the color disappears from the picture and shortly returns without any control adjustments. Properly solder all IC chroma pin leads with a low-wattage soldering iron (Fig. 9-15). Sometimes spraying coolant on the chroma IC may turn up the intermittent IC component.

The best method to check oscillation of the color 3.58 MHz oscillator is with the oscilloscope. Check the waveform at pin 17 going to the VCO color oscillator. The color crystal is found in a metal case quite close to the chroma IC and oscillator adjustment (Fig. 9-16). The color wave amplitude may change when a color picture is being received (Fig. 9-17).

A final measure

In TV sets similar to Figs. 9-1 and 9-3, replace the next likely offending tube. This is the oscillator control V20A, one-half of a 6GH8 in Fig. 9-3. Finally, in all types of TV set lineups the last stage is again the 3.58 MHz oscillator or the reference oscillator. If none of those steps seem to make an adequate improvement in color stability or normal picture the fault probably lies in one of a number of components in this general area, requiring troubleshooting by a professional using specialized instruments.

It should be added here that in a number of cases of mild color instability the fault lies outside the TV set. Occasionally, a color is transmitted under circumstances making

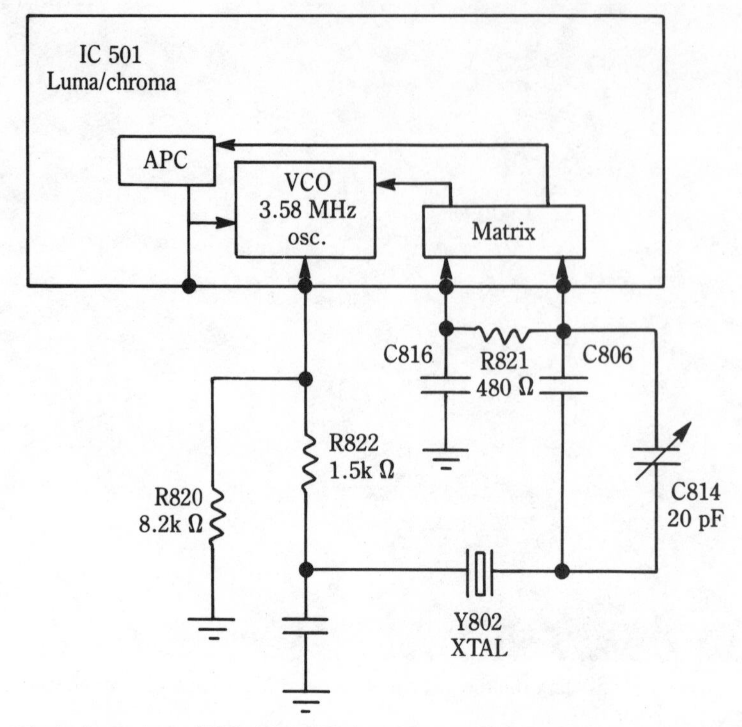

9-15 Block diagram of a VCO (3.58 MHz) crystal circuit in one of the latest TV sets.

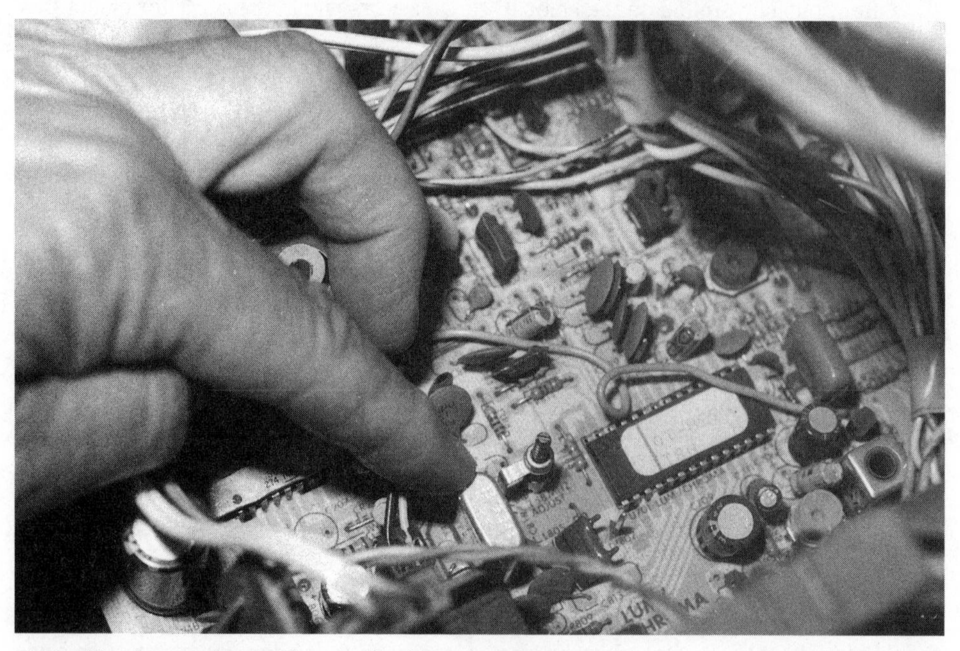

9-16 The finger points to the actual color crystal soldered directly into the PC board, next to frequency capacitor adjustment and color IC chip.

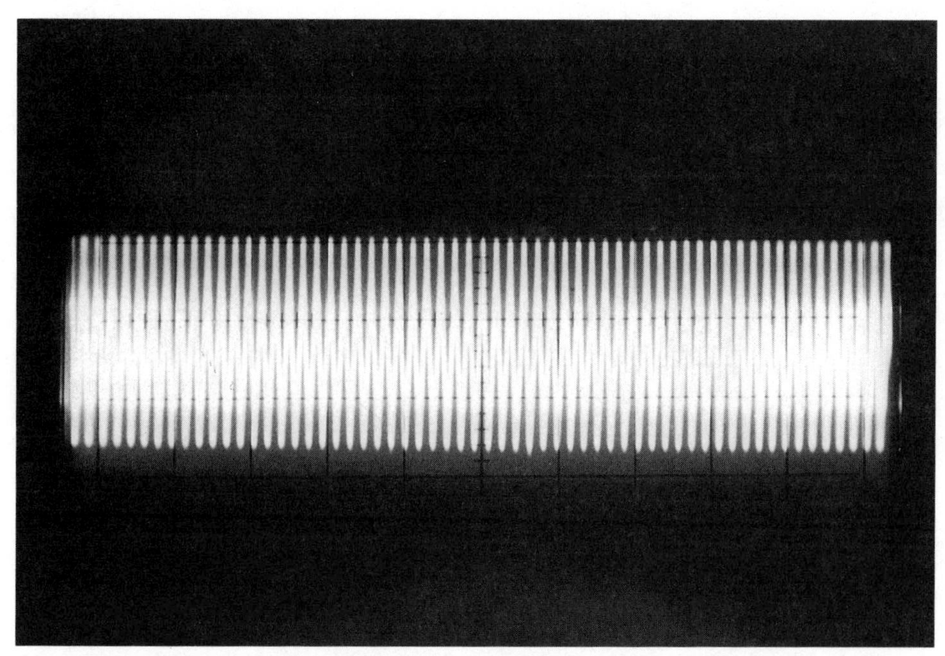

9-17 The color waveform taken from the 3.58 MHz circuit on pin 20 of Fig. 9-18.

perfect stability unrealizable. The owner can best verify this by switching to an adjacent color channel. If the problem disappears, the probabilities are that the station, not the TV set, is at fault.

Solid-state color problems

Intermittent and no color symptoms are the most common problems found in the color circuits. Drifting color bars may be caused by a defective component or misadjustment of APC alignment. When one color is missing, check the color output demodulator signal, color amp and color gun assembly of the picture tube. Recheck the spark gap of the missing color. Sometimes if a defective spark gap arcs continually, the picture tube will shut down the whole chassis.

Accurate voltage measurements on the chroma IC may turn up a leaky IC or component. Sometimes the chroma IC may be defective with fairly normal voltages and must be replaced. Scope the color input and output signals with the scope and color-dot bar generator connected to the TV antenna terminals. Remember, the color circuits must have a video and color input signal, usually from the comb filter, and an output color signal from the color demodulators to the color output amplifiers (Fig. 9-18). The VCO 3.58 MHz color oscillator must oscillate, and a horizontal pulse from the flyback circuits must be present before the color stages can operate in today's color TV chassis. Of course, some chroma circuits may have sand-castle input and BRM signals applied to the chroma processor. It's best to leave the color circuit problems up to the experienced electronic technician.

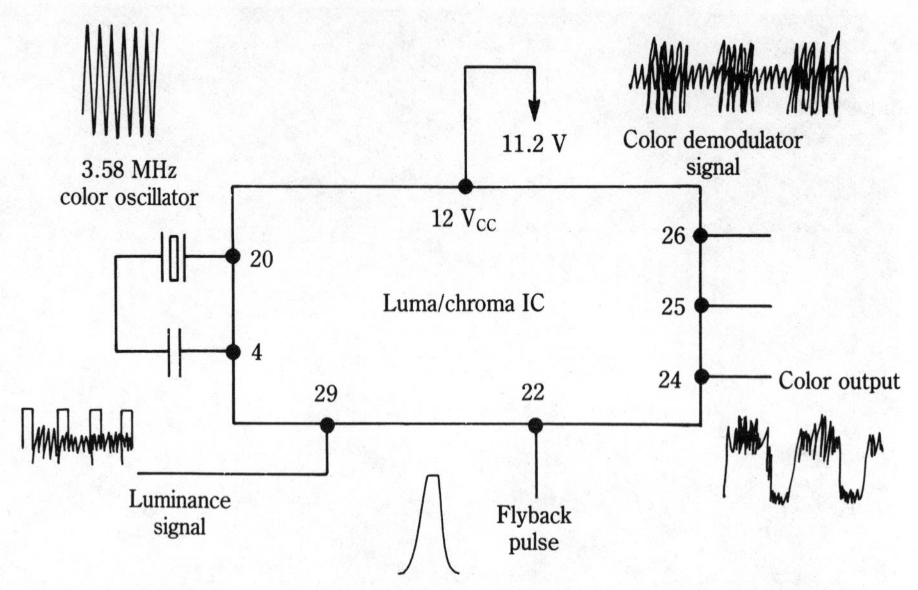

9-18 Signals and waveforms needed before getting color back into the picture.

Color TV antennas

Commercially available TV antennas vary all the way from a piece of ribbon lead-in type of wire cut to some approximate length, costing a little over a dollar, through various dipole versions, all the way to multielement, multibay structures, costing hundreds of dollars. The more elaborate the antenna, the higher the *gain* (a term indicating the strength of the signal delivered to your set), and vice versa. An important characteristic closely related to gain is the bandwidth.

The second factor, although must less scientific, is nevertheless just as important: viewer attitude. It is a fact that the average observer is far more tolerant when viewing a black-and-white picture than he or she is of color. This is due, in part at least, to the fact that even a faint picture can be seen and followed in monochrome, while the weak picture in color is intolerable and actually much more difficult to follow. Taking both of these into proper consideration I say the following about the color TV antenna requirements.

First, the satisfactory color antenna must have adequate gain. While almost any antenna, given a fair to poor signal, will provide an acceptable black-and-white picture, only a sufficiently high-gain antenna will serve as well for color reception under the sme circumstances. Second, the color antenna must have adequate bandwidth. That is, it must be able to intercept and receive a strong enough signal over the whole expanse of the TV frequency spectrum, from channels 1 through 13 and possibly channels 14 through 83. Of course, this does not mean that a narrow-band antenna will receive only part of the picture. It will receive a single channel or possibly two or three. Narrow-band antennas are usually *cut* for specific frequencies.

While it is beyond the ability of the TV owner, no matter how well technically informed, to determine the characteristics of an antenna he or she wishes to purchase, there are a number of ways this information may be obtained without great effort. First, there are a few consumer organizations who do impartial testing and reporting on manufactured products, antennas among them. Their reports are published and may be easily obtained for a minimum cost, or they may be seen in most libraries. Second, and without even any factual data, and contrary to other examples, the cost of the antenna is a good indication of its capability. As a matter of confidence, the purchaser may rely on the reputation of the manufacturer, be it a TV manufacturer who has antennas made under its label, or the antenna manufacturer itself. The only precaution one must observe is to steer clear of the sensational *amazing discovery* type of antenna bargains offered by (usually unknown) mail-order advertisers. The old cliché that good quality does not come cheap is very true here. In cases of color abnormalities, such as smear, color over-lap, and of course color snow, assuming that the set has otherwise been checked out, it is advisable to consider the antenna as the possible source of the trouble.

The foregoing observations are not intended to suggest that you should always buy the most expensive antenna. Not only is this unnecessary, but sometimes actually inadvisable for the following reasons.

The purpose of any antenna is to provide an *adequate* signal for the particular receiver location, not the maximum signal possible. A simple perusal of catalogs and advertisements will show that antennas are classified according to intended use; thus, there are antennas for metropolitan, suburban, near-fringe, fringe and deep-fringe applications. Advertisements often specify mileage ranges for each type, but these may be taken with a grain of salt since they are for *ideal* conditions, not usually realized. A practical guide would be to purchase an antenna one grade better than indicated by the advertised mileage; thus, if the TV is located in the near-fringe distance from the station, a fringe antenna would be adequate. Of course, local conditions, such as locations behind a hill, etc., will adversely affect the signal. In such a case, see what size and type of antenna that is used by others in the immediate vicinity. Having determined how big an antenna to get, it might further be advisable to get one from a reputable manufacturer.

Combination VHF-UHF antennas

In deciding on the type of antenna, consideration should be given to the reception requirements in the UHF portion of the TV band, both for the UHF reception as well as to the incidental effect on the more popular VHF stations. Since most network as well as independent TV stations of importance are concentrated in the VHF region, while UHF service at present is primarily of relatively minor scope, it behooves the TV owner to decide, before installing an antenna, where his or her primary interest lies. If it is in VHF only, then the best VHF *only* antenna will be best since almost all combination VHF-UHF antennas are but a compromise, with less than optimum reception in either range. If, however, the set owner is interested in the best reception of both types of stations, then a separate UHF antenna is desirable; even the very largest combination antennas give but mediocre to marginal UHF reception. The only exception to this situation is in areas of high-signal strength. Here a combination antenna will be adequate for both VHF and UHF.

Color smear

In a black-and-white picture, a condition similar to smear often results from an improperly installed or wrongly oriented antenna. The usual name for such an appearance is ghosts—a slight displacement of the picture edge as if two superimposed identical pictures were slightly out of register or slightly displaced from one another, either to the left or to the right. One check on the accuracy of this diagnosis in a color picture is to turn the chroma (or color gain) control all the way off so that a black-and-white picture remains. If evidence of ghosts exists the diagnosis was probably correct. Adjusting (rotating) the antenna, if it is not otherwise defective (broken elements, poor lead-in connections at the antenna, loose elements, etc.) should eliminate ghosts or at least reduce them to a minimum and effect a marked reduction in the color smear.

Degaussing problems

A fairly common cause of both color smear and color mixing or distortion, particularly around the edges of the picture, is accidental magnetization of some of the structural metal around the face of the picture tube by stray magnetic fields: ac lines in the wall, a heavy current appliance in the immediate vicinity of the TV set. Sometimes moving the TV set to a different wall causes improper purity by changing the picture tube angle to the north pole. Stopping or starting the electric sweeper in front of the TV set may magnetize a portion of the picture tube. Even placing strong magnets of the speaker column mounted next to the TV set may magnetize the front of the picture tube. Often, demagnetizing or degaussing the picture tube will remove any color impurities from the screen.

While most of the more recently manufactured TV sets have built-in degaussing circuits, the earlier color sets are not so equipped. With the built-in degaussing circuit, the front of the picture tube is degaussed each time the TV set is turned on. When power is turned on, the thermistor in series with the automatic degaussing coil (ADG) provides a low-resistance path to the degaussing coil (Fig. 9-19). The ac voltage is applied across the degaussing coil at the front side of the picture tube. The coil energizes and removes most color purity problems from the face of the picture tube. After a few seconds, the thermistor becomes warm and provides high resistance to the ac voltage shutting down the degaussing process.

If a section of the screen has a colored portion, shut the TV off and wait a few moments. Now, turn the TV on and see if the colored section is removed. If not, try degaussing the front of the picture tube with the external degaussing coil (Fig. 9-20). While the process of external degaussing is relatively simple and quick, it does require a degaussing device, a large (10- or 12-inch diameter) ring made up of many turns of insulated wire terminated in an ac plug.

Degauss the picture tube by plugging the ac cord into the receptacle and rotating the coil ring in a circular motion at the front of the picture tube. Make sure the coil passes

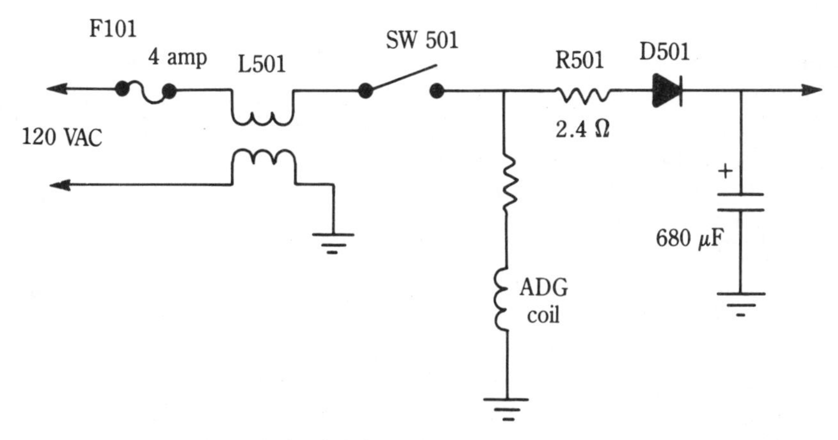

9-19 The degaussing coil circuit is in the low-voltage power supply and operates for only a few seconds when the power to the TV is turned on.

9-20 Use the external degaussing coil to remove stubborn color patches from the front of the picture tube.

around the outside edge of the picture tube. A few minutes are required to demagnetize the steel frame and rim of the color tube. To finish the degaussing process, bring the coil around to the outside edge of color tube, and while backing up, rotate the coil. Bring the coil away from the picture area by turning it perpendicular to the TV screen before unplugging the degaussing coil. Often, the electronic technician degausses the picture tube when finished with the TV repair.

Sometimes the degaussing coil or plug opens and no immediate degaussing occurs. Before long, a color section appears on the TV screen, usually at the corners. Check the thermistor resistance with the ohmmeter for open conditions, up to 25 ohms when cold. Examine the thermistor to see if it is open or has a ''popped off'' terminal caused by overheating. An open ADG coil or plug results in no degaussing. Also, the coil can be checked with the ohmmeter and should be below 100 ohms. In a few cases, the insulation of the degaussing coil has broken down and shorts against the metal framework. When the fuse keeps blowing, disconnect the degaussing coil for a possibly shorted condition to prevent the fuse from blowing.

10 Solid-state circuitry

BEFORE LISTING SOME OF THE MAJOR DIFFERENCES BETWEEN TUBES AND semiconductors it should be made clear that transistors and diodes perform the same functions as tubes; that is, they detect, amplify, oscillater, etc. The end result, whether picture or sound, is exactly the same, regardless of how it was achieved. Nevertheless, the structural and operational differences between tubes and transistors are very sharp. Ones that are important to the consumer are:

- Unlike tubes, transistors need no heaters for their operation. This is perhaps their greatest single advantage, resulting in greater stability, longer life and more economical power consumption.
- Transistors are low-voltage devices. Most operate on between 1.5 and 15 volts, although a few rare ones require as much as 100 volts or so. By contrast, tubes usually require a few hundred volts.
- Transistors are not fragile. Shock and vibration are far more harmful to tubes than transistors. This relative immunity to vibration combined with low-temperature operation makes for long life.
- Transistors are simpler to manufacturer and hence less expensive.

Effect on TV design

Solid-state technology has produced a number of changes in the TV industry. To the do-it-yourself consumer, most of these changes have been positive. And although there may be a few disadvantages, as far as repair by the owner is concerned, on the average the consumer is ahead. Some of the more obvious advantages are:

- *Smaller size*—With the proliferation of semiconductor use, a major miniaturization technique has evolved, not only with respect to transistors versus tubes, but also for almost all other components except, of course, the picture tube and the speaker—capacitors, resistors, hardware, wiring, etc. A by-product of the parts miniaturization has been a decided improvement in the quality of components.
- *Lower heat*—Not only has the power consumption been greatly reduced, but heat generated also has gone down correspondingly. Less heat means less waste, but far more important, less wear and longer life for components. This is almost entirely due to the replacement of tubes with transistors, although some little progress in this direction was independent of the transistor factor.
- *Less wiring*—One of the greatest and most obvious effects of the solid-state technology has been the tremendous reduction in wires and wiring. And although wires are not usually a source of trouble, their elimination wherever possible is a distinct plus for economy of space, labor and cost.
- *Modularization*—With parts being much smaller, it has been logical to group a larger number of them on a single common subassembly; thus, printed circuit modules or boards came into vogue. We shall say more about these devices when describing do-it-yourself techniques.
- *Repair standardization*—When a TV set consisted of almost 100 percent discrete parts, no two jobs were alike, that is, the manner of repair, and to some extent the efficacy of such a repair, was largely dependent on the individual TV servicer, professional or do-it-yourselfer. The PC board and module have eliminated a lot of haphazard repairs. For any particular repair, the replacement module connected exactly the same way as the original—the repairman has no choice in this matter—will produce essentially the same normal operation of the set, regardless by whom it was repaired.

At first glance, it might seem wasteful to replace a complete module when only one of its components is defective. Actually, this is not necessarily true. While it is sometimes possible to quickly locate the defective part and replace it only, thus saving the old board, in the majority of cases the replacement of the complete module is more economical. It saves time, money, and possibly even avoids the risk of damage to other components in the immediate vicinity of the defective part—a most unpleasant experience in view of the fact that the damage may not be discovered until after reassembly of the TV set.

Service techniques

There are a few precautions as well as some new techniques to be observed in working with solid-state circuits.

- Care must be exercised that nothing is disturbed in the vicinity of the part being removed.
- An *absolute minimum* of heat must be used in soldering components.
- Physical force where necessary must be applied very sparingly. Miniature parts simply cannot stand rough handling, by hand or with tools.

The following few tools are recommended as an essential minimum in servicing solid-state receivers.

- Needle-nose and long-nose pliers, small size.
- Diagonal cutting pliers (dikes), small size.
- Pencil-type soldering iron, not over 30 watts.
- Pair of tweezers, about 6 inches long.
- Dental probe—looking like a giant needle, with a flattened end for poking and scraping in tight quarters.
- Heatsink—a type of clamp which snaps shut in the manner of a clamp—to be used in absorbing any excess heat during soldering, thus preventing damage to parts.

One other tool, not exactly a must, but very handy, is a type of soldering tool which automatically removes solder by suction, incident to unsoldering small parts, transistors, etc. During such unsoldering, very tiny mounting holes used for parts mounting, unintentionally fill up with solder, preventing the new part from being inserted; however, with a bit of experience and care, the dental probe in conjunction with the soldering iron can accomplish the same job. Also, the unplugged hole can be carefully heated while a fine needle (not a wire!) is moved back and forth in the hole, thus dislodging the solder. Toothpicks work just as well if you can keep a supply on hand.

PC wiring and soldering

The printed circuit technique eliminates old-fashioned point to point wiring by electrochemically printing the wires on a phenolic or fiberglass board. First, a layer of copper is chemically deposited on one or both (as required) sides of the board. Next, an etching process removes the copper from all areas except those to be used as conductors or connecting points. On two-sided boards, the wiring of both sides may be connected together with metal feed through or griplets. Holes are then drilled for inserting the wire leads of the components to be mounted. Parts are then inserted in the proper holes and the board given a solder bath which solders all leads to their respective copper backs or areas (Fig. 10-1).

In the manufacturing process, a machine inserts the small components while a person may mount the larger parts. After all components are mounted, the board travels through the solder bath (just the bottom side of the board) and onto a testing board procedure. PC wiring boards that may have a defective connection are automatically rejected. Most of these discarded boards are repaired individually and passed through the testing station again. Sometimes, the human hand may never touch the printed board until it is mounted in the TV chassis.

When mounting a defective part, heat is applied with a fine-tipped soldering iron to the immediate area of the joint. Usually, a heatsink is not required when removing defective components. A heatsink *should* be used when installing solid-state components because excessive heat will damage their internal connections. Special heatsink clamps may be used or you can use a pair of needle-nose pliers. Grasp the transistor terminal you wish to solder with the pliers as you solder up the terminals.

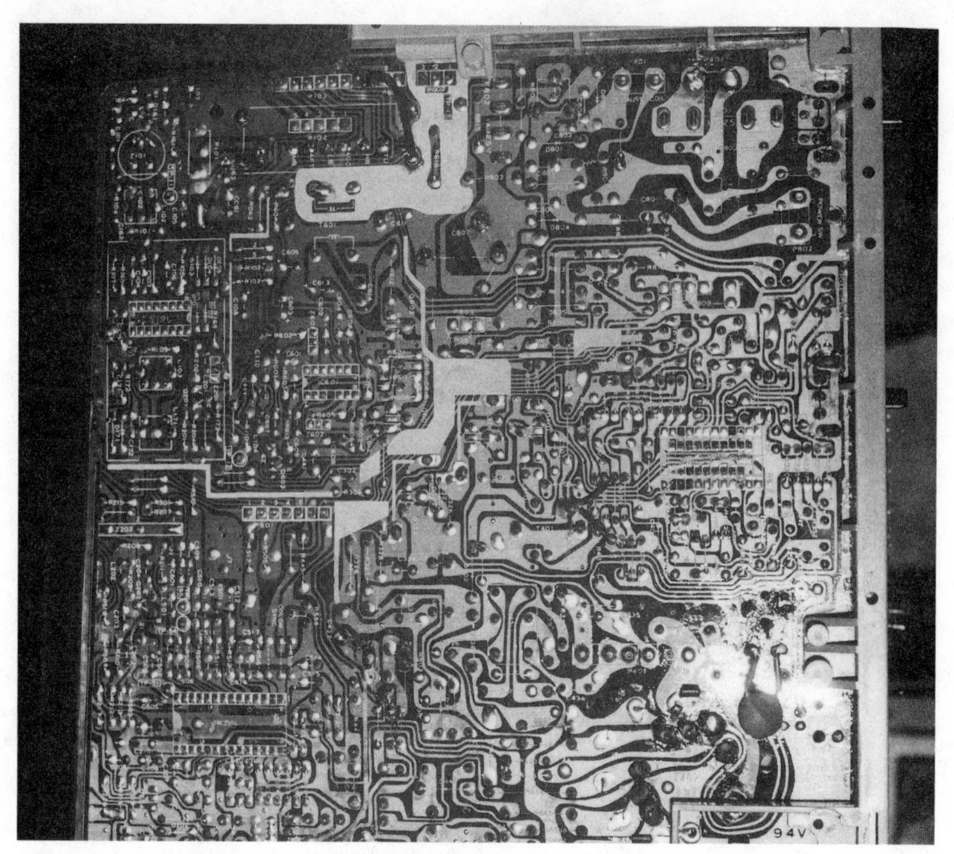

10-1 A printed circuit board chassis in a Japanese chassis.

As in all soldering and much more so here, time is the most important. The longer the soldering iron is applied to the component the greater the risk of overheating and damage. With care and a little experience, a good joint can be made by a touch-and-go application of the iron. This method requires:

- A shiny clean tip on the soldering tool.
- A corresponding clean spot on the work being soldered.
- A good contact between the soldering tool and the work.
- Choose a small iron (less than 40 watts) to solder transistors or IC circuits. A rechargeable battery type soldering iron is ideal (Fig. 10-2).

Place the point of the iron on the junction of the wiring and the component terminal, and flow the solder from the opposite side. It's best to use a small-diameter solder in transistor and IC soldering applications. A little practice on a few scrap pieces of material can soon make one a proficient solderer.

Choose the correct type of solder for electronic work. Never use acid core solder in the TV chassis. There is a great variety of solder on the market—from the coarse, lead-

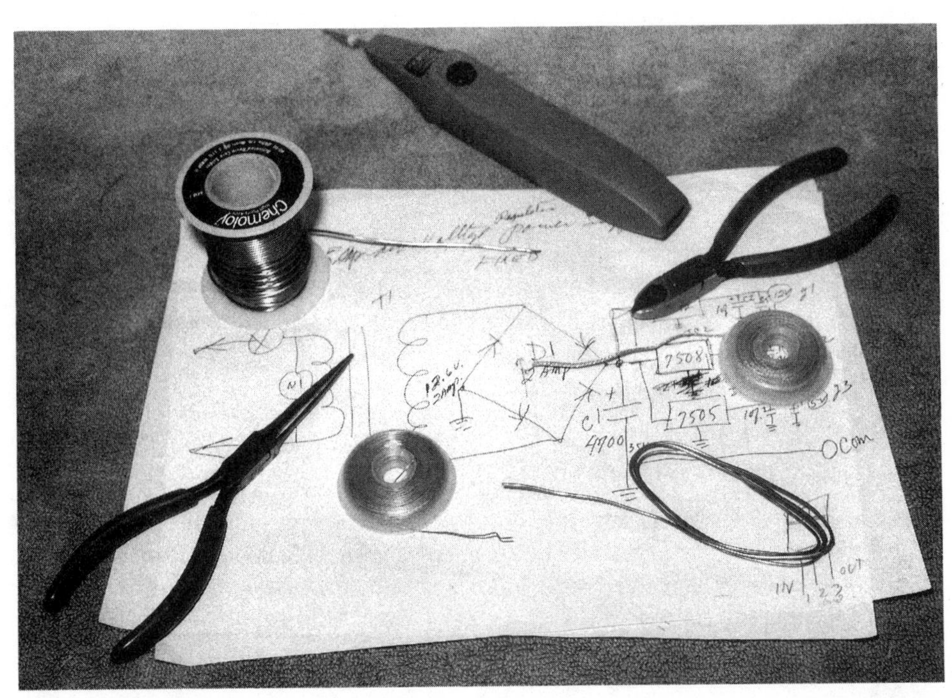

10-2 A rechargable-battery type of soldering iron is ideal to use in transistor and IC circuits. The small wattage prevents damaging the component from too much heat, yet it does a neat, efficient job.

like stuff used by plumbers to a very fine, high-grade type used in military and other electronic equipment. It is quite easy even for the novice to know which is which. Solder (especially the kind used in electrical work) is rated in the percentages of lead and tin used in the mixture. The higher the proportion of tin, the better the solder, and the shinier it looks. Thus, a 40-60 mixture is a relatively poor grade with only 40 percent tin. 50-50 is a commonly used grade in radio work, but is not recommended here. It is better to choose a 60-40 grade, that has a 60 percent tin content. Solder can be found in three different diameters or gauges. The two smaller diameter types are best for transistor and IC applications. A pound of this type can cost a few dollars, but it should last the do-it-yourselfers for a lifetime. This type of solder has a resin core (a resinous chemical in its center for cleaning the surface during soldering) so it is advisable to wipe the soldered area carefully and sparingly with alcohol to remove any resin residue, which could (in some cases) cause electrical leakage. Stubborn resin residue can be removed from the area with the blade of a pocket knife or with a small wire brush.

There are a number of factors to be considered by the TV owner before deciding whether or not to tackle a PC board. First and most important is the capability of the individual. This includes any previous experience, ability to work with small tools, as well as handling tiny parts, fine soldering, and general manual dexterity. Based on one's self-confidence with regard to the above, you may decide to go it alone or you may decide to seek the services of a professional.

The second factor is the nature of the suspected or failed part. For example, an encapsulated sub-module is determined to be defective, removal of the part and replacement is far preferable to the scrapping of a large PC board on which the submodule is mounted. On the other hand, when you suspect a relatively small circuit board that has parts soldered flat against the surface of the board with no accessible wires or leads, replacement of the whole unit is the wiser alternative. Finally, the cost of the replacement part is a factor. The purchase price of the part as well as the repair shop bill should be taken into account, the latter almost always being the larger of the two.

To summarize, if after careful examination of the defective part or parts and an equally careful study of the applicable service information, you think that you can do the job, you will probably be successful. If, however, after all these preliminaries, you decide that you are not up to the task, the job should be left to the professional TV servicer.

PC board parts layout

Today, most TV chassis have one large PC board with parts mounted on top. Often, the various components are found clustered together with the different circuits. Within RCA's 1990 13-inch TV chassis, parts are mounted on top of the board with miniature surface-mounted components under the same PC board (Fig. 10-3), although several tie wires may be found on top of the chassis to tie the various PC wiring together. In this particular chassis, PC wiring is found only underneath for the top-mounted regular parts

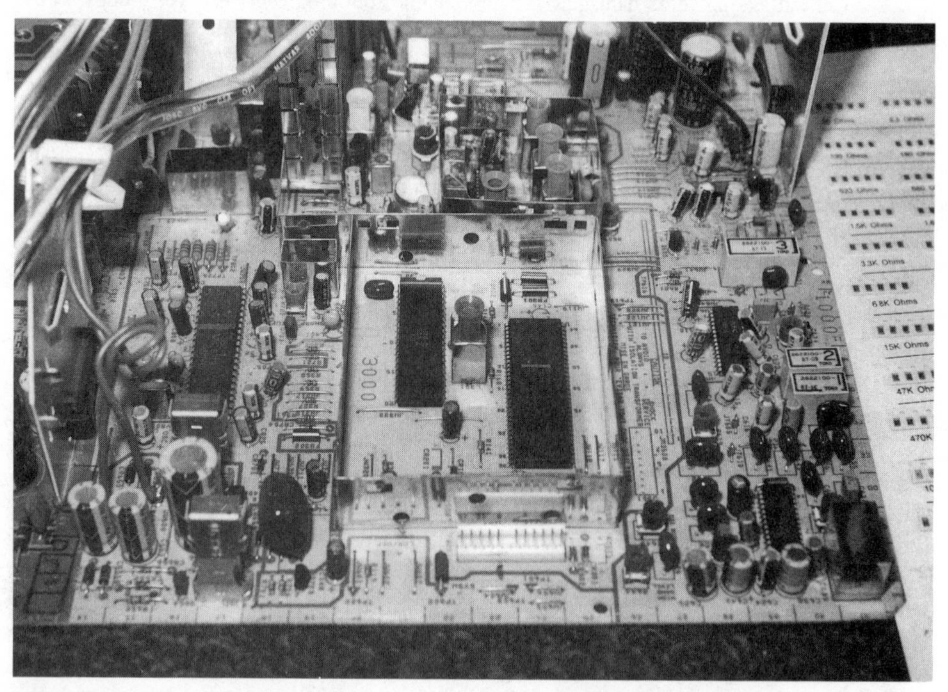

10-3 The regular parts are mounted on top in an RCA 1990 13-inch TV chassis with surface-mounted components underneath. At right is a card of surface-mount resistors.

and underneath with surface-mounted components, while in other TV chassis, you may find double-sided PC wiring both on top and on the bottom of the one board.

You may find it takes a little longer to locate the various components underneath and on top of the PC board. The manufacturer may have brief drawings indicating where parts are mounted. The manufacturer's service literature and schematics are must items in trying to service the solid-state chassis. Extreme care must be exercised when removing and installing the small components to prevent board wiring breakage plus miniature parts around the replacement. Try not to flex or bend the large PC board to prevent breakage connections of the small surface-mounted components.

Surface-mounted components

Surface-mounted components often appear as very small rectangular brown and black pieces with soldered ends. Bypass capacitors and resistors may appear with single or dual units within one component. The RCA CTC145E chassis (1990) has surface-mounted components underneath and on the same side as the PC wiring (Fig. 10-4). Standard components are mounted on the top side of the PC board. Here you will find thin lines of PC wiring connecting the surface-mounted and regular components together (Fig. 10-5).

Besides the TV chassis, surface-mounted components are found in CD players and camcorders. Surface-mounted ICs and microprocessors have wing-type soldered connections that solder directly onto the PC wiring. Most ICs are marked with a dot or

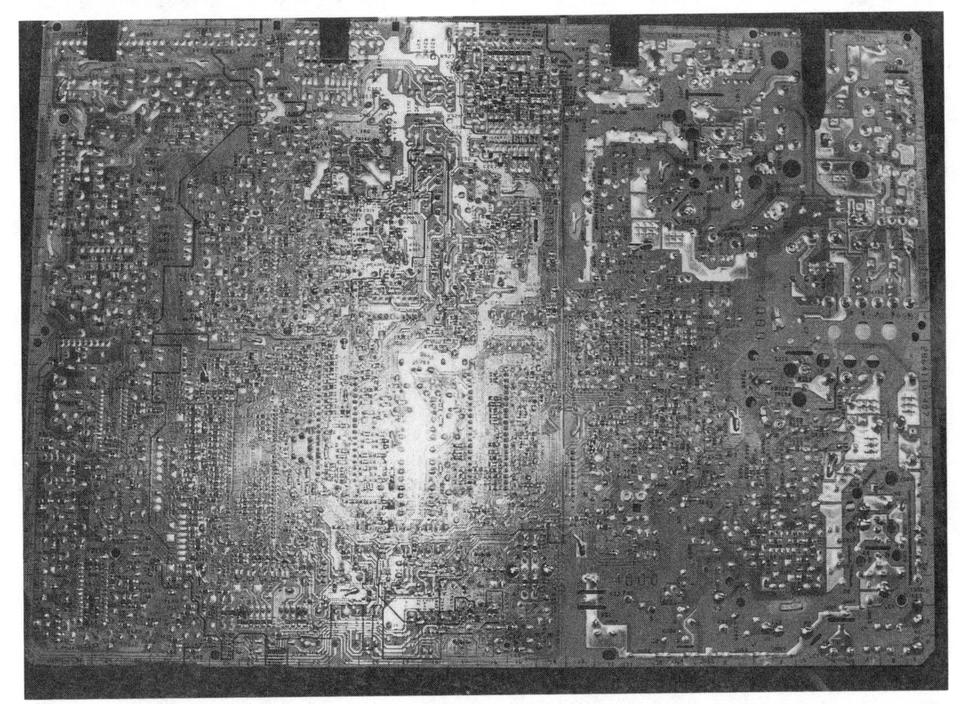

10-4 In the RCA CTC145E chassis (1990), the small brown and black components are actually surface-mounted resistors and capacitors.

10-5 A close-up view of the surface-mounted components have soldered ends that join the circuits together. Notice the thin lines of PC wiring.

number indicating the correct starting pin number. The small microprocessor may have up to 80 or more terminals (Fig. 10-6). Extreme care must be exercised in removing and replacing IC components. All surface-mounted components should be replaced with the exact part number. You need a pair of tweezers and a small-wattage pointed soldering iron for removing and replacing surface-mounted components.

10-6 The surface-mounted ICs and microprocessors have long, flat leads that solder directly to the PC wiring. These wing-type components can have more than 80 soldered terminals.

Flyback PC ring mounting holes

The flyback or horizontal output transformer may be mounted directly upon the PC board with ringed soldered eyelets. The pin terminals fit inside the ringed holes. To remove the defective flyback transformer, the excess solder must be removed on all eyelets before the transformer can be pried from the PC board (Fig. 10-7). If not, the board PC wiring or eyelet may crack and provide intermittent conditions.

When removing the excess solder with a 200- or 300-watt soldering gun and mesh solder wick, go over each eyelet three or four times to suck up the melted solder. Greater heat must be applied to the soldered terminal to remove all excess solder. Sometimes the small sucking type of iron will not clean up the soldered holes. After removing the transformer, clean up excess solder around the PC eyelet. Mount the new transformer and apply enough solder to fill each hole.

Brush and clean off all solder and rosin around the eyelets. Inspect the board for possibly poor connections. To make sure contacts are good, connect one ohmmeter probe to a flyback eyelet terminal and the other probe to where the same PC wiring connects in the circuit. Most PC wiring breaks occur right where it connects to the soldered eyelet.

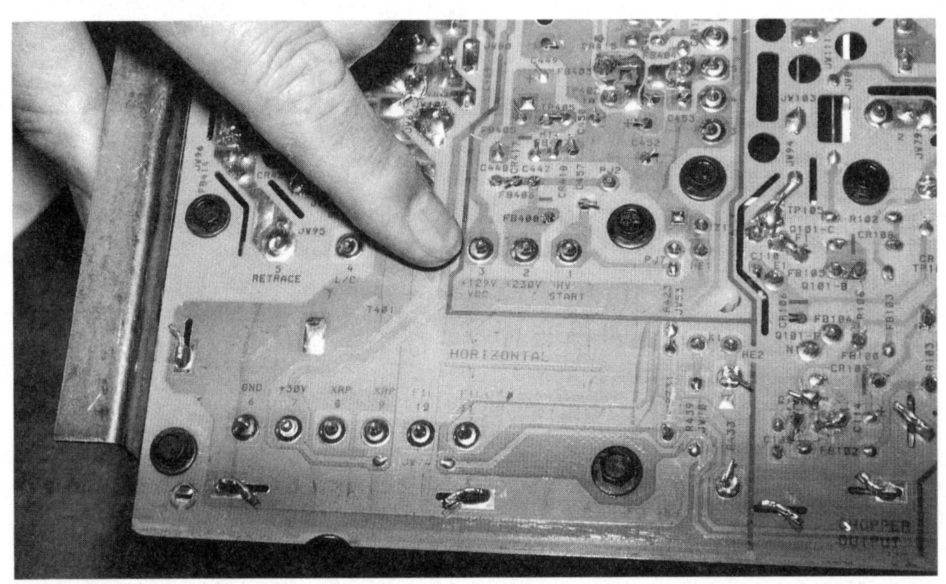

10-7 Round, ring-type holes on the PC wiring hold the contacts of a horizontal output transformer or flyback. All excessive solder must be removed before the transformer can be removed or replaced.

Broken PC wiring

If a TV set is dropped, inspect the board wiring with a magnifying glass. PC boards often break on sharp corners or where shields and heavy components are mounted (Fig. 10-8). If the board is broken in several places, it's wise to replace the whole board.

10-8 A cracked PC board was repaired with bare and encased hookup wire when the TV set was accidentally knocked to the floor.

Small breaks or cracked boards around parts can be repaired with bare or covered hookup wire. Locate all broken PC wiring. This may occur along the cracked area. Sometimes heavy components have a broken terminal or two. Scrape off the end of the PC wiring to be bonded together. Lacquer must be removed to make good contact. Scrape each broken end wiring with a sharp tool to get down to the copper PC wiring.

Do not try to solder the two broken ends together with just solder and the soldering iron. The crack will remain. Place hookup wire across the broken area and solder each end. If several components or bare, soldered joints are close by, use covered hookup wire. Double-check all connections along the cracked area. Make sure each broken PC wiring is on the exact wire. Bare hookup wire pieces may be used where it will not touch other connections. Now check each broken connection with the ohmmeter. Go from the first soldered connection to the next to ensure good bonds.

After the TV chassis is operating, lightly push down around the cracked area to see if set has an intermittent connection. Likewise, push down on small components nearby to find a broken connection or part. Sometimes it is worth it to try to repair the board because it could take some time to obtain a new board.

Modular chassis design

Soon after the all transistor chassis was designed, the modular system was adopted by about every TV manufacturer. Modules are easily removed from the chassis and

10-9 The plug-in module is easy to remove and replace. Just select the correct one and interchange it.

replaced with a new one (Fig. 10-9). The modules are constructed of lightweight components for use in practically any circuit. The large or heavy parts such as transformers, filter capacitors and picture tubes are mounted directly on the chassis. The modules can be plugged directly into a socket on the chassis or into a socket on a subassembly with wire leads connecting the module to the chassis.

Each TV manufacturer has its own set of modules. In other words, an RCA module will not interchange with a Zenith module and vice versa. Although, some modules with the same circuit function can be interchanged with another chassis of the same manufacturer. Most modules are replaced instead of trying to repair or locate the defective component in the module. In some cases, more than one module has to be replaced to correct the problem.

In some brands, the vertical and horizontal circuits are found in one module, but in other TV sets, the vertical and horizontal circuits are separate modules. The video, brightness, and sync circuits are located in one module in most TV chassis. The modules may be mounted flat or on edge in the chassis (Fig. 10-10).

The latest TV manufacturing trend is getting away from separate circuit modules, and you may find only the tuner, electronic tuning, and control circuit as separate modules in the TV chassis. All other circuit components are soldered and mounted on one large PC board. Although, these late model receivers cannot be classified as a modular TV chassis, the control and tuner modules are plugged into one another and are connected to the TV chassis with cables and plugs.

10-10 Some modules plug in flat in the chassis while others can be mounted on their edges. Always clean off the module contact points with rubber eraser of a pencil or cleaning fluid before inserting.

Module replacement

Before exchanging the various modules one must know which modules to interchange. A schematic with a chassis module layout is essential. A lot of the TV manufacturers have drawings or photo layouts of the different modules. Look for a module location chart inside the TV cabinet. Practically anyone can interchange a module if they know what to look for and where to look. Some TV owners purchase a complete package of modules so they can service their own TV chassis (Fig. 10-11). Check the following service suggestions before attempting to replace a possible defective module.

SAW filter component

The surface acoustic wave (SAW) filter utilizes the surface wave of a piezoelectric element of very compact design. The SAW filter is a finger-like electrode that is arranged in length and width to assure the desired bandwidth characteristic of the IF frequency. A surface acoustic wave filter device establishes the proper IF response. The SAW filter may be located after the tuner or between the tuner and IF amp and video detector (Fig. 10-12). The SAW filter is in a grounded, shielded container.

Very seldom does the SAW filter cause any service problems, but it may be checked with the DMM for leakage. Check each SAW filter terminal with the ground terminal for continuity. Any measurement between common ground and other elements indicates leakage. A normal filter has no continuity measurement. The unit may become cracked or open. You can check the SAW filter with a crystal checker in or out of the circuit (Fig. 10-13). Often, a lower measurement is found with in-circuit tests. A higher reading results when the SAW filter unit is out of the PC board. When the IF signal will not pass through the SAW filter device, suspect a defective unit.

10-11 Check the chassis and service literature layout of each module location. In the RCA chassis pictured, the five modules lay flat.

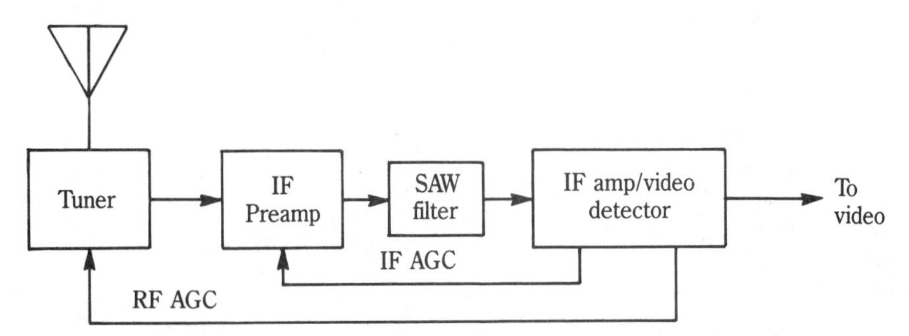

10-12 The block diagram of a SAW filter network between IF amp and IF amp/video detector.

10-13 The SAW filter is made up of a piezoelectric element. Notice that the small, round SAW filter unit is well shielded with soldered terminals.

Sand-castle generator

The sand-castle waveform is developed by combining the horizontal blanking, vertical blanking, and burst keying pulses applied to the luminance/chrominance processing IC (Fig. 10-14). The output of the horizontal oscillator is fed to the sand-castle generator. During the horizontal and vertical blanking period, these three pulses are combined to form the sand-castle waveform. The three-level signal waveform is fed from the sand-castle generator to the luminance/chroma IC. The sand-castle waveform can be observed at the luma/chroma IC input pin. The IC has an internal decoder network that decodes the three signals and sends them to the correct circuits. If the sand-castle generator is defective, there will not be any luminance or color from the luminance/chrominance processing IC.

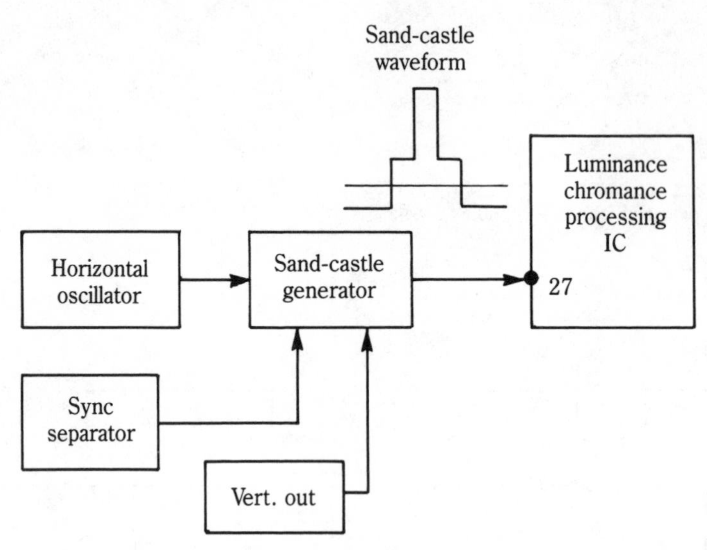

10-14 Block diagram of the sand-castle waveform generator.

Chopper power supply

The chopper-type power supply was first used in Japanese TV chassis. The chopper power supply may consist of a low-voltage bridge rectifier circuit providing voltage to the primary winding of the chopper transformer. The dc voltage from the chopper transformer is connected to the chopper transistor. Most chopper power supply circuits can be serviced with a DMM (Fig. 10-15).

The pulse-width-modulated (PWM) chopper regulated power supply works like the horizontal deflection systems in many of the later TV sets. The regulator control is a free-running oscillator with a frequency of about 15 kHz. This oscillator is triggered by a horizontal pulse derived from the secondary of the IHVT transformer, which locks the regulator to the scan frequency (Fig. 10-16).

The output of the PWM regulator circuit is applied to the regulator drive transformer. The output of the drive transformer is connected to the base terminal of the

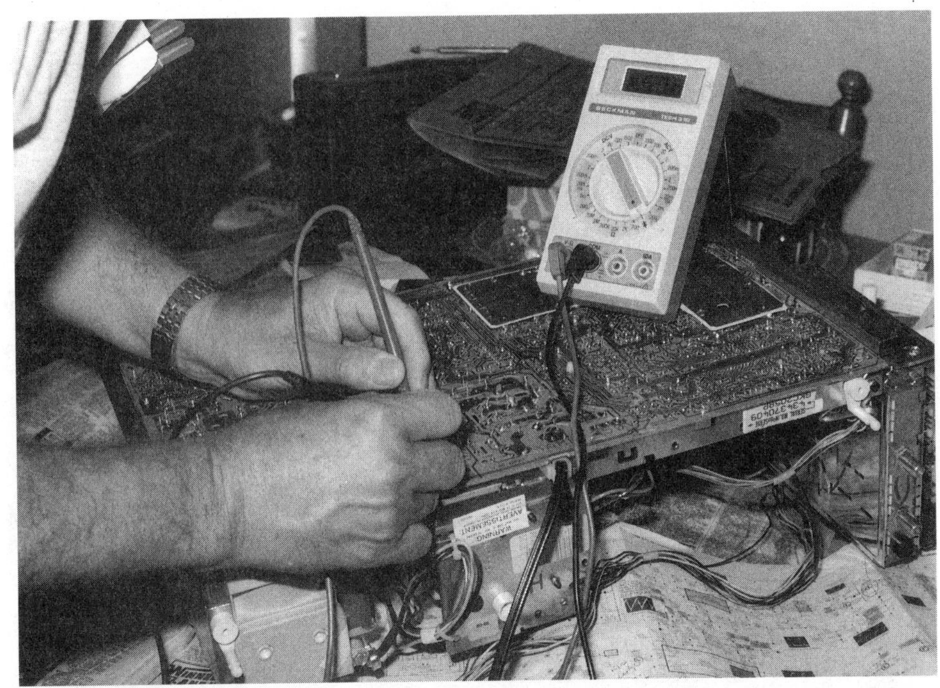

10-15 Checking transistors and components in the RCA chopper power supply with a DMM.

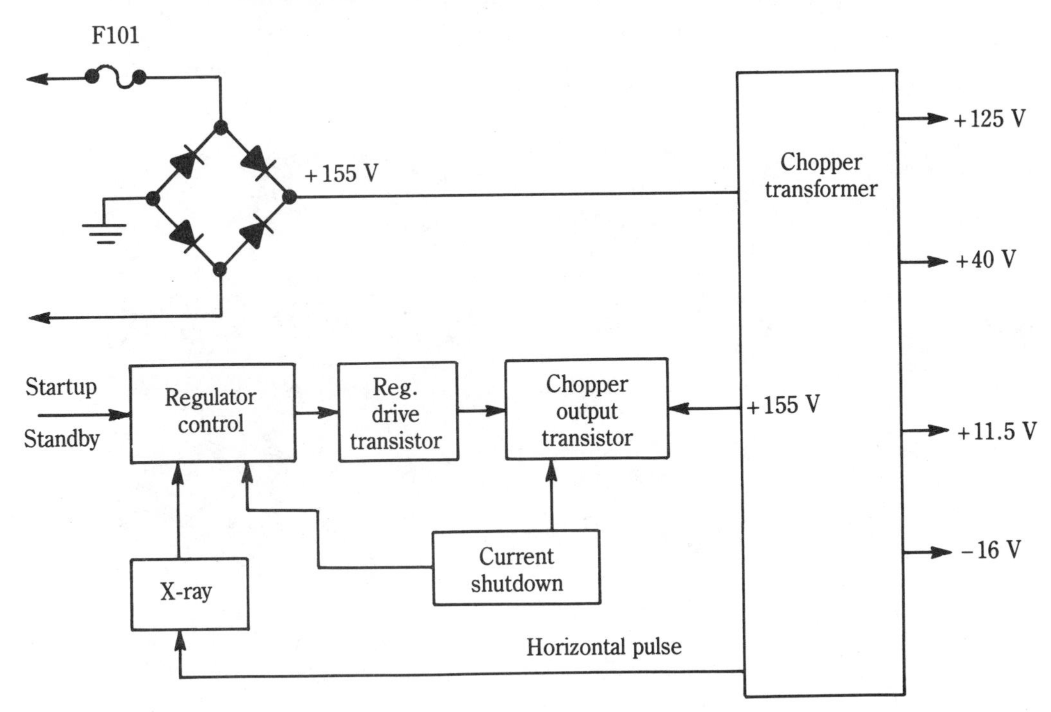

10-16 Block diagram of a chopper (PWM), regulated, low-voltage power supply.

chopper output transistor. This turns the chopper transistor on and off. The raw dc voltage from the bridge rectifier is applied through the primary winding of the chopper transformer to the collector of chopper output transistor (Fig. 10-17).

The on/off state of the chopper output circuit causes a pulsating dc voltage to occur in the primary winding of the chopper transformer. This induces pulses in the secondary winding of the chopper output transformer and is then rectified to provide voltage to several different voltage sources (Fig. 10-18).

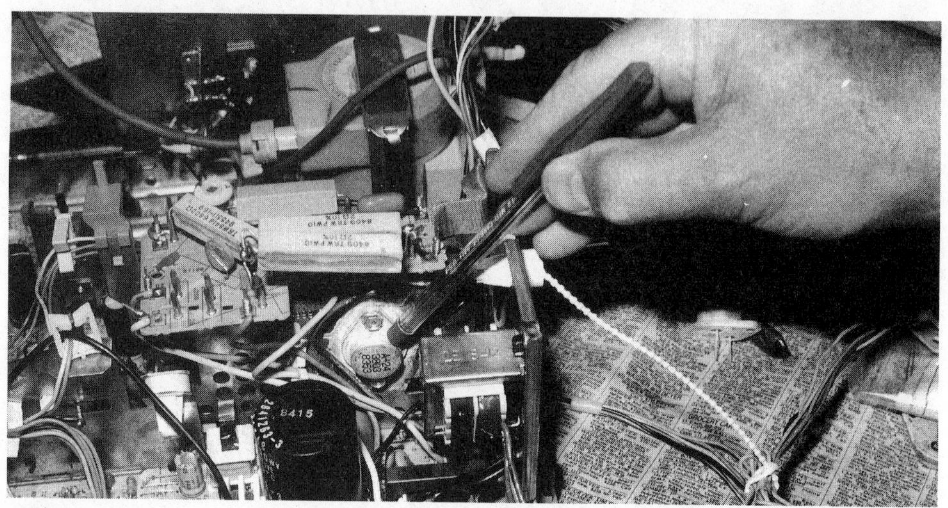

10-17 The pen points to the location of the chopper output transistor. The chopper transistor looks like a regular horizontal output transistor.

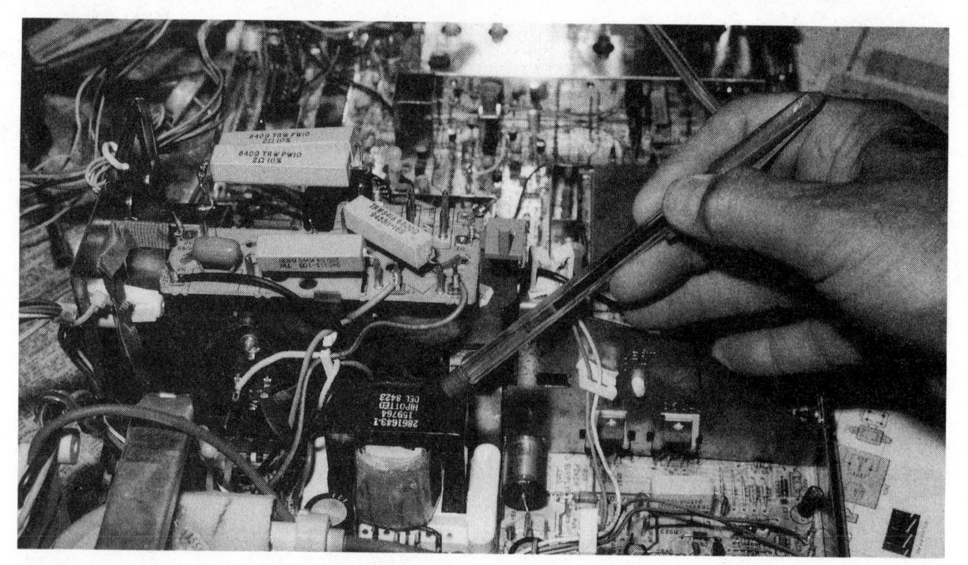

10-18 The pen points to the chopper power transformer. The chopper transistor is connected to the primary winding of the chopper transformer.

Scan-derived voltages

Scan-derived voltages, used in many TV chassis, supply different dc voltage sources from the flyback transformer windings. The vertical, horizontal, video, sound, and luminance/chroma circuits may be powered from the horizontal output transformer. Remember, when voltages are taken from the flyback, the horizontal circuits must be operating. Some chassis provide a voltage from the low-voltage bridge circuits to start up the horizontal circuits (Fig. 10-19), while other manufacturers have start up circuits that eventually run after the horizontal circuits are started.

10-19 The scan-derived voltage sources are taken from the new integrated (IHVT) horizontal output transformer.

The scan-derived voltage sources are taken directly from additional windings on the horizontal or flyback transformer. Each winding is rectified by a small silicon diode and capacitance filtering. Zener diodes may be used to provide additional regulated voltages (Fig. 10-20). Higher voltage sources for the color output transistor and picture tube circuits are taken from another winding. When a leaky component within one of these secondary voltage sources occurs, the overloaded circuit may shut down the horizontal output transistor and all voltage sources. To determine if the flyback transformer or overload voltage source is defective, slowly raise the ac line voltage with an isolated power transformer.

Switched-mode power supply

The advantage of the switched-mode power (SMP) supply is to slowly bring up the rectified power line voltage to prevent damage to critical IC processors and transistors within

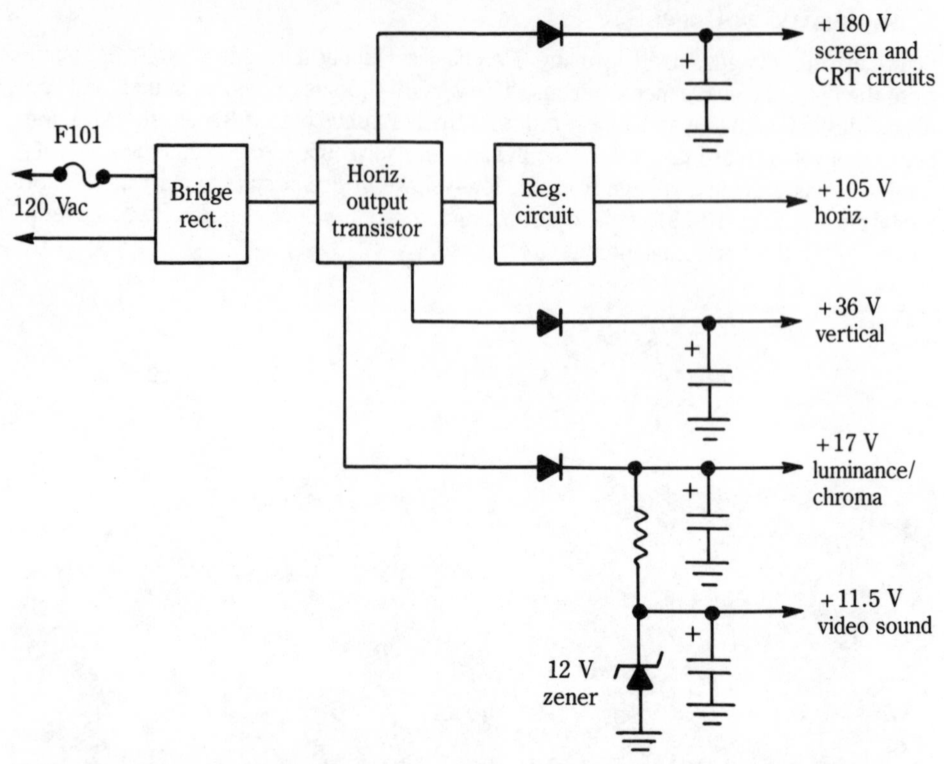

10-20 Block diagram of the bridge rectifier voltage source, which supplies voltages to the horizontal and many other circuits.

the TV circuits. The SMP supply contains a self-oscillating transformer and voltage regulator circuit (Fig. 10-21). The switched-mode power transformer operates from an ac-line-derived +155 volts dc. T402 also provides power-line isolation with the optoisolator component (IC 403).

Magnetic energy is stored in the transformer during the on time and then transferred to the secondary winding of switched-mode transformer during off time. The amount of energy transferred is controlled by the switched-mode regulator transistor. The frequency of the power supply, under normal load, may range from 20 kHz to 40 kHz.

The differential amplifier (Q 407) monitors the output voltage and compares this voltage to a zener diode. The connection voltage is fed to the main control amplifier transistor, Q 404, and fed back through the optoisolator (IC 403). The control circuitry varies the duty cycle and maintains correct voltage output.

No sound and no raster

First, check for a possible defect in the circuit breaker or fuse. Replace the fuse or check for continuity with the low range of the ohmmeter. Now, look on the module layout chart for the location of the low voltage module. Replace the low voltage module with a new

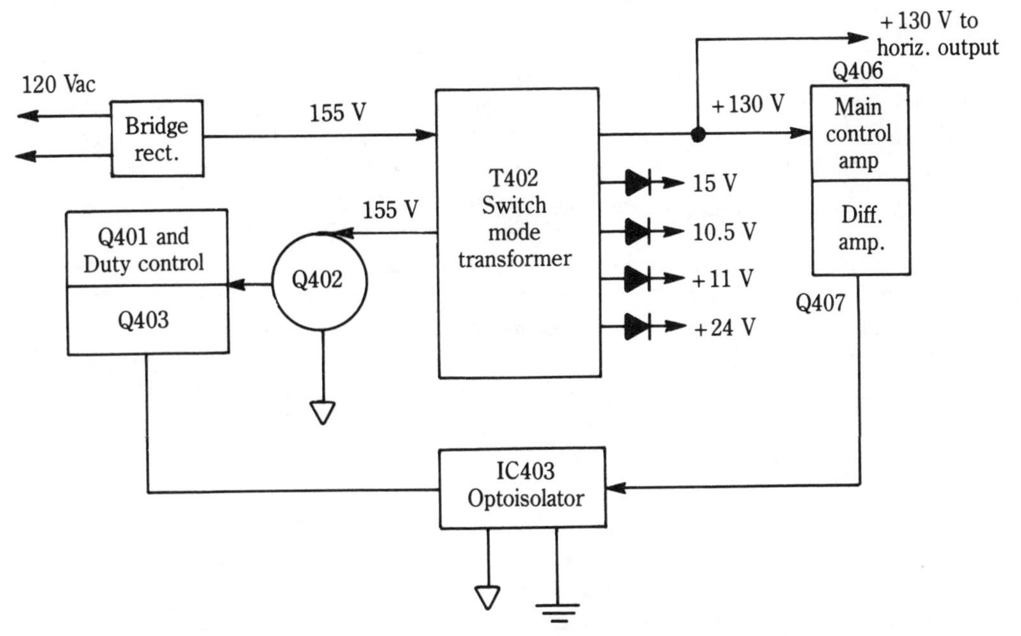

10-21 Block diagram of the switch-mode power supply circuits.

one. Plug in the TV chassis and listen for the sound to come on. If the sound is heard, this often means the horizontal and output circuits are functioning normally.

If no sound or raster is observed, proceed to the horizontal oscillator and driver module. Interchange the horizontal modules. Most problems found in any TV set are caused by the horizontal circuits. A defective horizontal output transistor might not be physically part of the suspected horizontal module and might be located on the TV chassis. In other modules, this transistor is located in the horizontal module. A defective horizontal output transformer and high voltage circuits can cause a no sound–no raster symptom.

Insufficient height or horizontal white line

A horizontal white line and insufficient vertical height indicates problems within the vertical circuits. Some modules have all the vertical components mounted on one module. In others, the vertical output transformers are mounted on the chassis. Usually, replacing the vertical module eliminates the vertical height problem.

If the raster almost fills out the screen after replacing the vertical module, readjust the vertical size control. Just one vertical control may be found in transistor chassis. The raster should completely fill out the screen after the adjustment. If not, other components could be defective, such as an open vertical coupling capacitor between the vertical circuit or the yoke might be open. The capacitor values may range between 470 μF and 4700 μF. A horizontal white line may also be caused from an open yoke winding or a poor vertical yoke socket.

No raster and normal HV

Suspect a defective video output circuit or the voltages applied to the picture tube to cause a no raster problem. You can detect high voltage in the picture tube by feeling the front of the TV tube or listening for the sound of the yoke expanding when you first turn the set on. Place your arm next to the TV screen; if a high voltage potential is present you will feel the hairs on your arm being attracted toward the screen. If there are no signs of high voltage check the filament of the tube and see if it is lit.

Now, replace the video—luminance module. You can find a separate matrix-color amplifier module or the luminance driver transistors are mounted on the CRT (cathode-ray tube) socket. Replace both the video luminance and video drive modules if necessary to achieve correct brightness. In some chassis, a kinnie module may be found for each color output circuit. Excessive or insufficient brightness can often originate in the matrix-power amplifier stages.

Snowy or weak picture

A snowy picture can be caused by a defective tuner, IF stage or outside antenna. Cleaning the tuner may solve a snowy or erratic picture and sound symptom. Replacing the tuner module in the latest color chassis can cure a snowy or weak picture. A snowy picture can be caused by a defective AGC circuit.

Check and see if there is one module which includes the entire IF and AGC circuits. In some TV sets, the sync and AGC circuits are found in the IF module. In other TV sets, the AGC and sync circuits are found in a separate module. Replace the IF and AGC modules after replacing the tuner or checking the tuner for possible defects.

Poor sync

Suspect a defective sync circuit if the picture drifts either vertically or horizontally and cannot be adjusted with either corresponding control. Notice if these sync circuits are located separately or are a part of the IF module. Sometimes just replacing the IF module will cure both the out of sync and snowy picture. Always adjust the AGC control after replacing the AGC module. The AGC control may not exist in some of the latest TV sets.

Color problem

Color problems originate in the color IF, demodulator stages, and color video-output circuits. Just replacing one color module may not solve the weak or no color problem. You may find two different color modules in the color circuit of one color chassis and all of the color circuits on one large PC module in another color chassis (Fig. 10-22).

Before attempting to replace the color modules, adjust the fine-tuning control. Tune in the best possible picture and sound from a local TV station. A defective or misadjusted color killer control can cause color rainbows or missing color. Check the manufacturer's literature for correct color killer adjustment. Replace the color module when no color killer control is found on the TV chassis.

10-22 Simply replacing one color module can solve the color problem. Some TV sets use two separate color modules. Replace each one until the color reappears.

Solid-state sound circuits

In the early solid-state TV chassis, the entire sound circuits were made up of transistors. Today, you may find one large IC may contain the whole audio output sound circuits (Fig. 10-23). Sometimes the sound circuits are included in one large IC that contains luminance/chroma, video, sync, AGC and deflection circuits. You may find a combination of sound IF and discriminator circuits in one IC and the audio output circuits with transistors.

Often, the sound takes off after the first video stage to a sound IF transformer. The audio IC processor amplifies the weak audio signal and discriminator coil tunes in the FM audio signal. The audio signal is controlled by the volume control and passed on to the audio output stage and eventually the speaker. A slight touch-up of the discriminator coil may clear up the muffled or distorted sound when the sound signal is off frequency (Fig. 10-24).

The audio signal can be signal-traced with a signal tracer or external audio amp from sound detector to the PM speaker. Loud popping sounds may be caused with the large audio output transistor. Intermittent sound may be caused with defective IC or electrolytic coupling capacitor. Weak sound may result from leaky or open transistor, open coupling capacitors, or a dried-up electrolytic speaker coupling capacitor.

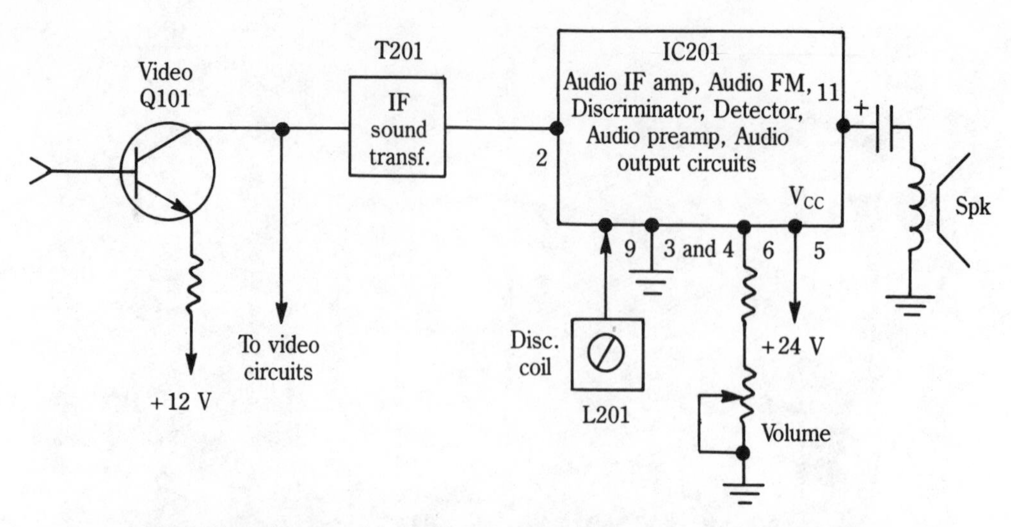

10-23 Block diagram of the solid-state audio circuits in one large IC component.

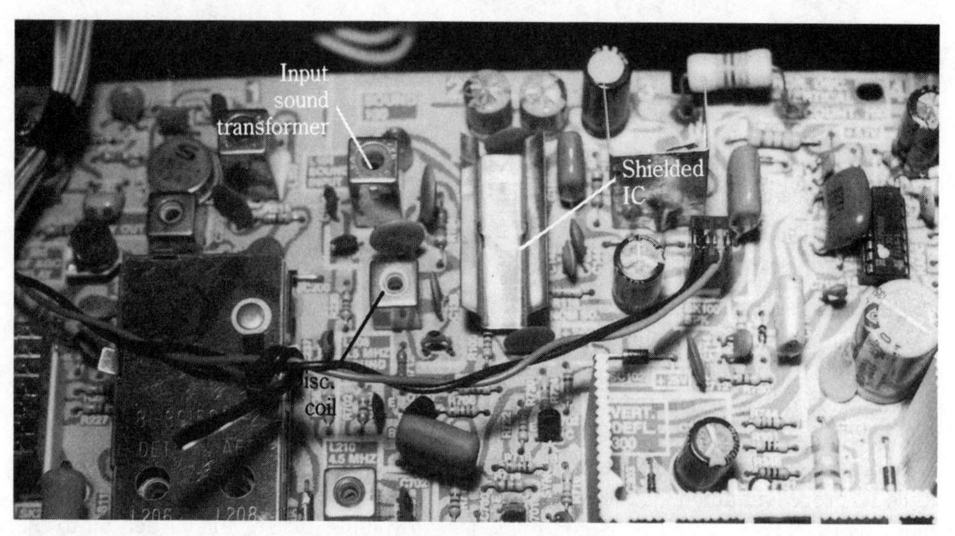

10-24 The location of the sound stages on a solid-state TV chassis. A slight adjustment of the discriminator coil could solve a problem of distorted sound.

Stereo audio

Most of the higher-priced TV sets have stereo sound. Today, some TV stations are broadcasting stereo audio and is noted when a red LED or light comes on with the stereo signal. All TV programs are not broadcast in audio stereo. When stereo sound is broadcast, the sound comes out of two separate amplifiers and speakers.

Within the stereo sound circuits, the stereo or mono sound signals are applied to the stereo demodulator circuits (Fig. 10-25). The baseband audio signal is developed by the system control IC and determines when a stereo signal is received. This stereo signal is

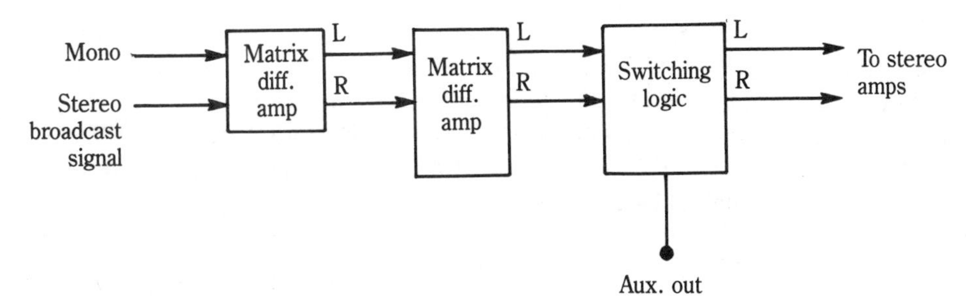

10-25 Block diagram of the stereo sound circuits in a late model TV set.

passed on to the stereo demodulator (IC1). The stereo audio signal is applied to a matrix and differential amp IC (IC2). The outputs of the differential amp and matrix are separate left and right audio channels. The L and R audio signal is switched by IC3. The switching IC applies mono or stereo signal to the audio amplifiers.

The stereo or monaural signal passes on to volume control IC4. Here, the signal controls the volume and may be muted by manual or remote control (Fig. 10-26). The left and right audio signal is applied to the dual audio IC amp (IC5). Here, the weak stereo audio signal is amplified and coupled to the stereo speakers with electrolytic coupling capacitors. The audio stereo signal can be signal-traced like any amplifier, starting at the audio sound output of the matrix IC2.

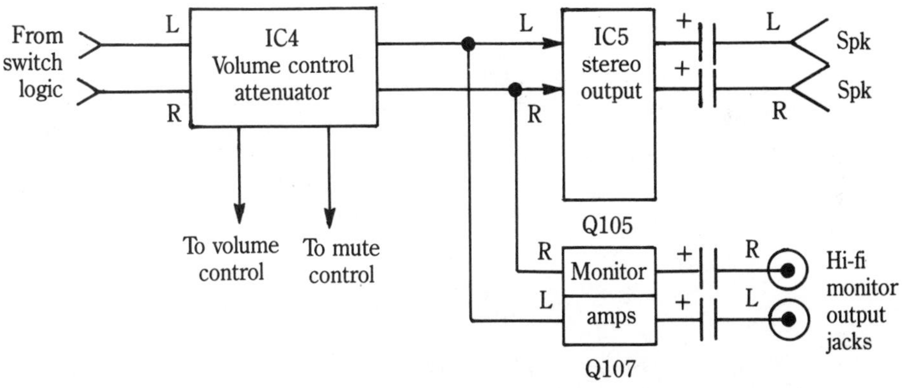

10-26 Block diagram of the stereo audio output circuits with stereo output jacks.

Weak or no sound

Replacement of the sound module can cure all sound problems. When modules first were found in the black-and-white or color chassis, the sound IF and preamp stages were mounted as a plug-in module. The sound output transistor was located on the chassis. In the modular chassis, the whole sound circuit can consist of one IC with its associated components located on one module.

If, after replacing the sound module, the sound is still weak, suspect an open or defective (speaker to module) electrolytic coupling capacitor. These small capacitors may produce weak, intermittent, and distorted sound. They are easy to locate by tracing the speaker wiring to the PC board and the PC wiring to the capacitor. The coupling capacitor is located between the output transistor and the speaker cable wires.

Be careful when replacing any TV module. You can easily crack the PC board by improper mounting or too much pressure. Make sure the module is properly seated. Double-check all external plugs connecting the chassis to the module. Check for any bent plugs or socket pins before mounting module. If several wire plugs are removed from the module before replacement, each plug should be marked so they can be properly connected. Do not clean the module contacts with a liquid cleaning solution. Clean off all the module contacts with a pencil eraser. A quick modular troubleshooting chart is found in Fig. 10-27. Replacing a suspected defective module can quickly solve the dead or erratic TV problem and save a few service dollars. You may want to leave module replacement up to the professional technician.

No sound and no raster:	Check for open fuse Reset circuit breaker Replace low voltage module Replace horizontal module
Sound and no raster:	Check for light of CRT heater No high voltage Replace video/luminance module Replace matrix and color output module
Snow picture:	Clean tuner Check antenna signal Replace IF AFT module Replace A AGC, IF, and RF module
Poor horizontal sync:	Adjust AGC control Replace AGC-sync module Replace horizontal oscillator module
Poor height and linearity:	Adjust vertical height or size control Replace vertical module
Poor vertical sync:	Adjust AGC control Replace AGC-sync module
Poor or weak color:	Check fine tuning and color-killer control Replace chroma module Replace matrix-luminance module
Weak or distorted sound:	Replace sound module Replace IF module Check speaker-to-circuit coupling capacitor Check for defective speaker

10-27 A quick modular troubleshooting chart.

Parts layout

Before you can find or replace a suspected or defective component, the part must be located in the TV chassis. Some parts are easy to locate, such as tubes, power transformer, flyback transformer, and picture tube. Others must be pointed out in a parts layout chart. Most manufacturers have a parts layout chart with photos and drawings labeling the different components. Sometimes critical part layout charts may be found glued to the side of the TV cabinet. Transistors and tube components can be located on a cabinet chart. Of course, the TV technician has original manufacturers schematic and layout charts for easy reference.

Other components

Besides transistors, you might have special diodes, IC circuits, SCRs, and tripler units in solid-state chassis. In many of the vacuum-tube chassis, silicon diodes were inserted in the low-voltage rectifier circuit. Today, special switching, zener, and damper diodes are found throughout a solid-state chassis. The silicon-controlled rectifier (SCR) can be found as a regulator in both the low- and high-voltage circuits. The latest color TV sets have high-voltage diodes molded right into the flyback transformer eliminating the need for separate regular tripler units (Fig. 10-28).

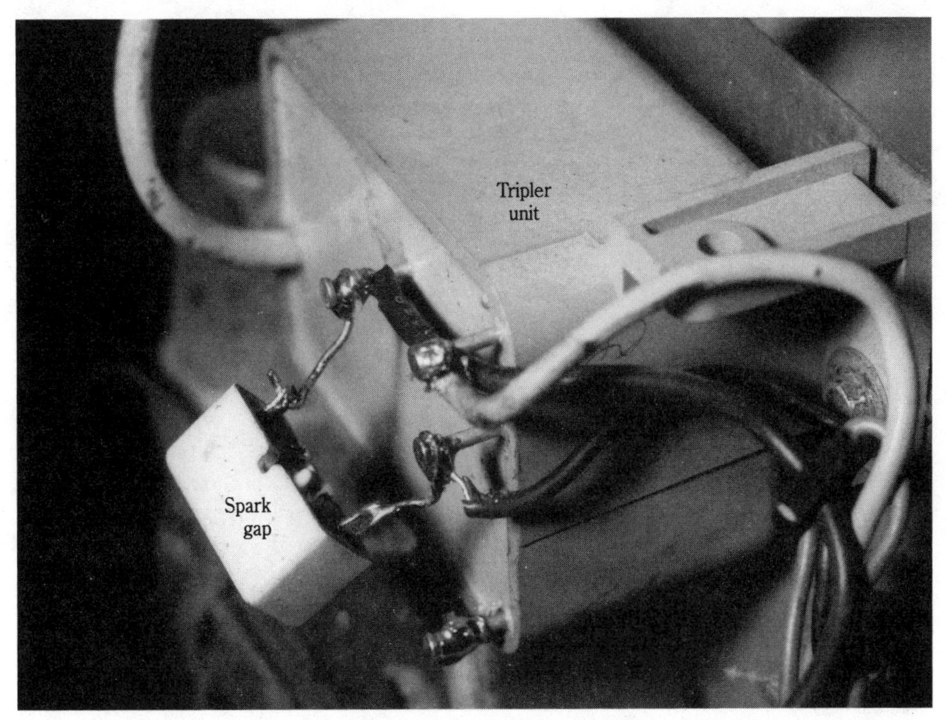

10-28 The tripler unit in the TV set is connected to the output of the flyback transformer to build up high voltage with diodes and capacitors in one component. A defective tripler could arc through the plastic cover to chassis ground.

The integrated high-voltage transformer (IHVT) has many different windings compared to the regular flyback or horizontal output transformer. The high-voltage winding that went to the high-voltage rectifier or tripler unit now has silicon diodes molded right inside the plastic molded case. A large high-voltage lead comes out of the flyback and goes directly to the anode connection of the CRT (Fig. 10-29) in addition to the primary windings that supply voltage sources to other TV circuits. Like the old flyback transformer, many of them become leaky, shorting HV diodes from internal acrover, mandating replacement.

10-29 The integrated horizontal output transformer or flyback has several additional windings for the various voltage sources. The silicon diode and electrolytic capacitor filter is in each voltage source.

Before the use of IHVT transformers, tripler units were used that contain high-voltage diodes and capacitors that build up high voltage from the flyback transformer. They too can break down and arc over. Sometimes the HV arcs over from the tripler body to the metal TV chassis. The tripler unit has input, ground, and focus terminals. The large HV insulated lead goes directly to the anode of the CRT.

The transistor comes in many sizes and shapes. High-powered transistors may be found in the sound output, horizontal and vertical output circuits, and low voltage regulator circuits. Medium-powered transistors are used in the vertical output stage, and in the sound stages. Low-powered transistors are found in the remaining solid-state circuits (Fig. 10-30).

The integrated circuit (IC) was first located in the sound preamp and driver stages of the tube chassis. Now one IC can combine the entire audio circuit. Some ICs were located in the color stages of the tube chassis, but now the entire color, luminance, and

10-30 Transistors come in various sizes and shapes. The larger transistor is the horizontal output transistor.

video circuits are contained in only one large IC. A separate IC package might house the oscillator and driver stages of either the vertical or the horizontal circuit or both of the circuits can be included in one IC package (Fig. 10-31). Replacing just one IC component can solve several problems in several different circuits.

Low-powered signal diodes are located in the RF, IF and video circuits of the TV chassis. In contrast, the damper diode is a special heavy-duty type with a very high voltage rating and is found in the collector circuit of the horizontal-output transistor. Always, be careful to observe polarity when installing a new damper diode and make sure it has the correct voltage rating. Zener diodes are found in the low-voltage circuits. They are used to start, stop, and shut down the regulated voltage circuits. The zener diode has a tendency to become leaky in circuit operation.

High-voltage shutdown

High-voltage shutdown may occur when the high voltage reaches the X-ray protection circuit and shuts the chassis down. Excessively high voltage may occur when the ac power line voltage raises above the 120 Vac range. Power-line voltages above 125 or 128

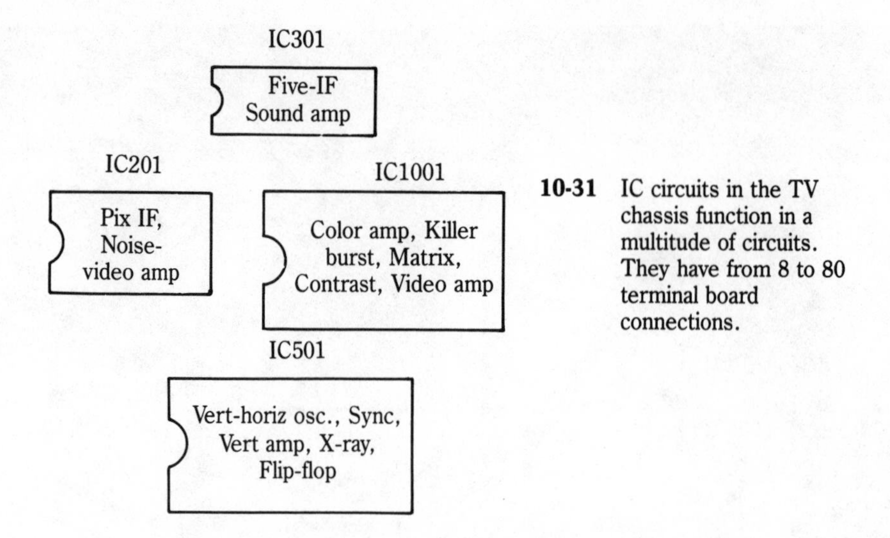

10-31 IC circuits in the TV chassis function in a multitude of circuits. They have from 8 to 80 terminal board connections.

volts ac may cause the chassis to shut down. An open or change in capacitance of bypass capacitor across the collector terminal of the horizontal output transistor may increase the high voltage (Fig. 10-32). A defective IHVT transformer may cause high voltage shutdown.

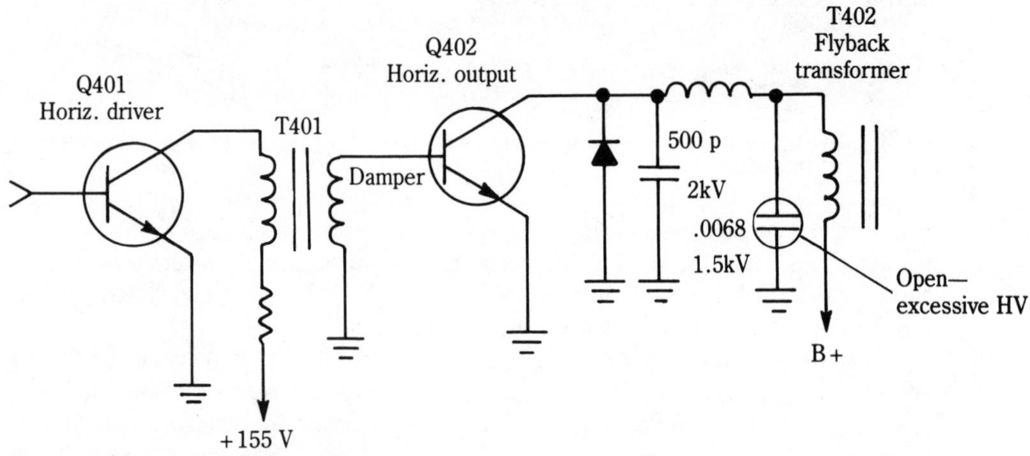

10-32 An open or change in capacitance of bypass or tuning capacitor across the collector terminal of the horizontal output transistor could increase the high voltage.

To determine if the chassis is shutting down from excessive HV, monitor the HV at the picture tube, the waveform at the base of the horizontal output transistor, and the dc supply voltage with the DMM (Fig. 10-33). Plug the TV into a variable isolation power line transformer and slowly increase the line voltage from 60 to 80 volts. Notice how the high voltage is measured at 80 volts ac. Keep advancing the transformer until the chassis shuts down. If excessively high voltage is noted before the 120 Vac voltage is reached, check the horizontal output and flyback for defective circuits.

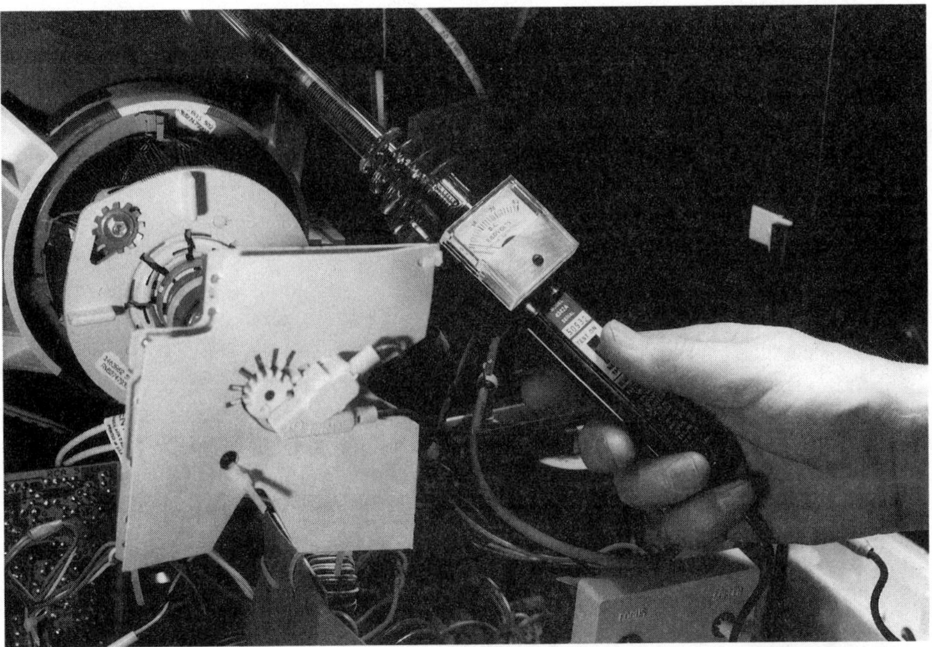

10-33 Monitor the high voltage at the anode connection of picture tube to determine if high voltage is shutting down the chassis.

X-ray protection circuits

The X-ray circuit protects the operator and chassis from the excessively high voltage applied to the picture tube. The shut-down circuit operates when the voltage or current has reached a certain limit and shuts down the horizontal output transformer and chassis. In the early chassis when the high voltage went over the manufacturers limit, the protection circuits would turn the horizontal oscillator stages off frequency. When high voltage was present, you could not successfully adjust the horizontal control and bring the picture into sync (Fig. 10-34).

A pulse from the flyback winding is applied to the X-ray shut-down circuits and applies rectified voltage to the shut-down circuits. As long as this voltage stays below the zener diode voltage, the diode will not conduct. When excessively high voltage is applied to the flyback, the pulse is larger and applies higher voltage to the zener diodes causing the diode to conduct. When the diode conducts, an SCR or diode provides bias applied to the horizontal oscillator circuits. The horizontal oscillator is disabled and the whole chassis shuts down. Check for pulse waveform from flyback transformer with a scope. Correct voltage and resistance measurements may reveal a defective shut-down circuit.

Power supply regulator circuits

In many of the lower-priced ac TV chassis, the power supply consists of a half-wave or bridge rectifier and output regulator. The ac voltage is fuse-protected (F101) and is

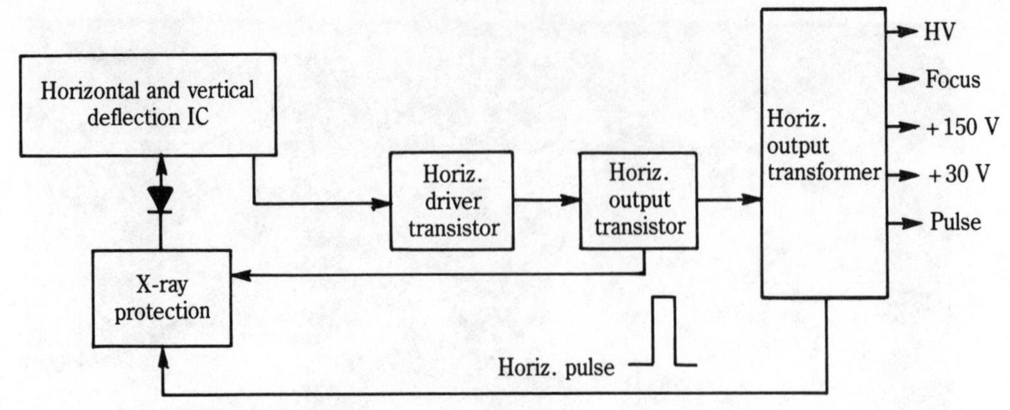

10-34 Block diagram of the x-ray protection circuit to shut down a TV chassis with excessive HV.

applied to the isolation resistor and silicon half-wave rectifier (Fig. 10-35). Again, the dc voltage is filtered with C1 and is fuse-protected with F102. The high dc voltage is applied to the input terminal of the low-voltage regulator.

The low-voltage rectifier may consist of an IC circuit that looks like a regulated horizontal output transistor, except it has three terminals instead of two, besides the metal body of the regulator. Often, the metal body is insulated away from the metal chassis and is the regulator output terminal. In other TV chassis, the IC regulator resembles a power output IC found in high-powered stereo audio output circuits. You may find low voltage regulators of 115, 120, 125, and 130 volts in the ac TV chassis. The correct voltage regulator must be replaced with the exact part or universal component number. The defective regulator may short, become leaky, or be open.

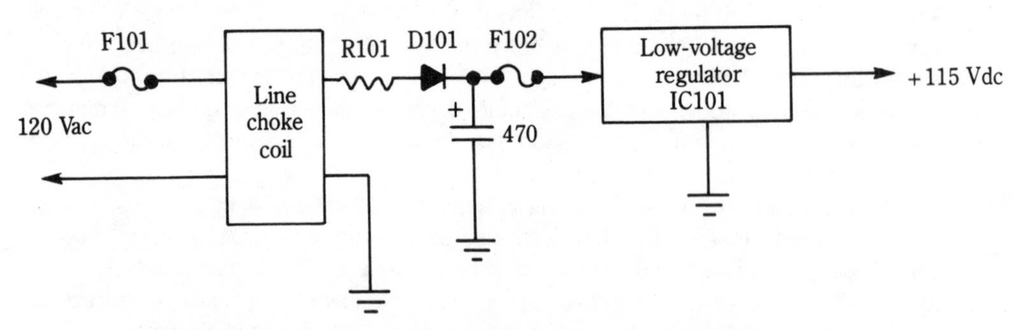

10-35 Block diagram of the low-voltage power supply regulator circuits.

Standby power circuits

The standby power supply provides voltage at all times to the IR receivers and start-up circuits of the horizontal output transformer. AC power is applied to the standby transformer (T401). The low ac voltage is rectified and applied to several standby voltage sources. Often, the standby power supply is responsible for standby voltage and regulation (Fig. 10-36).

In this circuit, the 21 volts is applied to the horizontal driver transistor during start-up. The 11.5-volt source supplies voltage to the infrared receiver and keyboard. A 5-volt standby source is fed to the system-control IC. The standby voltages are needed when the remote or keyboard is pressed to activate the TV chassis. Note that these standby circuits are on all the time. Of course, until the chassis is activated very little current is being used. Check the standby circuits when the remote will not turn on the TV chassis, providing the remote transmitter is normal.

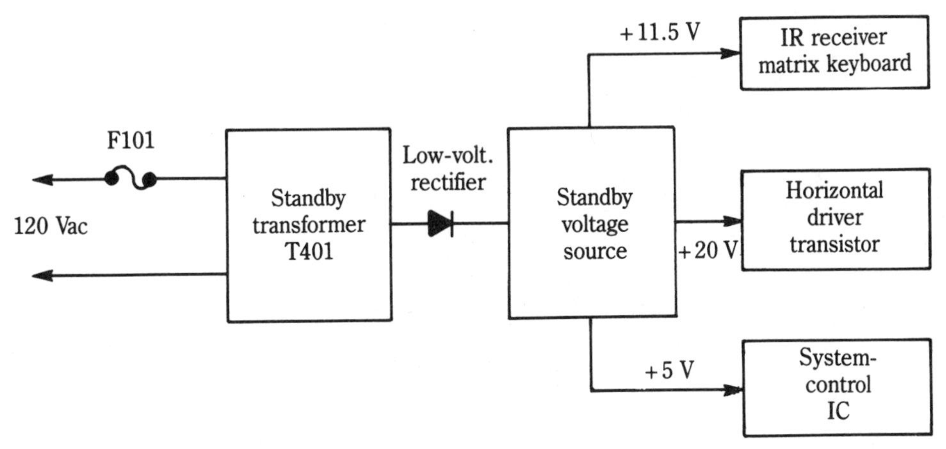

10-36 Block diagram of the standby circuits in the remote-control TV chassis.

Horizontal and vertical countdown circuits

The horizontal and vertical countdown circuit may be controlled with a fixed crystal oscillator. The horizontal and vertical oscillator circuits are found inside the processing IC. The frequency of the horizontal oscillator is determined by the crystal. In the RCA chassis, the output of the crystal is 32 times the horizontal frequency.

The crystal operates at 503 kHz. The output of the VCO is applied to the horizontal countdown circuit (Fig. 10-37). The countdown circuit divides the crystal frequency to 15,734 kHz. The output of the horizontal countdown circuit is amplified and a square-wave pulse is applied to the driving circuits.

The vertical circuits may operate from a countdown divider circuit. The vertical sync and separator circuits lock in the vertical countdown circuits. The 15,734 kHz frequency of the horizontal oscillator is applied to the vertical up/down counter and horizontal frequency doubler circuitry (Fig. 10-38). The output of the vertical drive (sawtooth waveform) is connected to the vertical output circuits, which scan the vertical deflection winding.

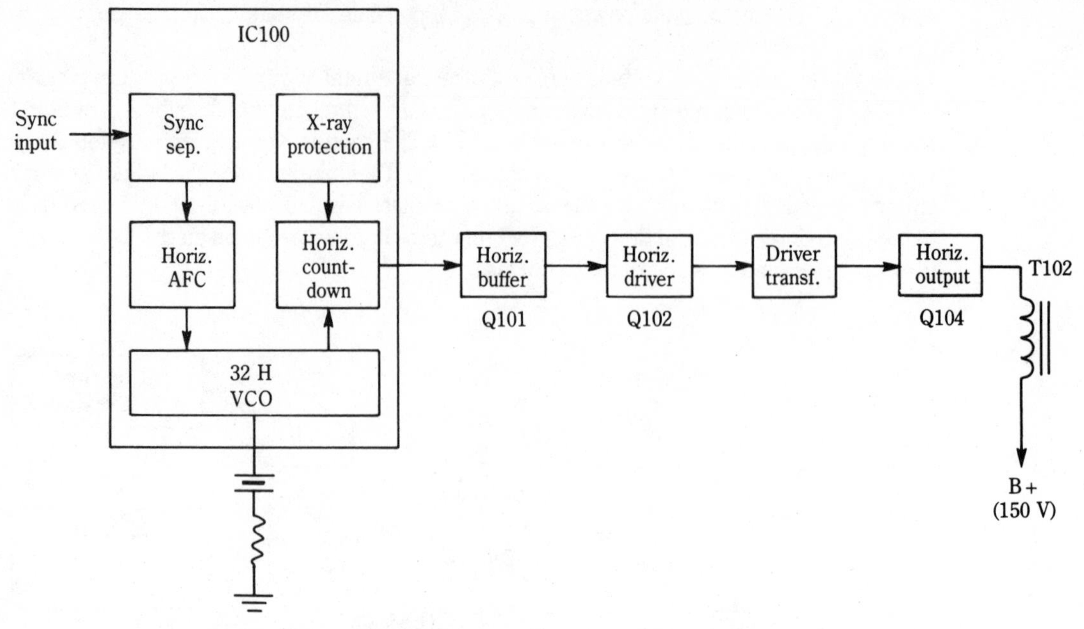

10-37 Block diagram of the horizontal countdown deflection circuits.

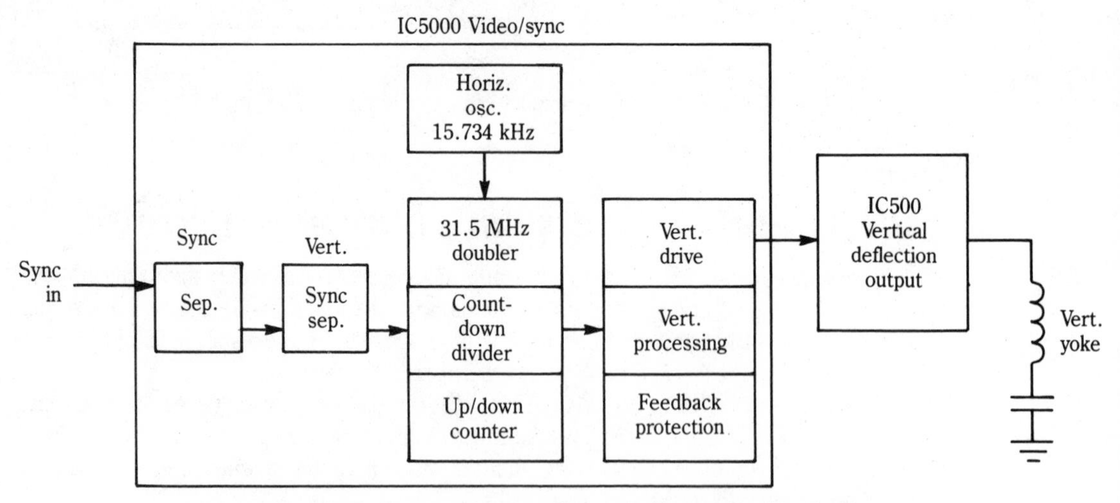

10-38 Block diagram of the vertical countdown deflection circuits.

IC deflection circuits

When the solid-state TV chassis was first designed, only transistors were used in the vertical and horizontal deflection circuits. Today, the horizontal and vertical oscillator circuits may be found in one large IC component. Transistors or IC components may be used in the vertical and horizontal driver circuits. The horizontal output transistor is found in most TV chassis, while the vertical output circuits may be transistor or IC components (Fig. 10-39).

In this vertical deflection system, the deflection IC 1001 pulse is applied to a vertical reset and sawtooth transistors. The error amp transistor provides a pulse to the input of the vertical output IC 103. Output terminal 7 applies a 55-volt p-p waveform to the vertical yoke winding.

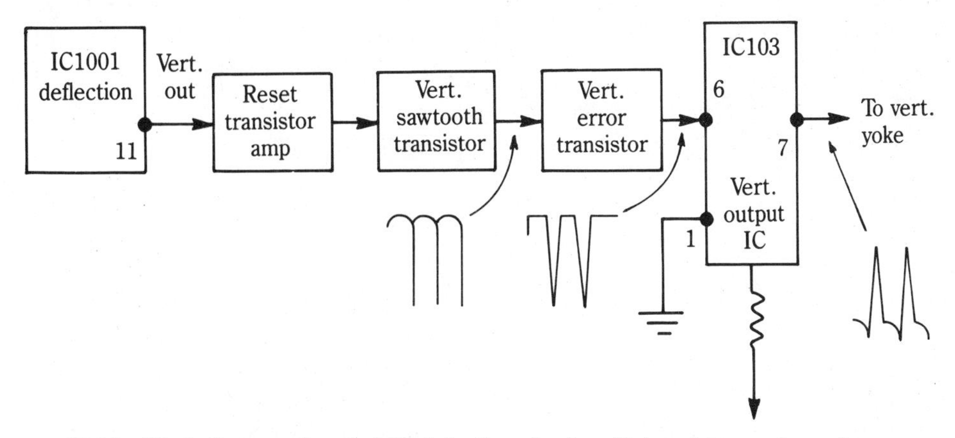

10-39 Block diagram of vertical IC deflection circuits with transistors and waveforms.

The horizontal deflection IC and amp supplies a 10 V p-p positive pulse to the horizontal driver transistor (Q 401). The collector of the driver transistor applies a pulse to the horizontal driver transistor (Fig. 10-40). The base of the horizontal output transistor is tied to the driver transformer. The collector terminal of the horizontal output transistor is tied into the primary winding of the yoke. Notice the damper diode is inside of the horizontal output transistor. The regular horizontal output transistor will not function in this circuit.

System control

The system control circuits found in the latest TV chassis are controlled with a large IC processor or microcomputer like those found in CD players and camcorders. The system control microcomputer generates the clock signal and data. The information transmitted on the serial data bus includes local and remote commands, AIU utilization data, and tuning commands (Fig. 10-41).

The system control microcomputer contains random-access memory (RAM) and read-only memory (ROM). The RAM portion stores information that frequently changes such as channel scan and customer settings such as picture and volume control adjustment. The AIU processor allows the set to return to the last channel viewed with the same volume and picture adjustments, even after the set is turned off.

The ROM portion of memory is used to store information that never changes such as program for the microcomputer, on-screen display, tuning, etc. Some manufacturers

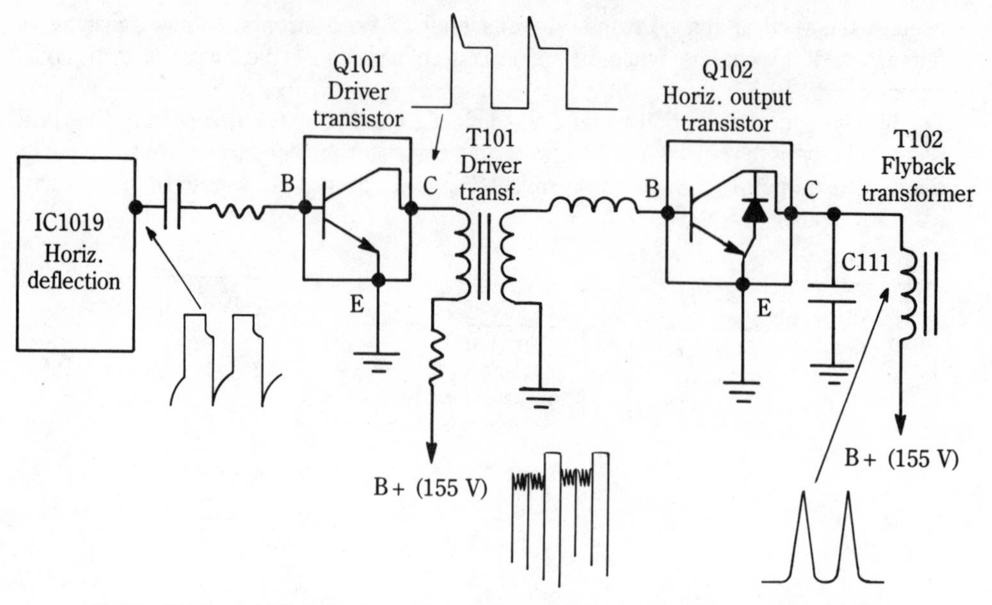

10-40 Horizontal IC deflection with transistor driver and output transistor circuits.

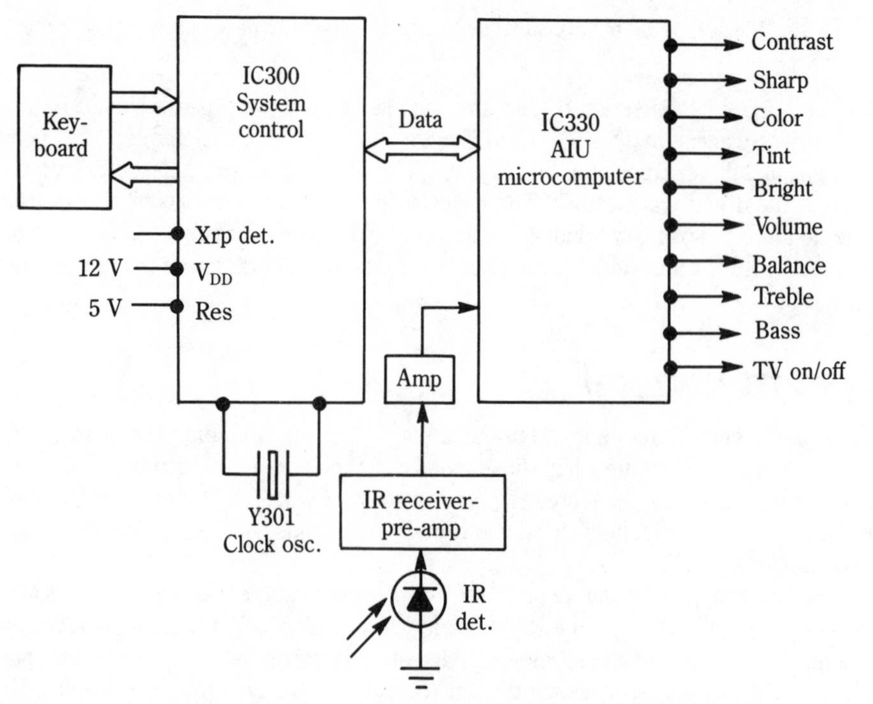

10-41 Block diagram of system or tuning control with IC and microcomputer operation.

call the system control the tuning system. The 5-volt power from the standby power supply keeps the microcomputer alive during power-off periods.

The tuning section selects channels by remote control, channel scan, and random access from the remote transmitter. The on-screen display and menus contain the audio, color, and picture, clock, and channel adjustments. Each menu can be selected by pressing a menu button on the remote control.

The system control may be serviced by determining what signals are present on an oscilloscope. The clock oscillator frequency can be checked with the scope. Critical voltage measurements on each IC or microcomputer terminal may indicate a defective circuit or IC component. The system control circuits are quite complex and should be serviced by the electronic technician.

11 Solid-state servicing

TRANSISTORS AND ICS ARE UTILIZED IN EVERY CIRCUIT IN THE TV CHASSIS. FIRST, the transistor was used in the sound circuit of the TV set. Next, the transistor was placed in the tuner, IF, and sync circuits of the hybrid chassis. Later, as more powerful transistors were developed they were installed in the horizontal output and regulator circuits, resulting in the all-transistor chassis (Fig. 11-1).

Both npn and pnp transistors are utilized in the circuits of a solid-state chassis. Although, in the older TV chassis a greater number of pnp germanium junction transistors were used, that trend has reversed and now more npn transistors are utilized in the circuits of a color receiver. The npn silicon transistor has a positive voltage applied to the collector terminal. The reverse is true of the pnp transistor, and a negative voltage is connected to its collector (Fig. 11-2).

A defective transistor may be located with an accurate voltage or resistance measurement of the device in the circuit. If the suspected transistor is removed, the transistor can be checked with a regular transistor tester or ohmmeter. The suspected transistor can be checked in or out of the circuit with a digital multimeter transistor diode tester. The suspected IC component can be located with accurate voltage and resistance measurements and signal in versus signal out tests.

There are many tests and adjustments the beginner can make in a solid-state TV chassis. Even you can test many components with the transistor tester and DMM (Fig. 11-3). Besides taking voltage and resistance measurements, you can check resistors, capacitors, coils, diodes, and transistors with the DMM. Careful voltage measurements on integrated circuits may indicate a defective IC. Today, you can purchase a portable DMM from $20 to $99. Pick up a low-priced digital multimeter (DMM) and give it a try.

Many different tests and adjustments for the beginner are given in this chapter along with several tests for the more experienced person. The intermediate or advanced person may often have acquired some experience from home or workshop facilities and is

11-1 Today's solid-state chassis are quite small compared with the tube chassis. Notice there is no power transformer in this ac-dc chassis.

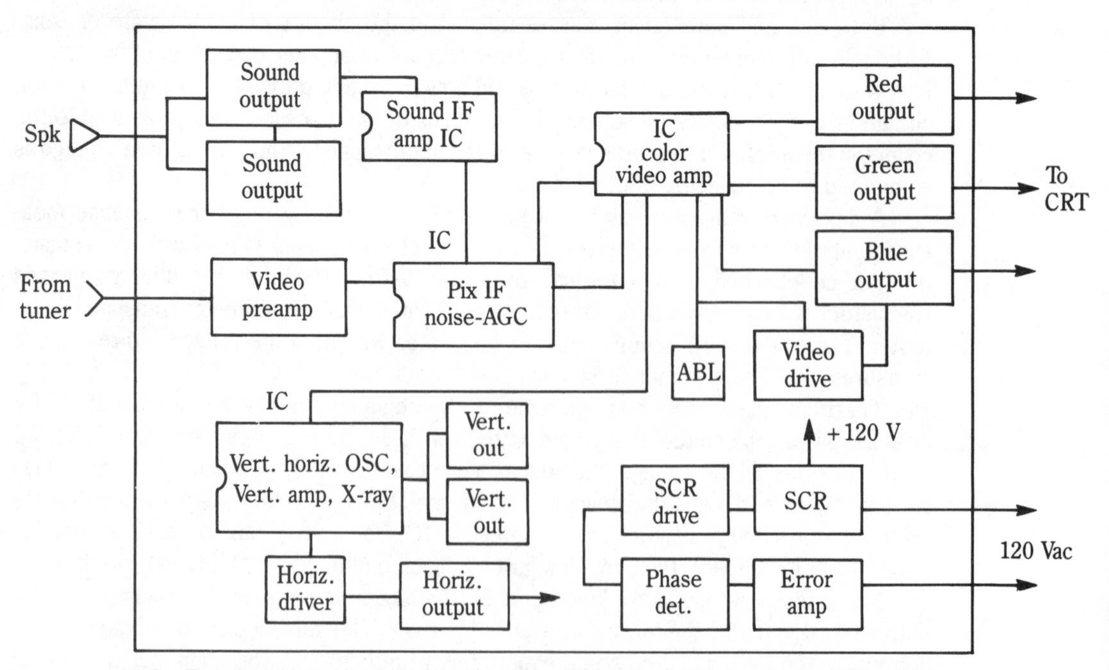

11-2 Transistorized chassis are now used in all black-and-white and color TV receivers.

11-3 Check transistors and IC components with the DMM.

able to operate some test instruments. Of course, the beginner should only go as far as he or she feels confident to go. A beginner should not attempt to make extensive repairs. It is better to call in a professional electronic technician.

Transistor resistance tests

There are many transistor testers on the market to help locate the defective transistor. Most transistors are checked for an open, leaky, or intermittent condition. These transistor testers may check the transistor in or out of the circuit. Test for an open base to collector and base to emitter. Testing may be done with an ohmmeter. In-circuit leakage tests within the circuit could result in an erroneous measurement from diodes, coils, or low bias resistors being connected across the transistor terminals. The transistor may test normal in or out of the circuit, then, break down under the circuit load and become intermittent. An intermittent transistor is one of the most difficult components to locate. Replacing the intermittent transistor with a new one is the best practice (Fig. 11-4).

The transistor may be checked with a pocket VOM in or out of the circuit with the low ohmmeter range. Remove the transistor from the circuit when a low resistance is measured between any two terminals. These resistance measurements should be made with the 2000-ohm scale and then repeated on a lower scale if a short or leakage is detected. After removing the suspected transistor, check the resistance between any two terminals. Actual resistance measurements of a leaky transistor are shown in Fig. 11-5.

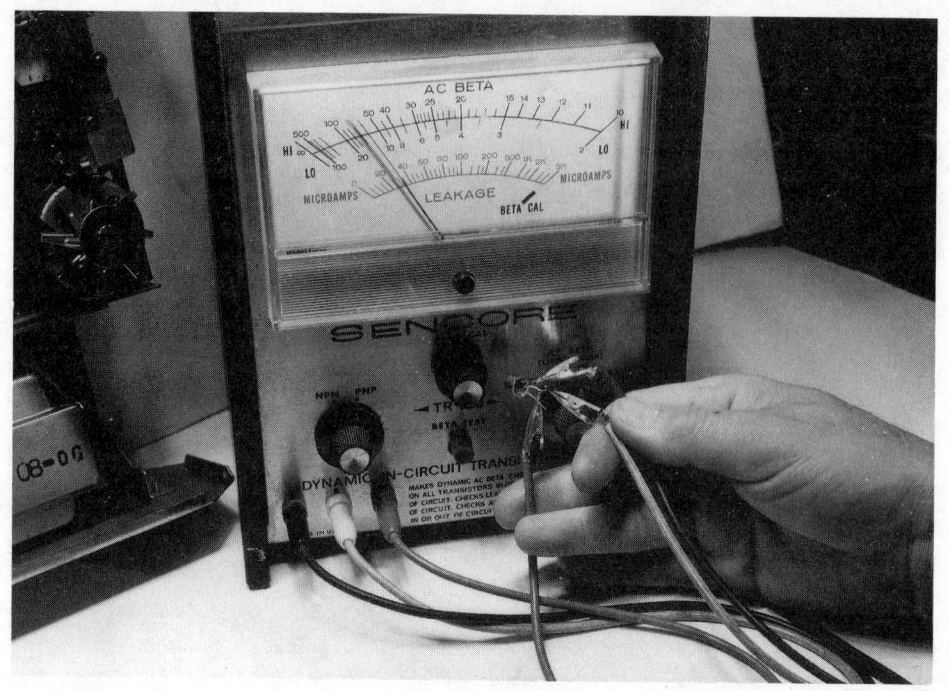

11-4 Transistors can be checked in and out of the circuit with a commercial tester.

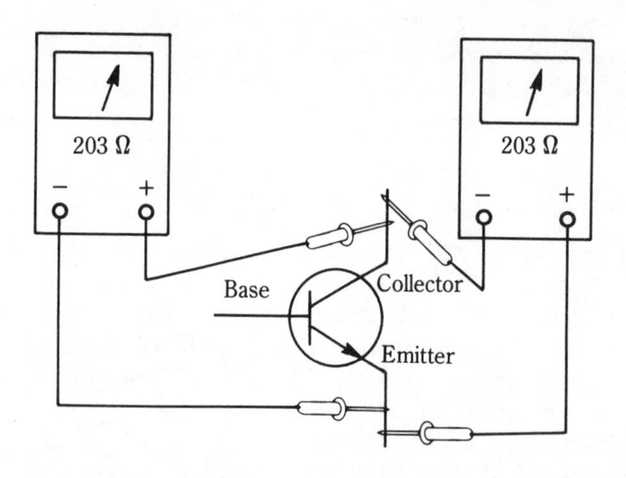

11-5 Actual resistance measurements between any two transistor terminals indicate the condition of a transistor. Very low resistance measurements in both directions point out a leaky transistor.

Besides providing very accurate voltage and resistance measurements, some DMM have a diode and transistor test capability that will quickly test all transistors in just a few minutes. Each transistor can be checked for an open, leaky, or normal condition right in the circuit. The digital multimeter (DMM) is a handy test instrument for checking a transistor or diode in or out of the circuit.

Rotate the function switch to the diode or transistor test position of the DMM. With npn type transistors, place the red probe (+) to the base terminal. Now place the black probe (−) to the collector terminal. No reading will be indicated for an open junction between the base and collector terminals. A very low resistance measurement may indicate a leaky transistor. Reverse the test leads and check for a low resistance reading in the other direction. Suspect a leaky transistor or diode when the meter indicates low resistance between the base and collector terminals. Likewise, place the black probe to the emitter terminal. A normal resistance measurement comparable to the emitter or collector terminal, indicates a good transistor (Fig. 11-6).

11-6 Check transistors on separate modules or after they are removed from the circuits to verify the in-circuit tests.

Now, check for leakage between any three terminal elements. Remove the transistor for out of the circuit tests, whenever a low resistance in two directions is found between any two terminals. Suspect a low-resistance component across the two suspected terminals if the transistor tests normal after it's removed from the circuit. Double check the resistance measurement with the DMM switched to the 2000-ohm range.

The pnp transistor can be checked in the very same manner. Place the black probe (−) to the base terminal. You may quickly check the polarity of any transistor with the base terminal probe in relation to the emitter and collector terminals. The transistor is a pnp type if you measure a low resistance with the black or negative probe of the DMM attached to the bast terminal of the transistor while the red probe is attached to the emitter or collector terminal. Likewise, the unknown transistor is an npn type if in order to get similar results you have to attach the red or positive probe of the DMM to the base terminal of the transistor (Fig. 11-7).

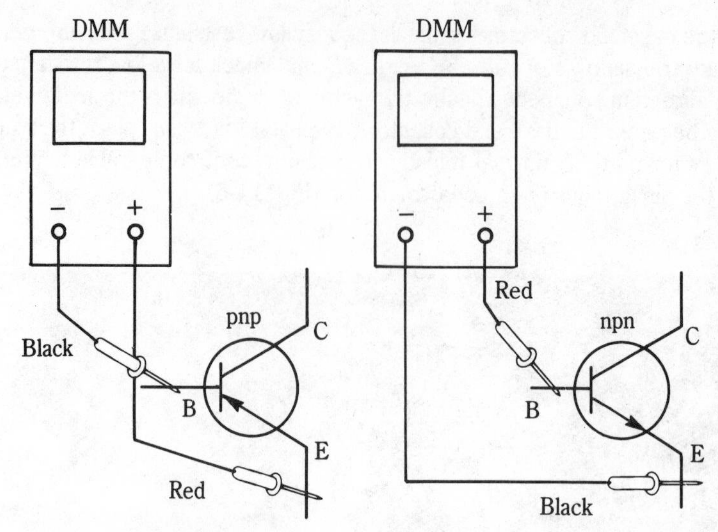

11-7 For npn-type transistors, place the red probe or positive lead to the base terminal of the transistor. The black probe should be placed at the base terminal to get similar measurements from a normal pnp transistor.

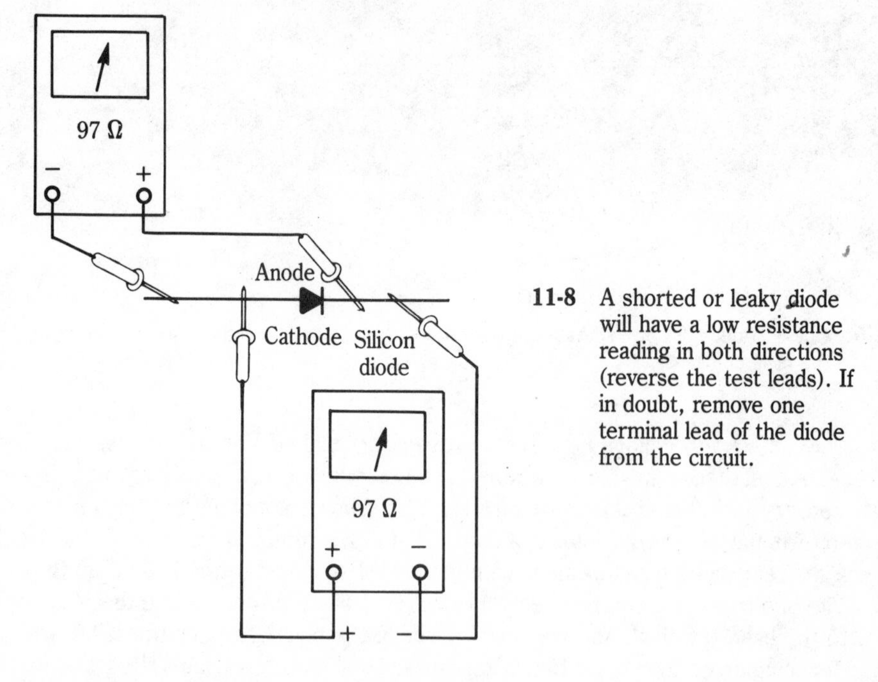

11-8 A shorted or leaky diode will have a low resistance reading in both directions (reverse the test leads). If in doubt, remove one terminal lead of the diode from the circuit.

The zener diode may be checked for open, leaky, or normal conditions with the diode scale of the DMM. For a quick normal test, place the red probe (+) to the positive terminal of the diode (anode) and the black probe (−) to the cathode lead. The normal diode will show a low resistance measurement. A shorted or leaky diode will have a low resistance measurement with reverse test leads (both directions) (Fig. 11-8). Double-check the resistance measurement by removing one terminal of the diode. The open diode will not show a reading in any direction. Most defective diodes have a shorted or leaky condition. The damper, boost, or high-voltage rectifiers may not give any indication on the diode test mode of the DMM.

Transistor voltage tests

The defective transistor may be located with accurate voltage measurements on each terminal. Accurate voltage readings may be taken with the VOM or VTVM. The low priced DMM is a very accurate voltage and resistance test instrument. First, take voltage and then resistance measurements.

Higher-than-normal voltage on the collector terminal may indicate the transistor is open. Always obtain a schematic diagram when taking voltage and resistance measurements. Check for no voltage or low voltage on the collector terminal. No voltage on the emitter terminal may also indicate the transistor is open (Fig. 11-9). Check the emitter bias resistor for correct resistance. An open emitter resistor may cause no voltage at the emitter and real high voltage at the collector terminal.

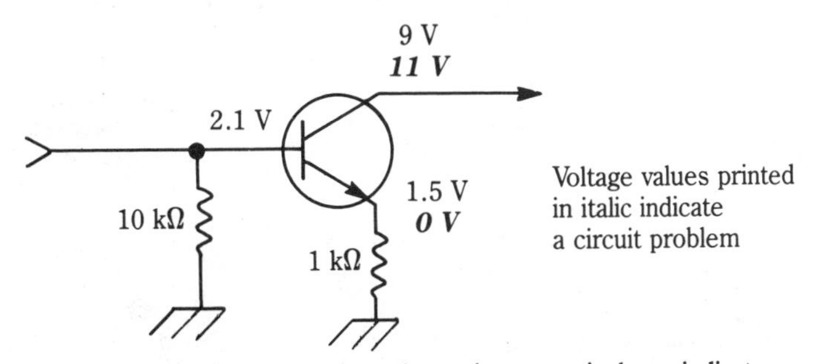

11-9 No voltage measured on the emitter terminal can indicate an open transistor, especially if the collector voltage measures higher than normal. Check for an open emitter bias resistor.

Low collector voltage may indicate a leaky transistor, increased collector resistance or improper supply voltage. Now check the emitter and base voltages. Suspect a leaky transistor when all three terminal voltages are about the same (Fig. 11-10). You might find some normal TV circuits where the voltages are quite close on all three terminals. A lower collector voltage with a higher emitter voltage may indicate leakage between the collector and the emitter elements. Make transistor in circuit tests or remove the transistor to test it out of circuit for leakage. Low collector voltage and a higher base voltage may indicate leakage between the collector and base terminals.

Low resistance measurements between any two elements may indicate a leaky transistor. Make sure outside components such as diodes, directly coupled transistors, coils and resistors are not making the low resistance path (Fig. 11-11). Measure the base and emitter resistors for correct resistance. Disconnect one of the two transistor terminals when low resistance or leakage is indicated. Desolder with a solder-wick or a special solder remover. Now, take another resistance measurement. Suspect a leaky transistor when a low resistance measurement is still present.

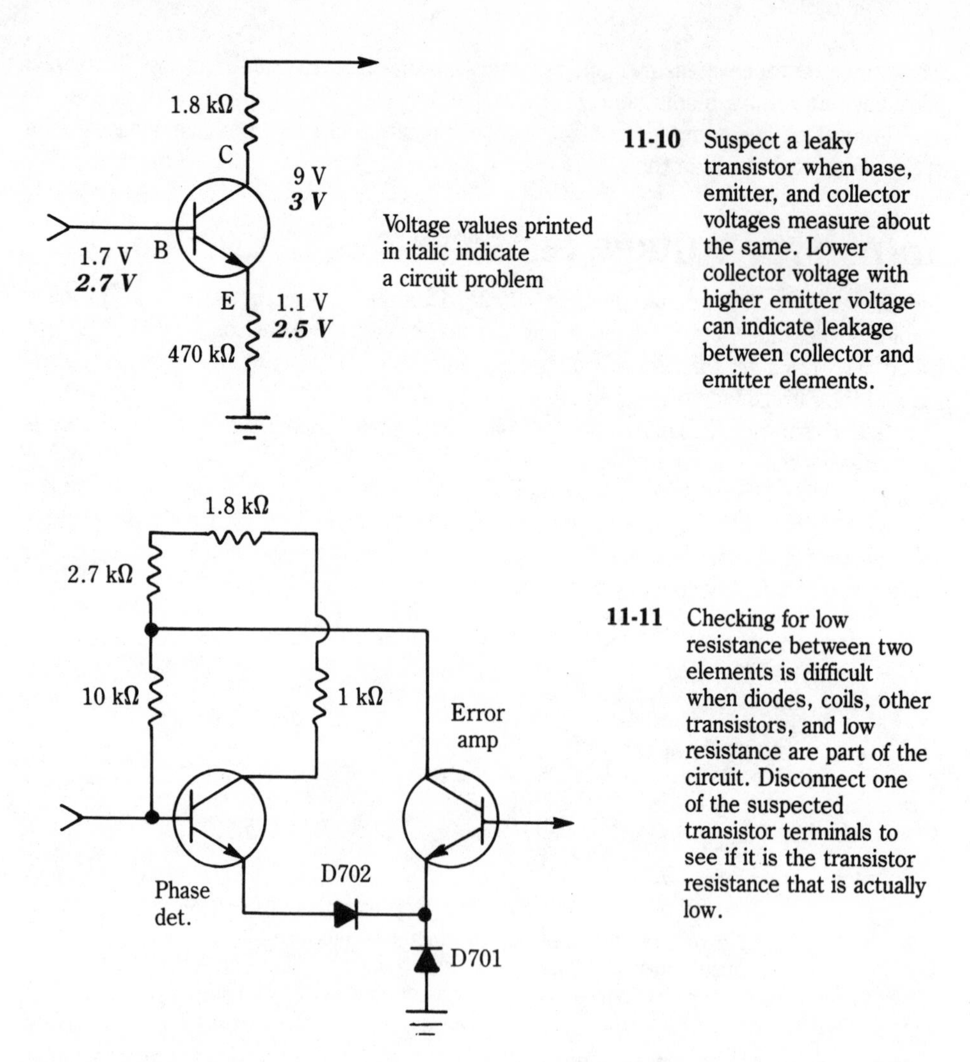

11-10 Suspect a leaky transistor when base, emitter, and collector voltages measure about the same. Lower collector voltage with higher emitter voltage can indicate leakage between collector and emitter elements.

11-11 Checking for low resistance between two elements is difficult when diodes, coils, other transistors, and low resistance are part of the circuit. Disconnect one of the suspected transistor terminals to see if it is the transistor resistance that is actually low.

Critical resistance and voltage measurements may be too much for the beginner, but may be easy for the intermediate or advanced person. The beginner can quickly learn electronics by taking voltage and resistance measurements of certain TV circuits. Voltage measurement on the high-voltage circuits should only be made by the advanced or professional electronic technician.

Checking integrated circuits

The IC component plays a large role in the operation of TV sets today. In the future the entire solid-state TV chassis might only contain IC chips with only a few discrete transistors. With digitally-operated circuits just around the corner, the TV set will be completely controlled by integrated circuits. The IC is a device whose components, resistors, capacitors, diodes, transistors, and connecting wires are all made by processing a piece (chip) of semiconductor material in such a manner as to produce a complex circuit within a single component (Fig. 11-12).

To locate a defective IC, first find out which circuit is malfunctioning, then compare the malfunction with the purpose of each IC. Sometimes more than one integrated circuit is involved. Most IC component failures in the TV chassis can be located by tracing the signal in and out of the IC package. Critical voltage and resistance measurements measured at the IC pins indicate a defective IC circuit. Always have the TV receiver schematic handy when trying to locate a defective IC.

The beginner might be able to locate a defective sound, AGC, or sync IC with voltage tests or with an IC replacement. Spray the suspected audio IC with a coolant when the sound is distorted, intermittent, or dead. Replace the defective IC if the sound appears normal after it is sprayed. Take voltage and resistance measurements at the IC terminals to verify whether or not the IC is defective. The AGC and sync ICs can be serviced in the same manner.

The more advanced person may be able to service the IF, horizontal and vertical, and video IC component by injecting a signal in and observing the output with an oscilloscope. No vertical output pulse from the vertical oscillator or drive section can be caused by a defective IC component. Improper horizontal oscillator signals can be caused by the same integrated circuit. Before removing the suspected IC, take accurate voltage and resistance measurement with a DMM. In many cases a leaky capacitor or a change in the resistance connected to an IC terminal can cause the IC circuit to not function properly.

11-12 IC components come in 8 and over 80 terminal pins in the regular IC and microprocessor. The large IC circuits are power output types.

Typical sound IC service procedure

The dead audio section should be checked by connected another PM (permanent magnet) speaker. Clip the new speaker across the TV speaker terminals. Turn the volume up halfway. Replace the defective speaker if sound is heard in the good speaker. If the sound is still dead, clip a 100 μF 25-volt electrolytic capacitor in series with the subspeaker (Fig. 11-13). In this circuit, attach the clip lead at pin 2 of IC 201, if the sound returns, replace the open speaker capacitor. Many electrolytic coupling capacitors become open after a few years of operation. These two sound tests can easily be made by the beginner in electronics.

For weaker or distorted sound conditions, spray the IC with coolant. Replace the defective IC if the sound returns. If not, the most critical voltage measurements are on pins 1, 2, 4, 10, and 11. Low voltage measurements on pins 1 and 2 may indicate a leaky IC as indicated by the small voltage in Fig. 11-13. Check the supply voltage at the 27-ohm supply resistor (26.1-V). Low supply voltage may indicate problems within the low-voltage power supply source or a leaky component.

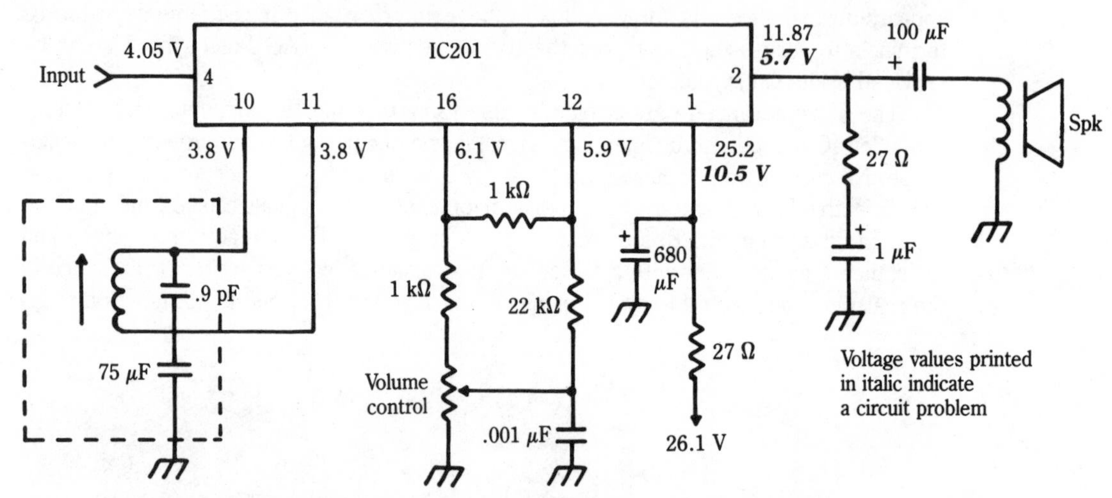

11-13 Check for improper voltage potentials on the IC terminals that indicate a defective IC. Low voltage on pins 1 and 2 indicates a leaky IC, as indicated with the voltages in italic.

A distorted sound IC stage may be caused by a defective speaker coupling capacitor, IC and improper alignment of L201. If after checking the speaker coupling capacitor and making the voltage test the distortion remains, take voltage measurements on the IC terminals. Touch up the sound coil (L201) if all voltage measurements are fairly normal. Select the proper alignment tool. If not, you may destroy the small metallic adjustment core. Turn the sound halfway up. Slightly rotate the core clockwise. Notice if the distortion clears up. If not, rotate the core counter-clockwise a full turn. Slowly rotate the core until the sound is best on all stations. It is usually only a slight adjustment. Replace the defective IC if voltages are normal and a quick sound adjustment does not clear up the distortion. These sound tests may be made by the beginner or intermediate person.

Signal tracing for sound distortion may be made with the use of an external audio amp. These tests should be made by the intermediate or more advanced person in electronics. Using the external audio amp, check the input sound signal at pin 15. This signal may be quite weak, so turn up the volume of the external amp. Now, check the sound at the output of IC 201 (pin 2). Suspect a defective IC when the sound is normal at pin 15 and distorted at terminal 2. All TV sound circuits may be checked for weak and distorted sound with the use of an external audio amp.

Removing and replacing the IC

After determining the IC is defective, install a new one. Locate pin 1 by the identification notch on one end of the IC. It is the first pin to the left of the notch when viewed from the top side of the IC (Fig. 11-14). Mark an X where pin 1 is located upon the chassis. Some manufacturers have the IC terminal numbers printed on the bottom side of the PC board. Desolder all pin terminals at the bottom side of the PC board.

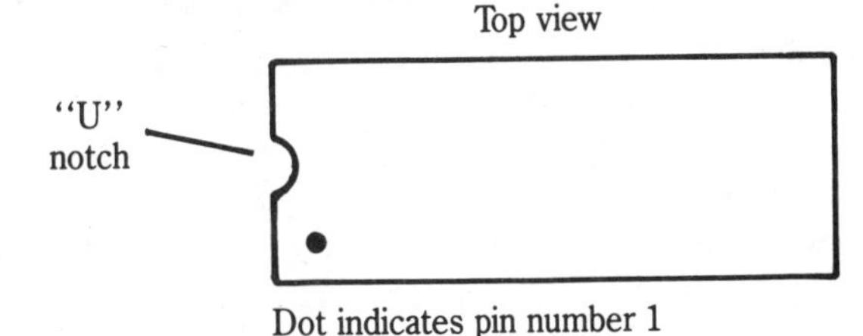

11-14 Locate pin 1 by the identification notch as viewed from the top of the IC. In some chassis, the IC pin numbers are marked on the bottom side of the PC board.

Select a piece of copper solder wick which is impregnated with soldering paste to aid you in removing the solder from around each terminal. This handy solder remover comes in a small roll, and is found at most radio and TV stores. It is used to lift solder off transistors, capacitors, resistors, and IC terminals. Hold the solder wick flat along side a row of IC terminals. Likewise, go to the other side and remove the excess solder. Proceed by removing solder from both rows of terminals. Flick each terminal with the blade of a pocket knife to make sure the connection is loose (Fig. 11-15).

11-15 Regular IC pins go through the PC board, but surface-mounted components are soldered on top of the PC wiring.

Lift the defective IC with a pocket knife or the blade of a small screwdriver. Check where pin 1 goes and insert the new IC into the PC board holes. Make sure all pins poke through each hole on the printed wiring side. Double check each one. These small pins have a tendency to bend over and not fit into their respective holes. Solder each terminal with a low-wattage soldering iron or a battery-cordless soldering iron. Place the soldering-iron tip to one side of the wiring and pin connection. Apply solder on the opposite side.

Do not leave the iron in contact with the connection too long to prevent overheating the IC component. Double check the soldering of each connection. Sometimes a magnifying glass is handy to check for good soldered connections. The solder may adhere to one side of the pin and not cover the whole connection. Retouch the terminals that appear to have poorly soldered joints. Run the back edge of a pocket knife blade between each pin connection. Besides removing excess rosin, the blade will clean out any solder bridges between the IC terminals that could damage the new IC replacement. If two connections are accidentally joined together, remove the excess solder with a piece of solder wick material. After replacing a few IC components, the beginner will soon acquire the experience to move into the intermediate class.

Solid-state trouble symptoms

This section is similar in content to chapter 10. Some of the headings indicating TV symptoms were discussed earlier. Nonetheless this is not a repeat of any earlier material. This section will deal with several common or typical symptoms that might cause problems in a solid-state TV receiver. Scan through the list of troubles listed and locate the desired symptoms; then follow the recommended action (Fig. 11-16).

11-16 Try to isolate the symptoms and section of TV before tearing into the chassis. Then take voltage and resistance measurements on the suspected component.

Dead set

First, check the circuit breaker or fuse for a no sound/no picture/no raster symptom. Replace a blown fuse with the exact current rating specified. The circuit breaker can be quickly checked by clipping a lead across its two terminals. Sometimes a fuse will blow or a circuit breaker open because of a brief overload in a transistor. Suspect a shorted or leaky silicon diode when a new fuse opens up (Fig. 11-17).

11-17 Try to locate the fuse on the chassis. The fuse might be soldered into the PC board. Sometimes the TV set has both a line fuse and a B+ fuse.

You may find four silicon diodes forming a bridge rectifier circuit, two individual diodes or one single diode in the power supply (Fig. 11-18). Each diode may be checked with the ohmmeter scale of a VOM or DMM. If you suspect any diode unsolder one of its leads and check it again. When lightning strikes a TV set, two or more diodes often become defective along with an open fuse and voltage isolation resistor.

If after replacing the open fuse the set remains dead, check the isolation resistor (R701) for an open condition. Check each diode for leakage with the ac power cord removed. Then plug in the power cord and measure the dc voltage at the output terminals of the power supply. Suspect problems within the horizontal output circuit if the low voltage power supply has a higher than normal low voltage.

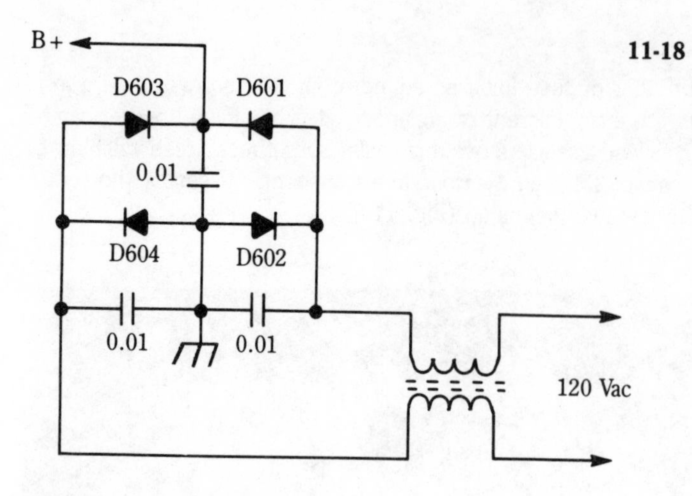

11-18 A bridge rectifier contains four separate diodes. These diodes can be discrete components mounted on the board or all four diodes can be molded into one component.

Keeps blowing the fuse

Check each silicon diode if the fuse keeps blowing. The bridge rectifier can be four silicon diodes incapsulated into one component. Check each diode for leakage with the VOM or DMM. You can save fuses and possible damage to other components by inserting a 100-watt light bulb in series with the load. Just clip the lamp across the two fuse terminals as illustrated in Fig. 11-19. The short is still present when the light has full brightness.

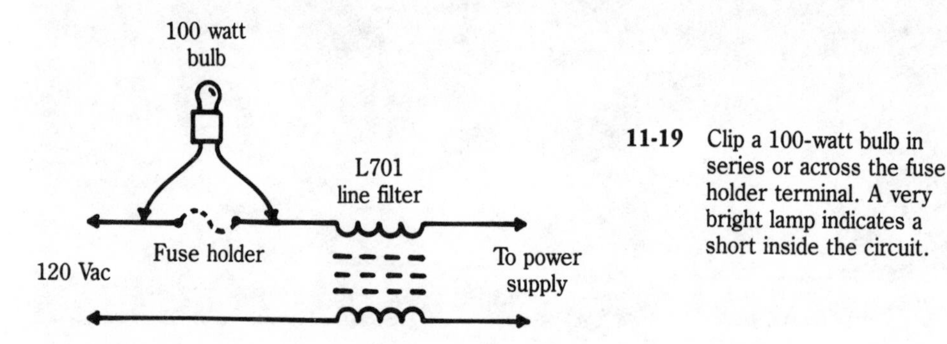

11-19 Clip a 100-watt bulb in series or across the fuse holder terminal. A very bright lamp indicates a short inside the circuit.

Often a leaky horizontal output transistor will cause most of the fuses in the TV chassis to open. Remove the output transistor or disconnect its collector lead. Then plug in the power cord and observe the light. If the light goes out, replace the leaky horizontal output transistor. If the light remains bright, suspect a defective flyback transformer. Call in a professional electronic technician for a transformer replacement.

A shorted or leaky filter capacitor may cause the fuse to open. If the fuse blows with the horizontal output transistor removed, suspect problems within the low voltage power supply. Either the voltage regulation transistor is leaky or the filter capacitor is defective. Remove the positive lead of the first filter capacitor and run the lamp brightness test (Fig. 11-20).

To check for a blown ac fuse caused by a shorted automatic degaussing coil (ADG), remove the voltage isolation resistor or the input lead of the silicon rectifier. This will isolate the ADG circuit and prevent it from opening the fuse or keeping the 100-watt bulb

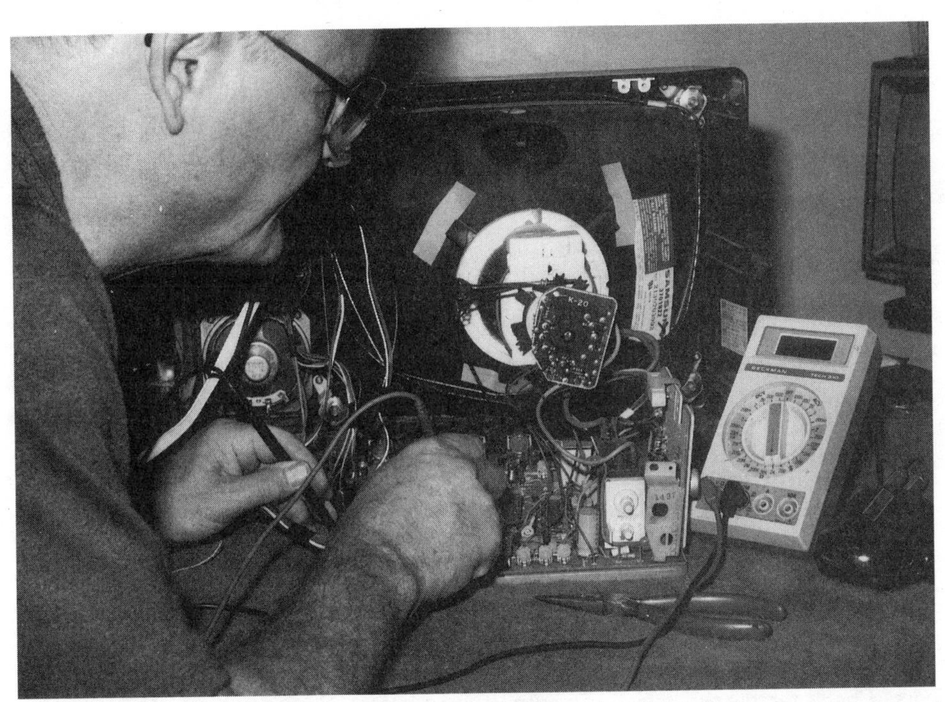

11-20 Check the fuse and diode continuity with the DMM. A good fuse will show continuity and a good diode will show a reading in only one direction.

real bright. Automatic degaussing coils will sometimes short against the metal shield of the picture tube assembly. Simply unplug the ADG coil assembly if it is not wired directly into the circuit. The light will go out if the coil is not shorted.

Circuit breaker problems

If the circuit breaker keeps opening up after resetting, suspect an overloaded condition or a defective circuit breaker. Clip another circuit breaker with the same amperage rating across the old breaker terminals. Reset the substitute circuit breaker several times. It will not hold if a true overloaded condition exists. Do not continue to push in on the circuit breaker if the unit opens immediately when reset.

Another method is to clip the 100-watt bulb test across the open circuit breaker. If the light remains bright, check for an overloaded condition in the power supply or horizontal output circuit. Check the low voltage power supply for the probable cause if the circuit breaker is in the ac input circuit (Fig. 11-21). If the circuit breaker is in the horizontal output circuit, check for a leaky horizontal output transistor or damper diode.

Hum bars

Poor filtering of the low-voltage power supply can cause vertical roll of two or more black hum bars. Improper adjustment of the B+ control can also cause hum bars in the raster. Try and eliminate the black bars by adjusting the B+ control. Dried-up filter capacitors

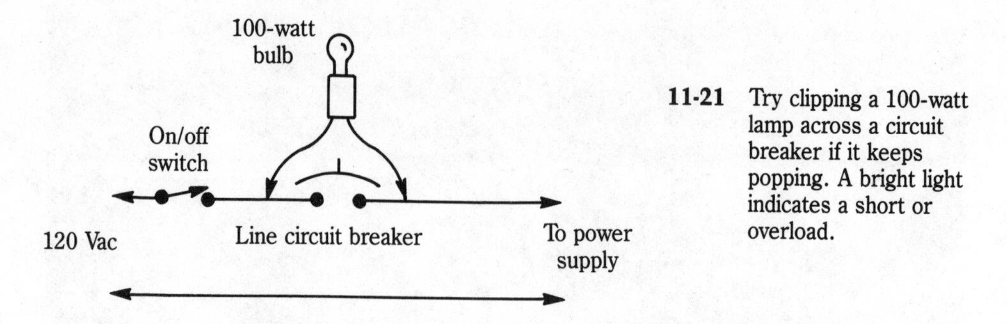

11-21 Try clipping a 100-watt lamp across a circuit breaker if it keeps popping. A bright light indicates a short or overload.

are a source of hum bars. To check, clip another equivalent value filter capacitor across the suspected one. Always, turn off the power and observe the correct capacitor polarity. Take the TV set to a professional when these two tests do not remove the black bars from the raster.

Regulator hum bars

When hum bars are found in the picture and cannot be eliminated with horizontal control or B+ control (found at rear of chassis), suspect trouble in the low-voltage power supply regulator circuits. Take voltage measurements on the regulator transistor and compare with the schematic (Fig. 11-22). Often, a leaky regulator transistor will produce hum bars. Check for burned resistors around the regulator. Shut down the TV and shunt small electroylytic capacitors with the same or higher values.

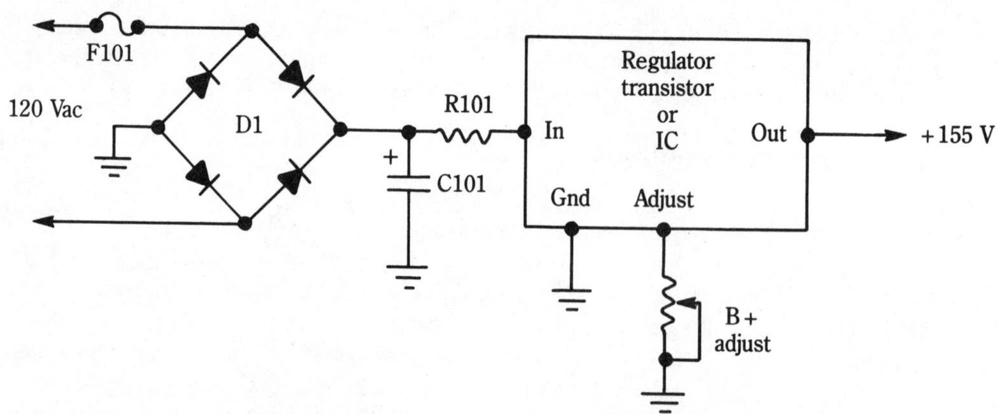

11-22 Hum bars in the picture can be caused by a defective low-voltage regulator. When the B+ voltage cannot be changed, expect a defective regulator transistor or IC.

No sound and no raster

A no sound, no raster problem may be caused by a defective low-voltage power supply or horizontal output circuit. The present-day solid-state chassis will not have sound until the horizontal circuits are operating. Measure for the correct dc voltage between the collector terminal of the horizontal output transistor and chassis ground. Higher-than-normal collector voltage can indicate a defective HV regulator or an open horizontal output transistor. High voltage shut-down can occur with too high of a voltage at the collector terminal. Low collector voltage can be caused by a leaky output transistor or a defective low voltage power supply.

A quick leakage test can be made between collector and chassis ground. Real low resistance may indicate a leaky transistor or damper diode. A low resistance reading between base and emitter terminals is normal for an in-circuit test since the secondary winding of the driving transformer and test it out of the circuit. Remove the horizontal output transistor and test it out of the circuit. Check for a leaky damper diode if the horizontal output transistor are normal. Sometimes the output transistor is open between base and emitter terminals. The transistor must be removed for this test. If in doubt, replace it with a new output transistor.

Chassis shut-down problems

Just about any component in the low-voltage power supply or horizontal circuit may cause chassis shut-down. Sometimes the chassis will not start up with a defective component. It's best to isolate the low-voltage power supply from the horizontal output circuits for chassis shut-down problems. Remove the B+ fuse going to the horizontal circuits. If a fuse is not used, remove the horizontal output transistor from the chassis.

Now, if the low voltage comes up to normal or goes higher, you may assume the low-voltage circuits are normal. Check the horizontal circuits for defective components. The electronic technician may check the drive voltage with the scope at the collector of driver transistor and base of horizontal output transistor. Check the horizontal output transistor with a transistor tester or with the DMM diode tests.

If the chassis shuts down with a horizontal white line before shut down, suspect the vertical circuits. Most problems caused in the vertical circuits are caused by the vertical output transistors. Be careful when checking them in the circuit. Sometimes they will pop back to normal and test good. Simply remove the vertical transistor that receives the operating voltage from the low-voltage power supply. When the chassis comes up with a horizontal white line and does not shut down at once, check the vertical output circuits and transistors.

With integrated IC vertical circuits, remove the V_{CC} operating voltage from that pin. Unsolder the V_{CC} pin or cut the foil of PC wiring feeding the supply voltage pin. Likewise, if the chassis does not shut down, suspect vertical output IC and circuits. Do not forget to bridge cut PC wiring with bare hookup wire.

Checking horizontal circuits

Since the horizontal output transistor fails more than any other transistor in the solid-state chassis, replacement of this transistor may solve the no sound, no raster symptom. The transistor can be removed and a new one temporarily installed or the original can be tested out of the circuit. Before the body of the transistor can be removed, it is necessary to remove the two screws that secure the base to the chassis and to desolder the base and emitter terminals from the circuit. Some horizontal output transistors are plugged into a separate transistor socket and can be removed without desoldering (Fig. 11-23).

11-23 Replace the horizontal output transistor after applying silicone grease and insulator under the transistor. Do not tighten the screws too tight so the transistor "bites" into the chassis, blowing the fuse.

Check the transistor with a transistor tester, or take resistance measurement with a VOM or a DMM. Low-resistance measurements in one direction are normal in a silicon horizontal output transistor (Fig. 11-24). A low resistance reading in both directions between any two elements indicates a leaky transistor. A transistor can be open between any two elements, resulting in a no raster, no sound symptom with higher than normal voltage on the collector.

Replacing horizontal output transistors When the horizontal output transistor is found leaky or suspected of breaking down with overload, replace it with a new one. Most power output transistors mounted on the metal chassis have a piece of insulation between transistor and chassis (Fig. 11-25). Remember, the horizontal transistor may have the highest voltage supplied to it besides the picture tube circuits. The metal body of the output transistor is the collector terminal with low voltage applied.

If the horizontal output transistor is mounted separately on a heatsink like in Fig. 11-1, no insulator is found. The insulated PC board keeps the metal heatsink and transistor above ground. Sometimes the heatsink will come off when the two transistor mounting screws are removed.

Place silicone grease on each side of the insulated piece and press it in place. Now, place the horizontal output transistor in the correct holes. You can reverse this transistor except the mounting holes will not match up. Remove it and turn 180 degrees and insert again. Make sure the insulator ends are not broken off where the screws go into it because the transistor body may short against the metal chassis. Lightly tighten up the two metal

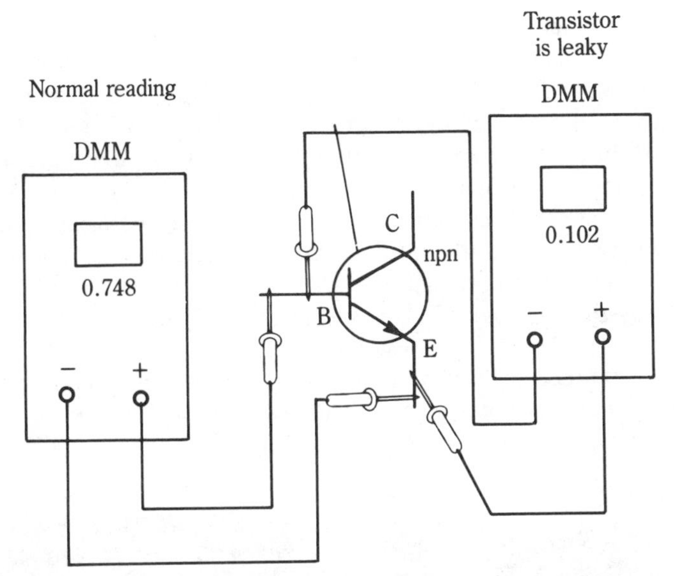

11-24 A low resistance measurement between any two elements could indicate a leaky transistor. Low resistance measurements in one direction is normal for a silicon horizontal output transistor.

11-25 Make sure the insulator is between the transistor and chassis before mounting. A coat of silicone grease on each side of the insulator will help keep the horizontal transistor from overheating.

screws. Do not overtighten because the transistor flange may bite into the chassis. Double-check with the low resistance range of ohmmeter between metal body of transistor and metal chassis. When replacing horizontal output transistors mounted on separate heatsinks, always place a thin layer of silicone grease on the bottom of transistor.

Horizontal lines Horizontal lines can be the result of improper adjustment of the horizontal hold control, horizontal circuit drifting, or a defective horizontal oscillator circuit. Check the circuit and chassis layout for the existence of a horizontal hold control or oscillator coil. In many of the lower priced models, the horizontal oscillator coil is also the horizontal hold control. Improper adjustment of this control will cause horizontal lines running diagonally to the left or right on the screen without a viewable picture (Fig. 11-26). Turn the control until the lines get wider apart. If the core of the coil begins to bind and cannot be turned any farther, remove the tab or stop around the plastic adjustment shaft. Now, rotate the core until the picture flops in. Don't go too far or the picture will form into horizontal lines of the opposite direction.

11-26 Improper adjustment of the horizontal hold control could cause horizontal lines in the picture.

Suspect a separate horizontal oscillator coil to be the cause of drifting when a regular carbon control is used as a horizontal hold control. Locate the horizontal oscillator coil which is usually located in a shielded container. The core could have a screwdriver adjustment or it may require a hexagon plastic tool to adjust it. If the core will not turn either way, heat the core by heating a metal hexagon tool that has been placed into the core. Use a soldering iron to heat the tool. Heat from a hair dryer or a regular heating tool can also help loosen the metal core area up. In many cases, the core is locked in place with excessive wax. Replacement of defective horizontal transistors and ICs should be left up to the professional servicer.

Insufficient width Look for a horizontal size or width control in the schematic TV parts layout. Adjust for a normal size picture. Improper adjustment of the B+ control can cause the sides to pull in. Measure the B+ voltage at the collector terminal of the horizontal output transistor. A defective horizontal output transistor can cause insufficient width.

No raster but sound OK

Suspect a defective picture tube or circuits when no raster or picture with normal sound and high voltage. If you have no high-voltage voltmeter, just listen for the yoke to expand when high voltage comes up. Hold the back of your hand near the screen of the picture tube and the hair should stand up, indicating high voltage is present.

Check the heater or filaments of the CRT for the light of each gun at the rear of the tube. No light may indicate an open filament of CRT or no heater voltage. Shut down the chassis and remove the socket of the picture tube. Check the schematic for correct picture tube heater pins. Any resistance measurement between 1 and 50 ohms may indicate the filament or heater is normal. If no measurement, the heaters are open. A new CRT must be installed.

When continuity of the picture tube filament is good, suspect a poor socket or poorly soldered connections. Inspect the socket for burned or poorly soldered connections from heater pins and socket to CRT board (Fig. 11-27). Take a continuity measurement across the two heater pins. No measurement indicates the heater winding or wire

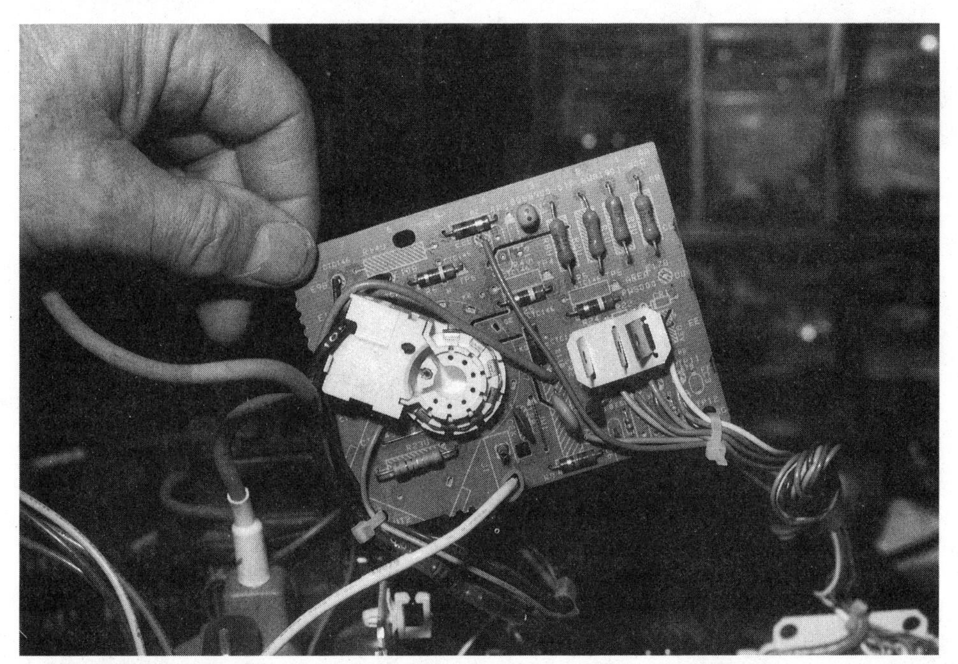

11-27 Inspect the CRT socket for poorly soldered connections or burned heater pins when the filament of the picture tube will not light up.

connections are open. Closely inspect the wiring and soldered connections. Simply soldering up the heater tube pin socket wires and chassis connecting wires may solve the open heater winding (Fig. 11-28).

Remember, in the tube chassis, the heater voltage comes from a power transformer winding. In today's solid-state chassis, the heater voltage is taken from a winding on the horizontal output transformer (flyback). Do not attempt to measure this ac voltage from the flyback transformer as you can with the tube-powered transformer winding.

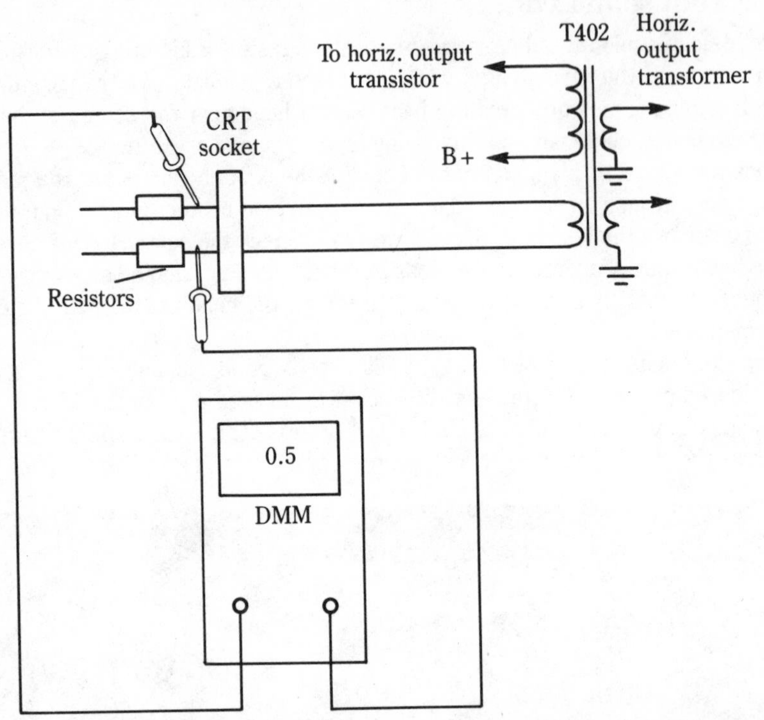

11-28 Make sure the heater or filament winding is not open by sticking the ends of resistors into each heater pin of the CRT socket and check for a low resistance across the filament pins with VOM or DMM.

Beware of high voltage

Be extremely careful when working around the horizontal output transformer, high-voltage lead or anode connections on the picture tube. You may receive a terrible burn or shock if you place your fingers or hand on the high-voltage connection (Fig. 11-29). The electronic technician always respects the high-voltage source of the TV chassis when the set is operating. If, for any reason, the chassis must be pulled out of the set for tests, the anode lead and socket must be discharged from anode to the black aqueduct coating of the picture tube.

Make sure the aqueduct coating of the CRT is properly grounded to the chassis. If not, arcing will develop and you can hear it in the TV's audio and see white dashes on the

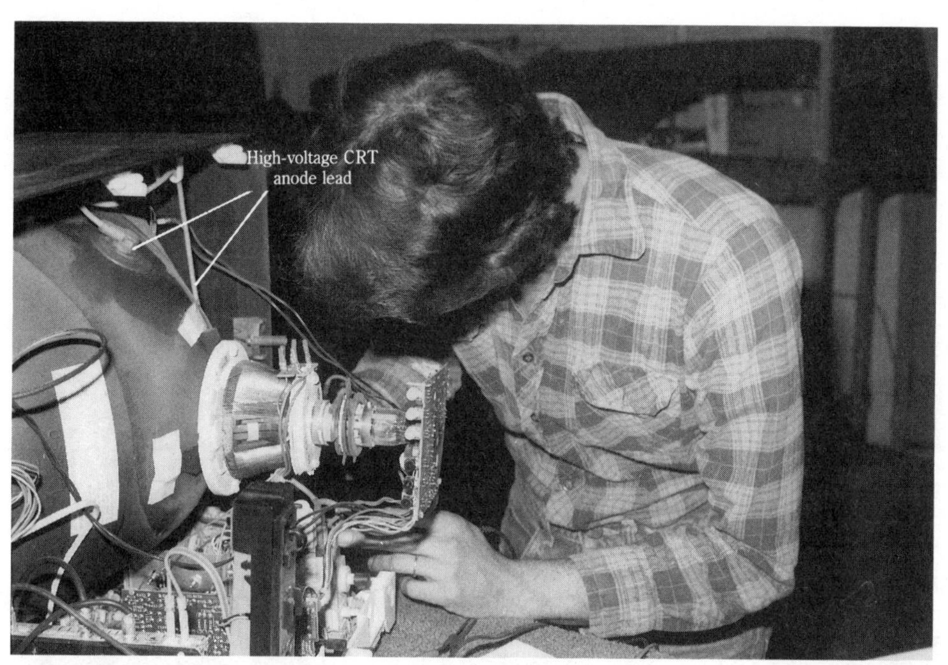

11-29 Beware of high voltage when the chassis is operating. Keep away from the large anode lead or socket of the picture tube.

screen. You may receive a shock when touching the picture tube with an ungrounded tube.

No raster but normal high voltage

A no-raster symptom with normal HV can be video or picture tube problems. Check for a filament light in the rear end of the picture tube. No visible light may indicate an open filament or improper voltage at the heater terminals. Pull the ac cord. Remove the picture tube socket. Measure for continuity across the heater terminals. No resistance indicates the heater is open internally.

Replacing the picture tube is the only answer. Check the heater wiring from picture tube socket to the PC board for continiuity. In older TV receivers,the picture tube heater received their voltage from a power transformer winding. Today, the heater voltage may be derived from a voltage in the flyback circuitry. Use the low ohm scale of a VOM to trace each heater wire back to its source. A broken PC wiring connection, tie-in wire, or an open isolation resistor may be keeping the heater from lighting up. Since video problems are very difficult to locate, leave them up to the professional TV technician.

Vertical circuits

Improper adjustment of the vertical height and linearity controls may cause the picture to roll. Adjust both controls with the vertical hold control to stop the picture from rolling. These controls are on the rear of newer black-and-white TVs and older color chassis. If the rolling does not lock in on the newer type of TVs without any vertical controls, check the vertical transistors and vertical sync circuits (Fig. 11-30). When the picture is rolling both horizontally and vertically, suspect the sync circuits. In the latest TV chassis, the sync circuits are usually found within the deflection IC.

Only a white horizontal line No vertical sweep may produce a single white horizontal line. Adjust the vertical hold, vertical size, and linearity controls to see if the sweep will return. You may find only one vertical adjustment control in a solid-state chassis. Vertical circuits are very tricky to service. Replacing both vertical output transistors can solve many vertical sweep problems, measure the voltage on the output transistor. An improper voltage can be produced by an open resistor in the low-voltage power supply.

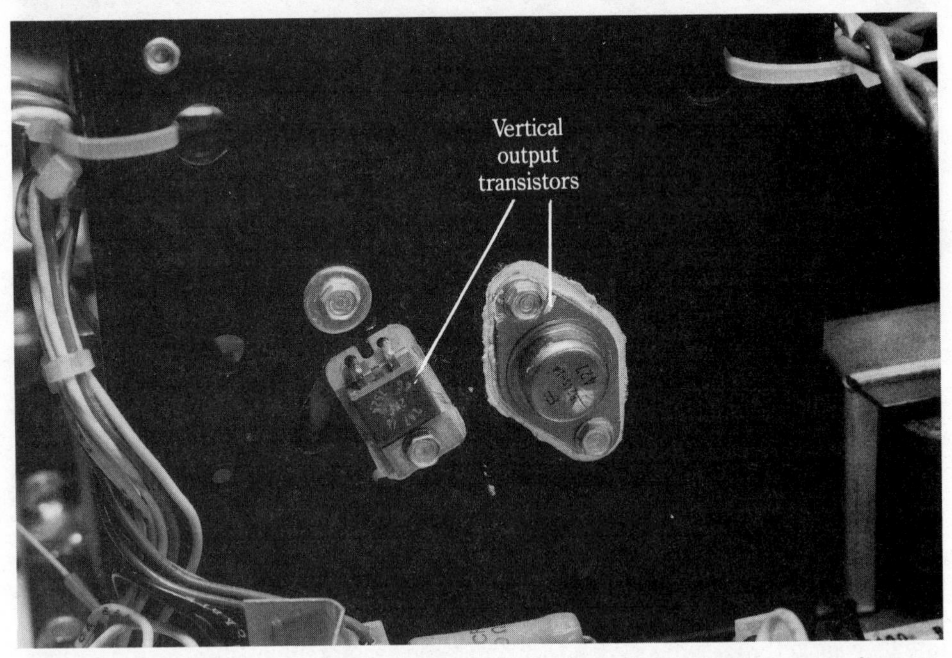

11-30 Try to adjust the vertical hold control on sets that have them to stop the picture from rolling. Suspect defective vertical output transistors or sync circuits when the control has no effect on the picture.

Vertical foldover Excessive vertical foldover cannot be adjusted up with the vertical height and linearity controls. Often, leaky or open vertical output transistors or a change in bias resistors may cause vertical foldover (Fig. 11-31). Improper voltages found upon the bottom vertical transistor may cause foldover. Shunt each electrolytic capacitor in the vertical circuits with a like value to see if foldover is caused by a dried-up capacitor. Sometimes, both of the vertical transistors must be replaced to cure excessive vertical foldover.

No sound but normal picture

Sound problems are fairly easy to service in a color TV chassis. The most troublesome components are the speaker, output transistors, ICs, speaker coupling capacitors, and

11-31 Excessive vertical foldover at the bottom of the picture tube could be caused by a defective bottom vertical output transistor or burned bias resistors.

bias resistors. A low hum in the speaker may indicate the speaker is normal with a defect in the sound circuit. No sound or hum in the speaker can indicate an open speaker or coupling capacitor.

To check, touch the center terminal of the volume control with your finger and listen for hum in the speaker. Place a finger on the base terminal of the AF or driver amp in a transistorized sound circuit and check for a loud hum when the volume control is turned wide open. Clip a substitute speaker across the speaker terminals. If no sound is heard from the speaker, take voltage and resistance measurement of the output transistor and IC circuit.

Distorted sound

A leaky transistor, burned bias resistors, leaky IC, damaged speaker, and improper sound alignment produce more distortion problems than all other components in the sound stages. Disconnect one speaker lead and temporarily clip another speaker into the circuit to verify whether the distortion is caused by a defective distorted speaker. Out of circuit transistor testing will provide more comprehensive results while in circuit testing may pinpoint any serious defects. Any marginal transistor operation could go undetected. Always check for a burned bias resistor after locating a leaky transistor.

A low and distorted sound stage can be signal traced with the aid of an external audio amplifier. This signal tracing method is especially helpful in checking input and output terminals of an IC sound circuit for distortion. Check for distortion at the input terminal of the IC. The signal at this point can be very weak. If the input signal is free of distortion, check the output terminal. Take voltage and resistance measurements on all the IC terminals. Replace the suspected IC if in doubt.

Chassis or HV shut-down

Component failure in low voltage power supply and horizontal circuits can cause HV or chassis shut-down. Shut-down circuits are circuits used in recent color TV chassis to prevent excessive radiation that otherwise could occur from excessive high voltage. Check the fuse and circuit breaker for the chassis shut-down circuit. An overloaded condition in the low voltage power supply or horizontal output circuits can cause chassis shut-down (Fig. 11-32).

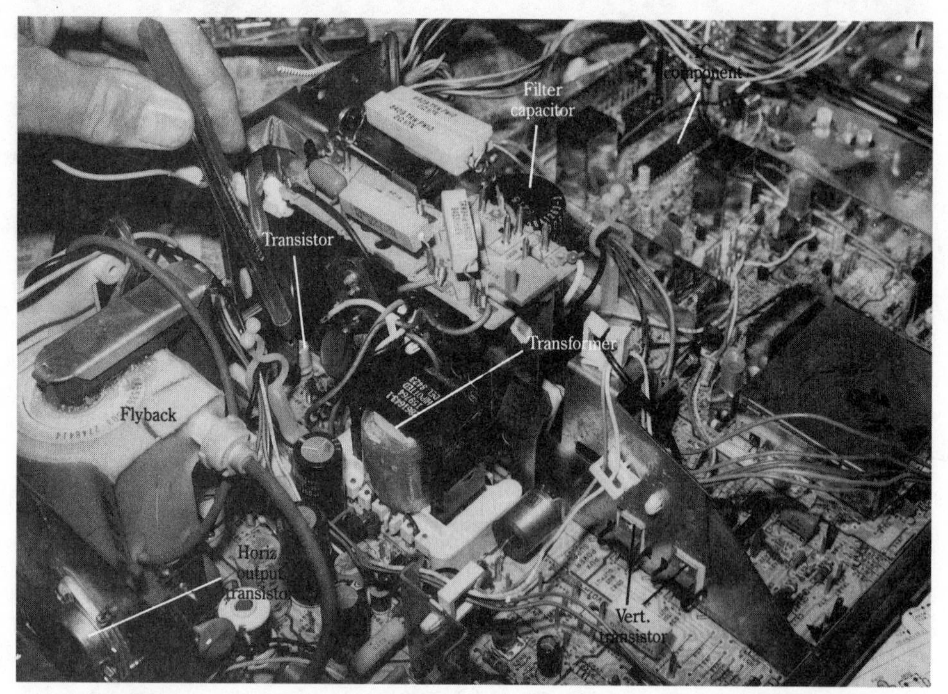

11-32 There are many components on the TV chassis that can overload and cause chassis shut-down.

High voltage shut-down may be caused from a higher than normal power line voltage, improper B+ from the power supply, or malfunction of components in the high voltage regulator and shut-down circuits. Locate the B+ or high voltage control located in the low voltage power circuits (Fig. 11-33). Readjustment of the B+ can sometimes solve a high voltage shut-down problem. Measure the voltage at the collector terminal of the horizontal output transistor and compare the result with the specified correct voltage. Reduce the voltage by adjusting the B+ control if the collector voltage measured slightly higher. Call in the professional technician when the measured voltage is real high compared to the schematic.

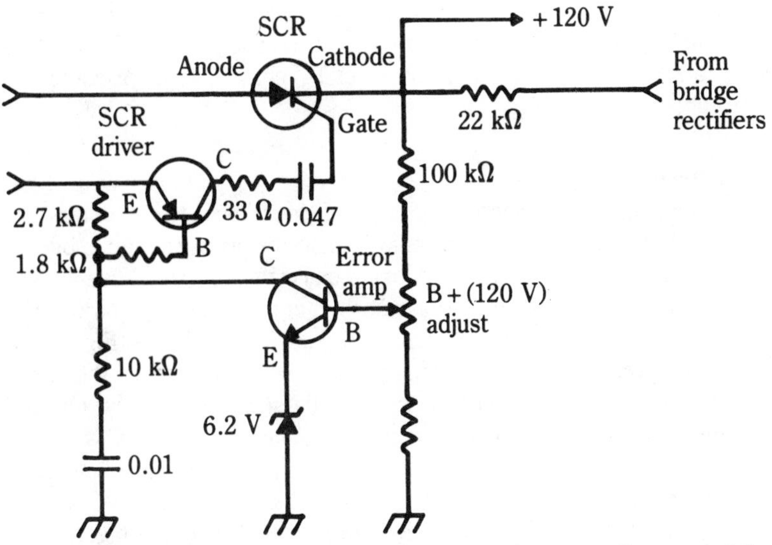

11-33 High-voltage shutdown can be caused by an improper adjustment of the B+ or high-voltage control. Lowering the B+ power supply voltage will lower the high voltage applied to the picture tube.

Only horizontal lines

When the horizontal lines cannot be eliminated with the horizontal oscillator control, suspect high-voltage shut-down. Measure the high voltage at the CRT anode socket with a high-voltage meter. Do not attempt to measure picture tube anode voltage with a regular VOM or DMM. This voltage may measure from 7.5 kV up to 31 kV voltage. You will also burn up the meter plus get a terrible shock from the high voltage.

In the earlier TV chassis when higher voltages were placed upon the picture tube due to increase in size and greater brightness, the HV shut-down circuit was designed to indicate higher voltage or shut down the TV chassis (Fig. 11-34). The X-ray protection

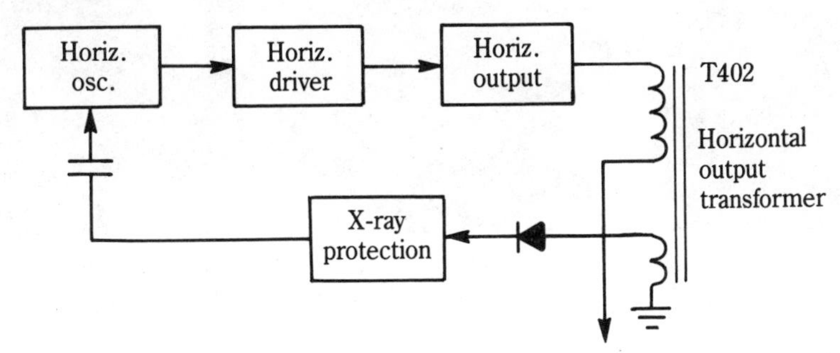

11-34 In early TV chassis, when horizontal lines cannot be adjusted out of the picture, suspect high-voltage shutdown. The x-ray or HV shut-down circuit disables the horizontal oscillator transistor.

circuit was placed in today's chassis to shut-down the chassis by disabling the horizontal oscillator or driver stage, when excessively high voltage is present. Remember, a defective X-ray protection circuit may shut down the TV chassis with normal high voltage at the picture tube.

Discharging the picture tube

The picture tube must be discharged before removing the chassis or in replacement of a defective tube. For tube chassis, electronic technicians discharge the anode connection to chassis ground with screwdrivers. Most technicians use two large or long screwdrivers. Place one screwdriver blade against the aqueduct coating and slip the blade of the other under the rubber cover of the anode high-voltage cable (Fig. 11-35). Crisscross the screwdrivers so they are touching when the blade reaches the anode button.

Do not try the screwdriver method with a solid-state chassis. You may accidentally destroy some transistors and IC components in the solid-state chassis with this method. Always discharge the anode cable connection to the grounded aqueduct coating on the bell of the picture tube. Remove the high-voltage cable after discharge. Short the anode button to the aqueduct coating to completely discharge the high voltage.

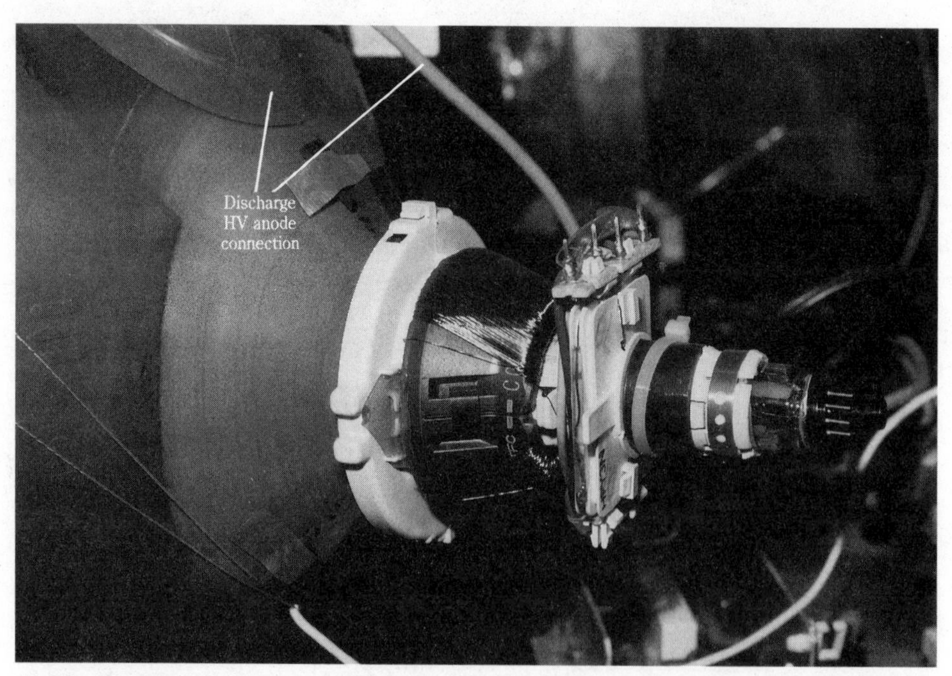

11-35 Discharge the high-voltage lead to the picture tube with two, long metal screwdrivers with well-insulated handles. Discharge HV to the aqueduct coating of the picture tube.

Snowy picture

Poor antenna signal, dirty or defective tuner, and AGC circuits all produce snowy pictures. The antenna signal can be checked with those of a second set. Notice if the stations come and go when the tuner knob is rotated. A dirty tuner can cause an intermittent or snowy color picture. First, clean up the tuner contact points with a good cleaning spray or solvent. Most defective tuners are sent to a tuner repair depot for any major cleaning and service.

Improper adjustment of the AGC control can cause a snowy picture. Some solid-state TV chassis have a regular AGC control and another delay AGC control Watch the TV screen with a mirror as you adjust both controls to eliminate the snow from a local TV broadcast. Tune in a fringe station broadcast and verify your adjustments. No obvious improvement in the picture quality after making the AGC adjustment may indicate a defective AGC circuit, tuner, or IF stage.

Dim and blotchy picture

If the picture remains dim after the screen brightness and subbright control are turned wide open suspect a bad picture tube. A blotchy close up appearance of a person's face is a good indication that the picture tube is defective. This condition can occur with the brightness control turned fully up then, when the contrast control is turned up a patch of brightness appears.

The picture tube can be given a new lease on life with a filament booster, or a rejuvenation charge. Rejuvenation is a cleaning process of the tube heater and cathode elements. Although these two methods of restoration of brightness very seldom last for more than six months to one year, life is added to the color TV chassis. The picture tube booster or brightner plugs on the end of the picture tube, but the rejuvenation of a picture tube must be done at the radio-TV shop with an expensive test instrument.

Poor color

Most color circuit problems must be serviced by a radio-TV technician. You may save a service call or a trip to the shop by rechecking the fine tuning adjustment of the tuner. The fine tuning control may be located right behind the tuner knob. Make sure the color control is turned up with the tint control set in the middle of rotation. Intermittent color can be caused by a dirty tuner. Readjust the color killer control if one is found on the back part of the chassis.

Color spots in the picture

Sometimes when the carpet sweeper is run close to the TV set and shut off, the TV screen becomes magnetized, and displays unwanted color spots. Placing large stereo speakers near the TV cabinet can do the very same thing. In most TV sets, each time the chassis is turned on, an automatic degaussing circuit functions for a few seconds. The screen is demagnetized once again, removing any patches of coloration in the raster.

Suspect a defective ADG circuit if stray spots of color are noted in several areas of the screen. The raster at the corners and the bottom half of the picture may have a discoloration. Check for an open degaussing coil or plug. An open or broken thermistor can cause the degaussing circuit not to function. The degaussing circuit is very simple and can be checked with the low ohm scale of the vom (Fig. 11-36).

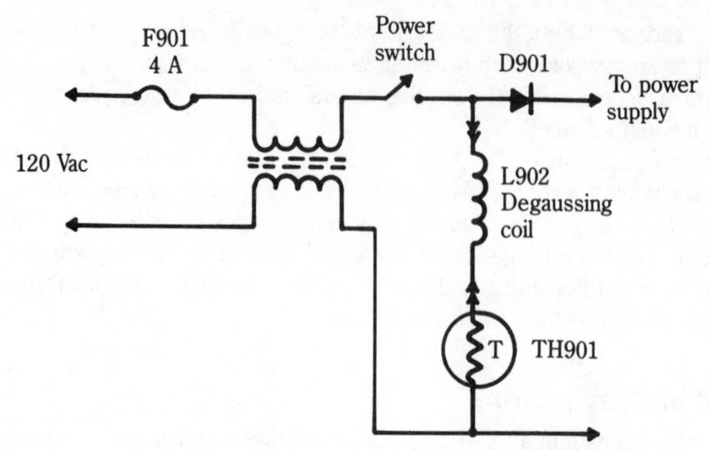

11-36 Check the degaussing circuit with the low resistance scale of a VOM. A broken or burned transistor can prevent the ADG circuits from functioning.

Arching

High voltage arcover is the most common of the arcing noises heard in a TV set. The arcing can occur at the anode connection of the picture tube, high-voltage flyback, transformer, high-voltage socket, and tripler unit. Be very careful when high voltage is arcing over the TV chassis. With the back cover removed, one can quickly see where the arcing occurs.

Before working around the high voltage, discharge the picture tube. Take a long and short screwdriver, touch one of the screwdriver blades to the metal chassis or shield of the picture tube and slip the blade of the other screwdriver under the rubber cover and touch the high-voltage anode connection of the picture tube. Touch both screwdriver blades together. Keep all hands on the insulated handles of the screwdriver. Hold the discharge for 30 seconds.

Cleaning off the high-voltage connection at the picture tube anode may solve arcing at the socket. Wipe off the anode connection with a clean cloth and isopropyl alcohol. Let the surface dry completely before replacing the anode socket. Make sure both wire clips are free from rust and securely caught inside the socket. A poor connection here may make the arcing worse. Arcing indications in the raster may be caused by a poor ground between the picture tube and the TV chassis. Clip an alligator lead from the metal chassis to picture tube shield or mounting bolts and notice if the arcing indications disappear. If so, install a shorting wire from the mounting bolt of the picture tube to a metal screw on the metal chassis.

Tripler arcover The high-voltage tripler unit may arc over between tripler and chassis ground. Often, diodes within the tripler will short out or high-voltage capacitors open and cause arcover. You can hear a loud arcing sound when the tripler is defective. Sometimes the defective tripler will arc internally and overload the flyback transformer. Shut the set down and feel the body of the tripler. These triplers should run cool, but if warm or hot, replace it. Replace the tripler unit when arcing is noticed around or beneath the unit (Fig. 11-37). Tripler units must be replaced with exact part numbers. Make note of where each lead goes before removing the old tripler.

11-37 Here the tripler unit has arced over through the plastic to chassis ground. A very loud noise is audible when arcing occurs.

You cannot repair these units with high-voltage arcover spray or putty. You are just wasting time and energy. At the moment, the arcover may quit, but it will resume before long.

What not to touch

There are many circuits within the TV chassis that the electronic beginner should not touch without the correct knowledge and test equipment. You can only make more problems and damage the TV set. Here are the primary circuits or procedures beginning electronic technicians should use caution with:

- Critical horizontal components
- High-voltage measurement
- High-voltage circuits
- High-voltage regulator circuits
- Replacement of color picture tube
- Critical chopper and switched mode power circuit
- System control circuit
- On-screen display
- TV receiver alignment
- Do not attempt to turn screws or coil cores to improve the picture without the proper test equipment
- Removing and replacing large IC components
- Removing and replacing microcomputer chips
- Video and sync circuits
- Stereo audio circuits

On the other hand, if you gain knowledge from reading this book or obtain on-the-job training, go as far as you can to put that TV set back into operation. Many of the repairs in this book you can do, but be careful not to damage any components or receive a shock when taking critical voltage measurements and working around the picture tube with the TV set operating.

Lightning damage

Lightning can damage the antenna lead-in, power transformer, or power supply circuit. First, check the fuse and silicon diodes for possible damage (Fig. 11-38). Look for a large voltage dropping resistor that has become open. Inspect the ac cord and plug for possible damage. Check for burned PC wiring in the power supply circuit.

Most newer TV sets have a capacitor in each antenna terminal to protect the balun coils (Fig. 11-39). In older TV chassis, the balun coils are wired right into the antenna coil circuit. Inspect the antenna lead-in and surrounding circuits for burned off terminal wires and components. After lightning damage has occurred suspect a defective tuner, if the picture is snowy.

Heavy lightning can strike the antenna and burn out sections of the lead-in wire. Cable TV can also be damaged by lightning. Inspect the cable where it connects to the antenna lead-in of the TV set. If there are signs of lightning damage replace the entire

11-38 Excessive lightning damage in the solid-state chassis could be too expensive to repair. Notice the black area on the bottom side of TV chassis.

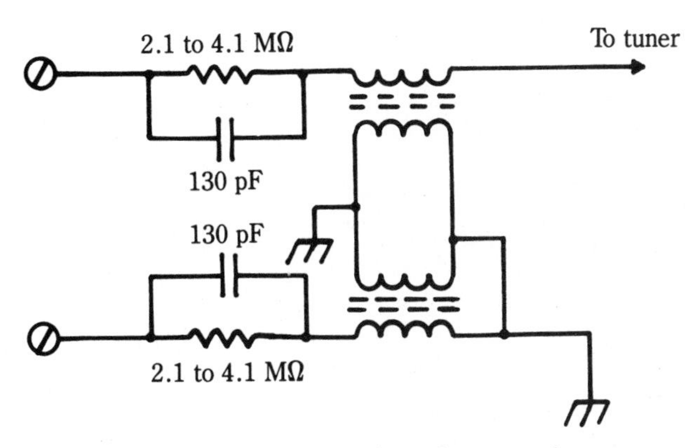

11-39 A small capacitor placed in each antenna leg can prevent lightning from entering the balun coils. Check the connecting capacitors for burned wires or leads.

lead-in wire whether it is flat 300 ohm type or the shielded 72 ohm cable type. Check all matching transformers and booster's connected to the antenna lead-in.

Remove the ac plug and disconnect the antenna cable when you go on vacation or when the TV will not be in use for several days at a time. Many of the new chassis have a surge voltage protector wired across the ac line. These line surge varistors can be wired directly into the power line circuit or a separate protector unit can plug directly into the ac power outlet. Then the TV's line cord is plugged into the protection device.

12
Symptoms and causes

THIS CHAPTER LISTS SEVERAL COMMON OR TYPICAL TROUBLE SYMPTOMS ALONG with things that might cause such a malfunction. Bear in mind that these are *typical*. They are intended to *guide* you in the right direction so that you can possibly pinpoint the exact trouble after checking over your TV.

You have seen some of the headings indicating TV symptoms earlier in this book. Nonetheless, they are presented here as an aid to refresh your memory after once reading this book. Just thumb through these pages to locate the trouble at hand. Then read through the listed possible causes, one by one, testing your receiver for defects.

When troubleshooting a TV receiver, always keep a mental image of the block diagram in your mind. Or better yet, keep the actual diagram at your side. You'll find that by referring to the block diagram you can isolate the defective area much easier.

If you pull several tubes to test at your local supermarket or electronic parts house, be sure to mark the location of each tube unless you have a tube layout diagram on hand. Even professional TV servicers have a difficult time remembering which tubes go where. Just outline the image of your particular TV chassis. Draw circles for the locations of the tubes with the type indicated inside each one. Also use care when handling the tubes. Besides being fragile, the numbers rub off rather easily. If you can't tell what kind of tube it is, it is of no use to you.

It's best to place a piece of masking tape on the tube and double check the tube location with the tube layout inside the TV cabinet. Then mark an X on the tape if a weak or defective tube is located. Do not throw the suspected defective tube away until you have installed a new one (Fig. 12-1). Tube substitution is the best replacement method, especially with power output tubes or ones that have become intermittent.

Likewise, when removing several components to get at the defective one, mark down how they were removed in succession and reverse the procedure. When removing and replacing IC components, mark on the PC board where terminal one is located.

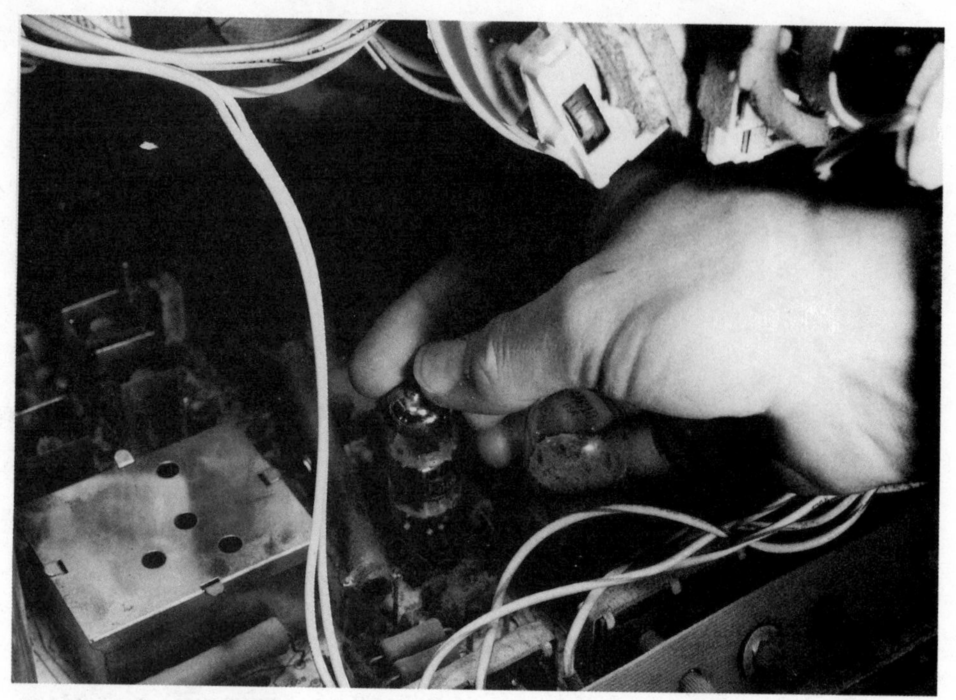

12-1 When removing tubes to be tested, place a piece of masking tape on the tube to identify tube type and location.

Mark all terminals of the suspected transistor on a piece of paper before removing it from the board. The same applies to diodes and electrolytic capacitors, to watch for correct polarity. Replace damper and HV diodes with exact voltage replacements.

Always replace capacitors with the same working voltage or higher. Electrolytic capacitors may be replaced with higher capacitance and voltage if the exact value is not available. When one filter capacitor is found leaky in with several others in one large container, replace the whole can. When one goes, they all will eventually go within a few months. Do not solder capacitors into the circuit with long leads because it will flop around and get in the way.

Do not replace a $^{1}/_{2}$-watt resistor with a $^{1}/_{4}$-watt one. It's best to use a 1-watt size if available. Try to replace all fuses with the same amperage rating. Do not wire across the fuse without providing protection to the TV chassis. Just use a little common sense and servicing precautions while working on TV chassis.

Set completely dead

- Check ac outlet for power with a table lamp
- Check line cord and plug for possible defects
- Check line cord interlock at rear of TV
- Check circuit breaker or main fuse in TV
- See if dial lights or tubes light; if so, check B+ fuse or circuit breaker
- Check on/off switch for proper operation

Solid-state TV: set completely dead

- Check the ac power cord
- Check the line fuse
- Check the B+ fuse
- Check power switch with DMM
- Check half-wave silicon diode or bridge rectifier
- Measure for dc voltage at low-voltage power supply

Raster OK but no sound or video

- Check tuner tubes if applicable
- Check shielded cable from tuner to main chassis
- Check tuner with spray lubricant
- Check IF amplifier tubes if applicable
- Check AGC pot setting
- Check noise gate pot setting if applicable
- Check video detector diode
- Check lead-in from back of receiver to tuner
- Check AGC tube if applicable

Solid-state TV: keeps blowing fuses (see Fig. 12-2)

- Place a 100-watt bulb across open fuse for indicator
- Remove horizontal output transistor or B+ fuse to see if horizontal circuit is blowing fuse
- Check for correct dc voltage at power supply after removing horizontal output tube
- Check diodes, electrolytic capacitors, and regulator in low-voltage supply if fuse still opens

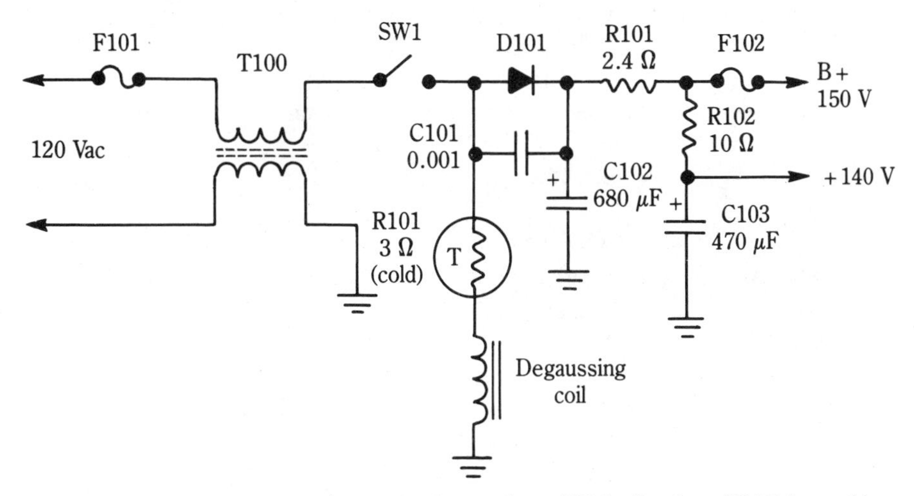

12-2 Check F101, D101, C102, and the degaussing coil if the line fuse (F101) keeps blowing. Suspect horizontal circuits when F101 and F102 open.

- Check for leaky damper diode or horizontal output transistor if low-voltage power supply is normal
- Check for shorted ADG coil against the metal frame of CRT

Solid-state TV: no sound or video but normal raster

- Check for snow in the raster
- Check the antenna connection on tuner
- Clean tuner
- If one or two channels are not present, suspect dirty tuner contacts
- Adjust the AGC control

Sound OK but no raster

- Check for illuminated picture tube filament
- Check high-voltage rectifier tube if applicable
- Check high-voltage regulator tube if applicable
- Check damper tube if applicable
- Check horizontal amplifier tube if applicable
- Check horizontal oscillator tube if applicable
- Check horizontal circuitry fuse or circuit breaker if applicable
- Check video output tube if applicable
- Check for connection of high-voltage wire to the picture tube

Solid-state TV: Sound OK but no raster (see Fig. 12-3)

- Notice if filaments are lit in end of CRT
- Check for open B+ fuse or resistor
- Check if high voltage comes up when you can hear the yoke expand and hair will stand up on your arm when placed near the TV screen
- Check horizontal hold control adjustment
- Check dc voltage at low-voltage power supply with DMM
- Check from horizontal output transistor to ground for leakage
- Check the horizontal output transistor out of circuit for leakage
- Measure dc voltage to collector terminal of horizontal output transistor while out of the circuit
- Check for leaky damper diode

Raster and video OK but no sound

- Check speaker by substitution
- Check audio output tube if applicable
- Check audio amplifier if applicable
- Check connection of wires between speaker and chassis
- Check connection of wires to and from volume control
- Check sound IF tubes if applicable
- Check sound discriminator/detector tube if applicable

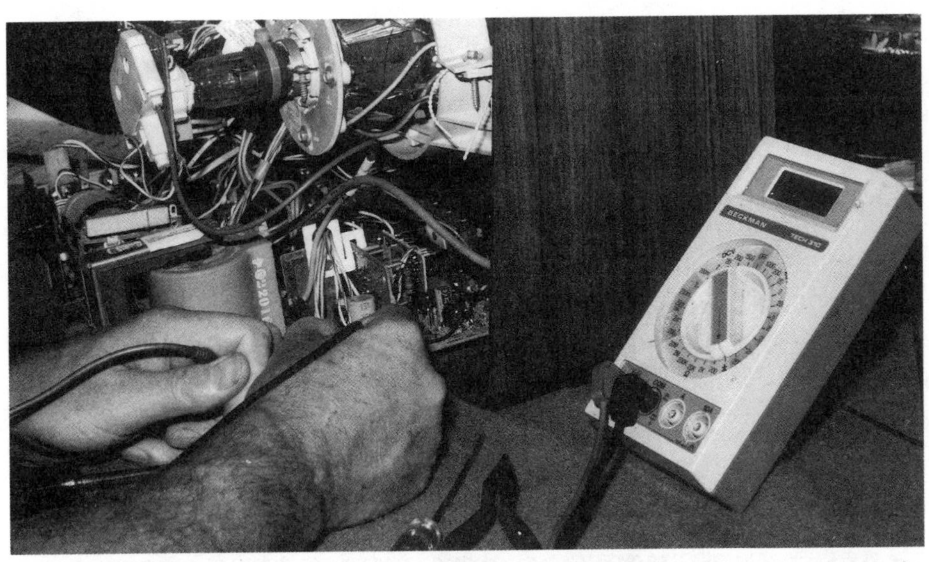

12-3 When there is no raster, check for an open B+ fuse and defective horizontal output transistor with the DMM. The filament of the CRT could be open.

Solid-state TV: raster and video OK but no sound (see Fig. 12-4)

- Substitute a new pm speaker or solder wires to speaker
- Rapidly rotate volume control back and forth to see if you can hear a scratching noise or hum sound
- Check voltages on the sound output transistors or IC circuit
- Test suspected audio transistors with diode test of DMM
- Replace defective transistor or audio IC with original part number

12-4 Clip another speaker across the suspected one. Solder all speaker wires and check the plug where wires connect in the chassis.

Loud arcing or popping noise

- Check connection of high-voltage lead to picture tube
- Check for defective rubber cup at picture tube where high voltage lead connects
- Check lead inside high-voltage gauge
- Check neck of picture for blue arc (defective tube)
- Check ground strap around picture tube

Solid-state TV: loud arcing or popping noise (see Fig. 12-5)

- See if arcing occurs around the tripler unit
- Replace tripler unit with exact part number
- Inspect high-voltage arcing around anode lead
- Check for arcing of ground on picture tube aqueduct wires
- If firing occurs in the picture tube gun assembly, replace defective CRT
- Check for defective audio output transistor when popping noise is heard in the speaker

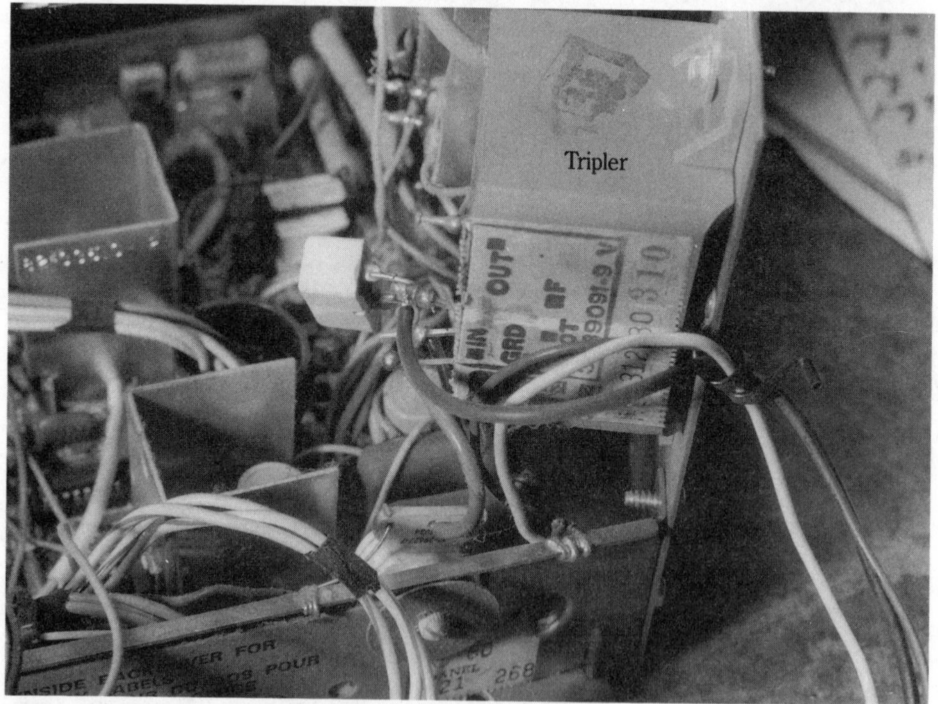

12-5 Check the tripler unit if you hear a loud popping or cracking noise. Sometimes the picture tube fires inside the gun assembly if the CRT is defective.

No red, and hue control has no effect

- Defective 3.58 MHz crystal
- Check color oscillator tube if applicable

Solid-state TV: no color control

- Check for defective 3.58 MHz crystal
- Make sure fine tuning, color, and tint controls are properly adjusted
- If one or two channels do not have color, suspect tuner or tuner adjustment

Picture is reduced overall

- Check low-voltage rectifiers (tube)
- Defective deflection yoke if raster has keystone appearance

Solid-state TV: picture is reduced over all

- Check dc voltage of low-voltage power supply
- Readjust B+ control, if available
- Check voltage in horizontal circuits if low voltage is normal

Horizontal line only

- Check vertical oscillator and amplifier tubes if applicable
- Check vertical size and linearity controls for "dead" spots
- Check vertical leads to and from deflection yoke
- Check vertical hold control for damage

Solid-state TV: Horizontal line only (see Fig. 12-6)

- Readjust vertical hold, linearity, and height controls
- Check vertical yoke socket and winding
- Check voltage in vertical output transistors and IC components

12-6 Locate the vertical output transistors and take voltage measurements. Test the vertical output transistors in the circuit with diode test of DMM.

- Check vertical control for breakage
- Remove vertical transistors for accurate tests

Raster reduced top and bottom

- Check vertical size and linearity adjustments
- Check vertical amplifier tube if applicable
- Check vertical centering control setting if applicable

Solid-state TV: raster reduced top and bottom (see Fig. 12-7)

- Readjust vertical hold, height, and linearity controls
- Check burned spot in any of the vertical controls
- Check for low dc voltage at vertical transistors or output IC
- Test vertical output transistors for leaky or open conditions
- Check for burned or open vertical transistor bias resistors
- Shunt large electrolytic capacitors in vertical circuits
- Poor width only: check horizontal output circuits
- Keystone slanted raster only: defective yoke
- Black at top and bottom of raster only: readjust vertical height and linearity
- Check for poor high voltage at CRT
- Leave high voltage and horizontal output problems to the electronic technician

12-7 This picture is not caused by vertical trouble, although the raster is dark at the bottom. Excessive noise blocked out the picture on channel 5.

Raster appears as a small circle

- Check position of deflection yoke or neck of picture tube

Lack of width

- Check horizontal amplifier tube if applicable

- Check damper tube if applicable
- Adjust horizontal width device

Solid-state TV: lack of width

- Readjust B+ control
- Readjust width control, if found
- Check horizontal output transistor; if very warm, replace
- Check open tuning capacitor in collector terminal of horizontal output transistor

Snow in picture

- Check tuber tubes if applicable
- Clean tuner with spray lubricant
- Check AGC pot setting
- Check noise gate pot setting if applicable

Solid-state TV: snow in picture (see Fig. 12-8)

- Check VHF or UHF tuner connections
- Check for poor antenna wire connection in back cover
- Check for outside antenna turned by the wind (in wrong direction)
- Burned or open balun coils
- Readjust AGC control
- Push down IF cable into socket

Horizontal weave with color bars

- Defective degaussing circuit: call a professional
- Defective low-voltage power supply: call a professional
- Shorted tube: check all tubes if applicable

12-8 Check for broken antenna leads on back of set and tuner. Inspect tuner input wires and cables for snowy conditions.

Solid-state TV: horizontal weave and bars (see Fig. 12-9)

- Defective degaussing circuit
- Open resistor in degaussing and low-voltage power supply circuits
- Poor filtering in power supply: shunt large filter capacitor
- Readjust B+ supply
- Defective low-voltage regulator and/or circuit

Movable ghost when fine tuning

- Check IF amplifier tubes if applicable
- IF alignment required: call a professional

Solid-state TV: ghost in picture

- Make sure ghosty picture exists on all channels
- If fine tuning will not adjust, suspect defective tuner
- Poor connection of IF cable from tuner
- Check direction of outside antenna
- Check for shielded lead-in wire
- Check for damaged outside booster (lightning or wind damage)
- Defective cable reception: call your cable company

Vertical roll

- Check vertical oscillator and amplifier tubes if applicable
- Check sync tube if applicable

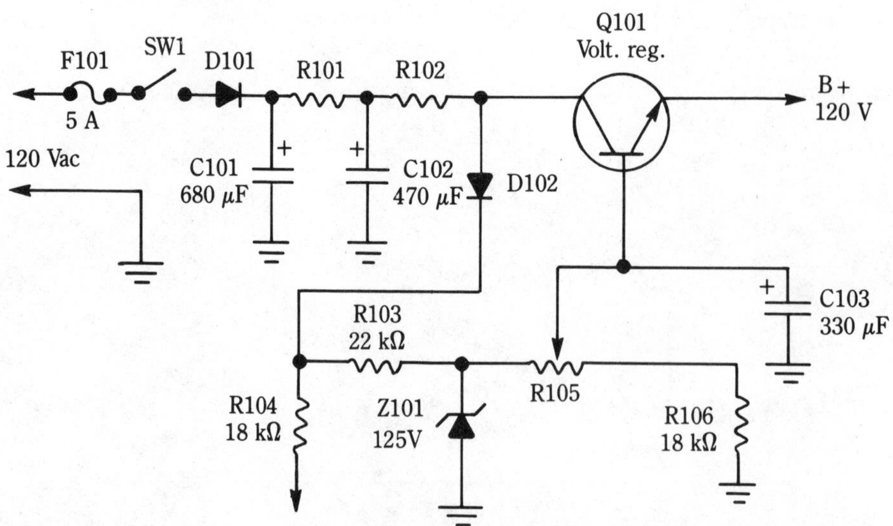

12-9 Weavy bars could be caused with improper B+ settings or defective regulator circuits. Suspect voltage-regulator transistor Q101 when the output voltage cannot be adjusted.

Solid-state TV: vertical roll

- Check vertical hold control
- Check adjustment of vertical height and linearity
- Check for excessive noise in the controls of picture (low channels only)
- Check antenna and lead-in of signal on all channels
- Check vertical sync and input circuits (refer to electronic technician)
- Check video amplifier and IF amplifier tubes if applicable
- Check AGC pot setting
- Clean tuner with spray lubricant
- Properly adjust vertical size and linearity controls

Smoke or fire

- Unplug receiver and look for smoking component or call a professional

Solid-state TV: smoking chassis (see Fig. 12-10)

- Try to locate smoking component after pulling ac cord
- Locate smoking flyback transformer (refer to electronic technician)
- Locate smoking power transformer (refer to electronic technician)
- Locate burning deflection yoke; replace with exact part number
- Burning resistors on chassis: find the cause and replace transistor
- Hole arcing between components: liquid spilled in TV set. Replace components, cut out burned parts, and restore PC wiring with hookup wire
- Replace arcing power transistor socket
- Replace defective tripler unit

12-10 If the chassis is smoking, it could be caused by a shorted power transformer, flyback transformer, or deflection yoke. Replace with original part number.

Poor black-and-white on color receiver

- Check screen control settings
- Check drive control settings
- Check color amplifier tubes if applicable

Picture pulling horizontally at top

- Check horizontal AFC tube if applicable
- Check AGC pot setting
- Check noise gate pot setting if applicable
- Check horizontal AFC diodes by substitution if applicable

Barber-poling or poor color sync

- Check color oscillator tube if applicable
- Check burst amplifier tube if applicable
- Check 3.58 MHz crystal by substitution
- Check color killer control setting

Poor focus in color receiver

- Check focus rectifier tube/diode by substitution
- Check focus control setting
- Check focus lead (heavy black wire) to picture tube at socket
- Check for excessive picture tube current by reducing brightness
- Check high-voltage rectifier tube/diode by substitution
- Check high-voltage regulator tube if applicable
- Focus divider assembly defective where applicable

Solid-state TV: poor focus (see Fig. 12-11)

- Readjust focus control
- Replace focus control if firing occurs in control itself
- Check for corroded focus wire to CRT socket
- Replace corroded picture tube socket
- Clean off focus pin on picture tube
- Replace defective picture tube (refer to electronic technician)

Misconvergence

- Check for connection of convergence assembly plug on chassis
- Static convergence magnets displaced

Reduction in sound after warm-up

- Check audio amplifier and output tubes if applicable
- Defective audio output transistor where applicable
- Dirty volume control
- Check sound IF amplifier tubes if applicable

12-11 Readjust focus control and note when rotated if it fires internally and causes firing lines in the raster.

Solid-state TV: sound distortion (see Fig. 12-12)

- Substitute another pm speaker
- Replace damaged speaker
- Check for dried-up or leaky speaker coupling capacitor
- Check for leaky output transistor or IC
- Measure supply voltage to sound circuits
- Replace burned bias resistor of transistor and IC
- Replace entire sound IC output component

Improper tint control range

- Alignment problem: call a professional
- Check burst amplifier tube if applicable
- Check leads to and from tint control

Sound beat pattern in picture

- Improper fine-tuning adjustment
- Improper operation of cable TV system where applicable
- Alignment problem: call a professional

Ac hum bar floating in picture

- Check *all* tubes if applicable
- Power supply or picture tube trouble: call a professional
- Antenna lead-in too close to power lines

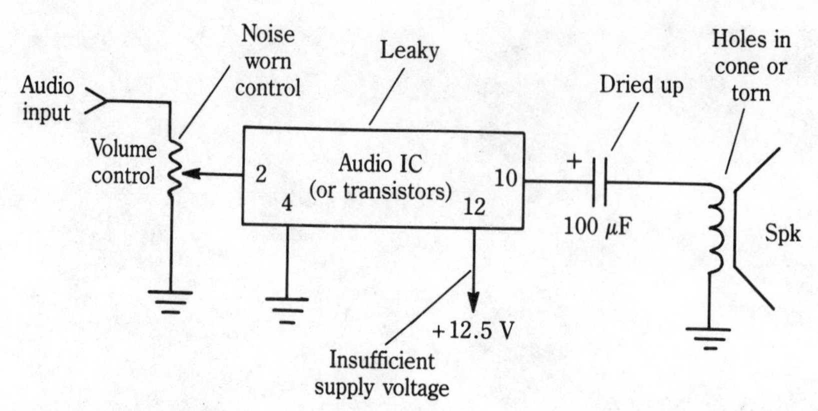

12-12 Check the indicated components when you hear distortion in the speakers.

Solid-state TV: hum bars

- Check defective degaussing circuits
- Check defective regulator circuits
- Readjust B+ control
- Shunt filter capacitor in low-voltage power supply
- Leaky horizontal driver and output transistors
- Check for high-voltage power lines near the outside antenna

One color missing from picture

- Check color amplifier tubes/transistors by substitution
- Check screen and drive control settings

Solid-state TV: missing one color

- Inspect picture tube heater with one dead or no light of filament
- Remove and replace the picture tube socket
- Check dc voltage on color output transistor on CRT board
- Check color bias and output transistor of color missing

Solid-state TV: all-red screen

- Defective picture tube; leaky red gun assembly
- Leaky red bias or color output transistor
- Missing voltage on red output transistor
- Check the above corresponding conditions if the screen is all green or blue

Solid-state TV: HV shut-down

- Check power line with DMM with over 125 volts
- Readjust B+ control
- Check low-voltage power supply for high voltage
- Check horizontal output circuits with variable isolation line transformer (refer to electronic technician)

Glossary

ac Alternating current. The type of electricity normally used in homes and most industries. Its contrasting opposite is direct current, or dc, now obsolete except for certain specialized applications. All batteries supply dc.

ACC Automatic color control A circuit similar in function and purpose to AGC, except that it is supplied exclusively to the color bandpass amplifiers to maintain constant signals.

ac hum A low-pitch sound heard whenever ac power is converted into sound, intentionally or accidentally. The common ac hum is a 60-hertz note.

AFC Automatic frequency control. A method of maintaining the frequency or timing of an electrical signal in precise agreement with some standard. In FM receivers, AFC keeps the receiver tuned exactly to the desired station. In TV, horizontal AFC keeps the individual elements or particles of the picture information in precise register with the picture transmitted by the TV station.

AGC Automatic gain control. A system that automatically holds the level or strength of a signal (picture or sound) at a predetermined level, compensating for variations due to fading, etc.

amplifier As applied to electronics, a magnifier. A simple tube or transistor or a complete assembly of tubes or transistors and other components may function as an amplifier of either electric voltage or current.

antenna A self-contained dipole or outside device to collect the broadcast signal from the TV station. The collected signal is fed to the TV set with a shielded or unshielded lead-in wire.

anode The positive (+) element of a two-element device such as a vacuum tube or a semiconductor diode. In a television tube, an anode is an element having a relatively high positive voltage applied to it.

aperature mask An opaque disk behind the faceplate of a color picture tube; it has a precise pattern of holes through which the electron beams are directed to the color dots on the screen.

arc An electric spark that jumps (usually due to a defect) between two points in a circuit which are supposed to be insulated from each other, but not adequately so.

aspect ratio The relation or proportion between the width and height of a transmitted TV scene. The standard aspect ratio is 4:3, meaning that the picture is 3 inches high for every 4 inches of width, or four-thirds as wide as it is high.

audio Any sound (mechanical) or sound frequency (electrical) that is capable of being heard is considered as audio. Generally, this includes frequencies between about 20 and 20,000 hertz.

B+ Supply voltage, as low as 1 volt dc in transistor circuits and as high as hundreds of volts in tube circuits, which is essential to normal operation of these devices. The plus sign indicates the polarity.

B+ boost A circuit in TV receivers which adds to, or boosts, the basic B+ voltage. The boost source is a by-product of the horizontal deflection system. Also see *damper.*

balun coil A set of balun coils are found between antenna connection and tuner to match the input of 300 ohms to 75 ohms at the tuner input.

bandpass amplifier In a color TV set, one or two color signal amplifiers located at the beginning of the color portion of the TV set; they are designed to amplify only the required color frequencies. They *pass* a certain *band* of frequencies.

Barkhausen A term applied to a display of one or two black vertical lines on the left side of the picture tube due to some spurious behavior (oscillation) in the circuit. These lines are usually seen best when there is no picture on the screen (just a blank raster).

blanking A term used to describe the process which prevents certain lines and symbols, which are required for keeping the picture in step with the transmitter, from being seen on the TV screen.

bridge rectifier Four diodes are wired in a series circuit to provide full wave rectification of a two lead power transformer. The ac-dc TV chassis may use a bridge rectifier after the line fuse.

brightness Refers to both the amount of illumination on the screen (other than picture strength) and the control that is used to adjust the brightness level.

burst In color TV, a precise timing signal. It is not continuous, but comes in spaced *bursts*. It is transmitted for controlling the 3.58 MHz oscillator essential for color reception.

burst oscillator The precision 3.58 MHz oscillator vital to color reception. It is kept in step (sync) by the burst.

buzz This is sometimes called *intercarrier buzz*, a raspy version of ac hum, usually caused by improper adjustment of some IF circuits.

B-Y The blue component of a color picture minus the monochrome.

capacitance A measure of a capacitor's ability to store electrical energy. The capacitor was called a *condenser* at one time. Bypass and electrolytic filter capacitors are found in many TV circuits. The unit of capacitance is the farad.

carrier The radio signal that carries the sound or picture information from the transmitter to the receiver. The carrier frequency is the identifying frequency of the station (e.g., 880 kHz, 93.1 MHz, etc.)

cathode The negative or minus element of a two-element tube or semiconductor. The cathode and the anode combine to form a diode (two-element device). The cathode is also the source of electrons in such devices.

cathode-ray tube A tube in which electrical energy is converted to light. An electron beam (or beams), originating at the cathode, impinges upon a phosphor light-emitting screen. TV picture tubes, radar tubes, tuning eyes in some FM sets and many similar types are basically cathode-ray tubes.

chassis The base where the majority of electronic components are mounted. The metal chassis may be common ground. Today, in the solid-state chassis, the PC board wiring is the main chassis.

cheater cord An ac line cord for operating the TV set without the back cover or the cabinet when troubleshooting and repairing. The original cord is attached to the back of the cabinet as a safety measure.

chroma Another term for *color*. Color amplifiers are often called chroma amplifiers. The term is also used to denote the control used to increase or reduce the color content of a picture.

chopper circuit The chopper power supply is a pulse-width-modulated (PWM), regulated power supply. The chopper supply circuits may be quite similar to the horizontal deflection system.

circuit breaker The circuit breaker may work in place of the line fuse to open when an overload is found in the TV circuits. Some horizontal output tubes have a separate circuit breaker in the cathode circuit.

clipper A term describing the operation of one of the sync circuits in a TV set. It is the stage (tube or transistor) that separates the sync (timing) signals from the picture information.

color bar generator The color bar generator provides patterns for color alignment and color TV adjustments. Some of the NTSC generators have from 8 to 10 different patterns.

color killer A special circuit whose function is to turn off the color amplifier circuits when a black-and-white signal is being received. Also the control used to adjust the operation of the circuit.

comb filter The comb filter circuit separates the luminance (brightness) and chroma (color) video information, eliminating cross-color that may occur in other chassis.

contrast The depth of difference between light and dark portions of a TV scene. Also the name given to the control for adjusting the contrast level.

convergence The system that brings the three electron beams together in a color picture tube so they all pass through the same hole in the shadow mask and strike the correct dots on the screen.

converter A stage in the tuner or front end of a TV set or any radio receiver which converts an incoming signal to a predetermined frequency called the IF or *intermediate frequency*. All incoming signals are converted to the same IF.

corona Similar to an electric arc, except that this is a characteristic of much higher voltages (thousands). Corona occurs as a continuous, fine electrical path through air between two points, sometimes accompanied by a faint violet glow, usually near the picture tube.

crystal A quartz of synthetic mineral-like slab or wafer having the property of vibrating at a precise rate or frequency. Each crystal is cut to vibrate at the desired frequency. Such a crystal is used in the 3.58 MHz oscillator to control its frequency.

CRT Cathode-ray tube; another name for the color picture tube.

damper A diode, tube or semiconductor used in horizontal amplifier circuits to suppress certain electrical activity. It incidentally provides B+ boost voltage.

dc Abbreviation for direct current.

deflection The orderly movement of the electron beam in a picture (cathode-ray) tube. Horizontal deflection pertains to the left-right movement, vertical deflection the up-down movement of the beam.

deflection IC Today, the deflection circuits may have both the vertical and horizontal oscillator and amplifier circuits in one IC component. You may find the deflection circuits in one large IC with many different circuits.

degaussing Demagnetizing. In color TV sets, an internal or external circuit device that prevents or corrects any stray magnetization of the iron in the picture tube faceplate structure. Magnetization results in color distortion.

demodulator A demodulator separates or extracts the desired signal, such as sound energy or picture information from its carrier.

detector Same as *demodulator*.

digital multimeter (DMM) The digital multimeter will measure voltage, resistance, current, and test diodes. Most have an LCD display. Today, you may find the DMM also measures capacity, frequency, tests transistors and is a frequency counter besides the regular testing features.

diode A two-element electron device—a tube or semiconductor. The simplest and most common application of a diode is in the conversion of ac to dc (rectification).

discriminator An audio detector in an FM receiver or TV sound circuits. Also, a detector performing a similar function in other frequency control circuits, such as horizontal frequency control.

electrolytic capacitor These capacitors may be used as filter or decoupling capacitors in the TV chassis. Large filter capacitors are found in the low-voltage power supply.

faceplate The front assembly of a picture tube. In a color tube, it includes the tricolor phosphor and the aperture mask.

field One scanning of the scene on the face of the picture tube, in which every alternate line is (temporarily) left blank. The scan duration of a field is 1/60 second. Two fields, the second one filling in the blank lines left by the first one, make up a frame, or a complete picture. A frame duration is 1/30 second.

filter The electrolytic filter capacitor is found in the low-voltage power supply. Always replace with same voltage and capacitance or higher (never lower values).

flyback, retrace Name given to return movement of the electron beam in a picture tube after completing each line and each field. You don't see flyback or retrace lines (normally) on the picture tube because they are blanked out.

flyback transformer Another name for the horizontal output transformer. The flyback transformer takes the sweep signal from the horizontal output transistor and builds up the high voltage to be rectified for the HV of CRT. The flyback provides horizontal sweep for the yoke circuits.

focus Some picture tubes are constructed internally with self-focusing elements, while in other TV sets, a focus control varies the voltage applied to the picture tube focus element. This voltage may vary from 4 to 5.3 kV.

frame The combination of two interlaced fields is called a *frame*. Since it consists of two fields each of 1/60 second duration, the frame duration is 1/30 second.

frequency The number of recurring alternations in an electrical wave, such as ac, radio waves, etc. Frequency is specified by the number of alternations occurring during 1 second and given in hertz (cycles per second), kilohertz (1000 cycles) and megahertz (million cycles).

frequency counter Actually, the frequency counter test instruments counts the frequency of various circuits. The frequency range may vary from 2 Hz up to 100 MHz.

gain Relative amplification. The number of times a signal increases in size (level) due to the action of one or more amplifiers. The overall gain of a signal often is millions of times.

gas Refers to the presence (undesirable) of a trace of gas inside a vacuum tube. A gassy tube is a defective tube.

ghost Most commonly a double-exposure type of a scene on the TV screen. Usually a fainter picture appears somewhat offset to the right of the main image caused by the reception of two signals from the same station; one signal is delayed in time.

G-Y The green color signal minus the monochrome.

high voltage Generally refers to the multithousand picture tube voltage, but it can be used to mean any potential of a few hundred volts or more.

high-voltage probe The high-voltage probe is a test instrument that will check the anode and focus high voltage at the CRT. The new probes may measure up to 40,000 Vdc.

hold-down capacitors The hold-down or tuning capacitors are found from collector terminal of horizontal output transistor to common ground. When the hold-down capacitor opens or dries up, the high voltage will increase, causing HV shut-down in the latest TV chassis.

horizontal Pertaining to any of the functions associated with left-to-right scanning in a picture tube including the horizontal amplifier, oscillator, frequency, drive, lock, AFC, etc.

H.O.T. Horizontal output transformer, which steps up the low oscillator voltage with usually a driver and horizontal output transistor between. This voltage is rectified by the HV rectifier and applied to the anode terminal of the picture tube.

hue In color TV, the basic color characteristic that distinguishes red from green from blue, etc.

hum Same as *ac hum*.

IC Integrated circuit. A structure similar to a *module*, in which a number of parts required for the performance of a complete function are prewired and sealed. It is not repairable.

IF Intermediate frequency. In the tuner of a TV or radio receiver the incoming from the desired station is mixed with a locally generated signal to produce an intermediate signal, usually lower than the frequency of the incoming signal. The IF is the same for all stations. The tuner changes to accommodate each incoming signal.

IHVT The integrated horizontal or high-voltage output transformer has HV diodes and capacitors molded inside the flyback winding area. The new IHVT transformers may also provide several different voltage sources for the TV circuits.

in-line picture tubes A more recent development in color tube structure that produces the three basic colors in adjacent strips or bars instead of the earlier types which produced three-dot or triad groups. Improved color quality as well as simplified design and maintenance is claimed for this type of design.

isolation transformer The isolation transformer may be a variable type that raises or lowers the power line voltage to the TV set. Always use an isolation transformer with ac/dc powered TV chassis.

intercarrier A term describing the current system of TV receiver design in which a common IF system is used both for picture and sound information. In older TV sets, the split-sound design was employed, in which separate IF channels for the picture and sound were used.

ion trap *See* trap.

kinnie The name often referred to as the picture tube.

leakage Undesired current flow through a component.

linearity Picture symmetry. Horizontal linearity pertains to symmetry between the right and left sides of the picture, best observed with a standard test pattern. Also, an adjustment for achieving such linearity. Vertical linearity refers to symmetry between upper and lower halves of a picture.

line filter A device sometimes employed between the ac wall outlet and a radio or TV set to reduce or eliminate electrical noises.

line, transmission The antenna lead-in wire or cable.

lock, horizontal An adjustment in some TV sets for setting the automatic frequency operation on the horizontal sweep oscillator.

loss Usually refers to the amount of signal loss in the antenna lead-in (transmission line). This is particularly serious on UHF.

low-voltage regulator The low-voltage regulator is found in the low-voltage power supply. The regulator may be transistors or an IC component. The fixed regulator supplies a well-filtered, regulated, constant voltage.

microcompressor The microcompressor or microcomputer chip is built like a regular IC with 8 to 80 (or more) separate terminals. The microcompressor IC may be found with surface-mounted terminals.

modulation The process of combining (by superimposition) a sound or picture signal with a carrier signal for purposes of efficient transmission through air. The carrier's only function is to piggyback the intelligence.

module A subassembly of a number of parts, usually including transistors and diodes. It is encapsulated and not repairable. *See* IC.

modular chassis A TV chassis that is made up entirely of separate modules for each circuit in the TV set.

motor boating A "putt-putt" sound caused in the audio sound input and output circuits. Motor-boating may be caused by poorly grounded or poorly filtered circuits.

oscillator Generator of a signal, such as the 3.58 MHz color subcarrier signal, the RF oscillator in the tuner, the horizontal oscillator signal (15,750 hertz) and vertical oscillator signal (60 hertz).

oscilloscope A test instrument that can show exact waveforms throughout the TV circuits to help troubleshooting and locate defective components for the electronic technician.

PC board Printed circuit board. A subassembly of various parts, not necessarily all for one and the same function, on a phenolic or fiberglass board on which the interconnections are printed on metal veins or paths. No conventional wiring, except external interconnections, is used.

parallel A method of circuit component connection where all components involved connect to common points so that each component is independent of all other components. For example, all light bulbs in your house are connected in parallel.

phosphor The coating on the interior of the *faceplate* of a picture tube, which gives off light when struck by an electron beam. The chemical composition of the phosphor determines the color of the light it will emit.

picture projection Three small projection color tubes are used to project the TV image on the front or rear of a large screen TV receiver. The projection tubes are found inside the cabinet of a rear projection color set.

picture tube The picture tube receives the video color signal that displays the picture upon the picture tube raster. The new picture tube sizes are 27- and 31-inch.

power supply That portion of a piece of electronic equipment which provides operating voltages for its tubes, transistors, etc.

preamplifier A high-gain amplifier used to build up a signal so it is strong enough to present to the normal level amplifiers, for example, an antenna preamplifier for fringe area reception.

pulse A single signal of very short duration used for timing and sync purposes. Sync pulses are the best example of this type of signal. Pulses occur in precisely measured bursts.

purity, color The display of the various true colors without any accidental or unwanted contamination of one color by any of the others. Color purity is largely dependent on correct convergence adjustments.

raster The illuminated picture tube screen fully scanned with or without video.

regulator A transistor or IC component that regulates the voltage for a given circuit in the low-voltage or HV power supplies.

remote control A hand-held transmitter that controls the function of the TV set by the operator from a distance. Today, the stations are tuned in electronically instead of the old method of rotating the tuner with a motor.

resistance Electrical friction represented by the letter R. The ohm is the unit of resistance. Resistance limits current flow.

RF Abbreviation for radio frequency.

retrace The return movement of the scanning electron beam from the extreme right to the extreme left and from the bottom to the top of the raster. *Also see* flyback.

retrace blanking The extinction or darkening of the light on the face of the picture tube during retrace time in order to make these lines invisible. Should retrace blanking fail white lines sloping downward from right to left would be seen on the screen.

R-Y The red color component of the overall color picture signal minus the monochrome.

sand castle The sand-castle generator is a three-level signal pulse that includes horizontal and vertical blanking and burst keying pulses.

saturation Pertains to the full depth of a color, in contrast to a faint, feeble color. Saturated colors are strong colors.

scanning lines The horizontal lines that you can see up close in the picture or raster. The scanning lines make up the picture from left to right looking at the front of the TV screen.

SCR The silicon-controlled rectifier is used in the low- and high-voltage regulator power supply circuits. In some TV chassis an SCR can be used as the horizontal output transistor.

semiconductor A general name given to transistors, diodes and similar devices in differentiation from vacuum tube devices.

series A connection between a number of components or tubes in chain fashion; i.e., one component follows the other. If any one component opens or burns out it breaks the series circuit.

shadow mask Same as *aperture mask*.

shield A metallic enclosure or container surrounding a component, tube, cable, etc. *Also see* tube shield.

shielded cable A wire with a metal casing on the outside to prevent unwanted electrical energy from reaching the inner conductor.

signal Electrical energy containing intelligence such as speech, music, pictures, etc.

signal-to-noise ratio A mathematic expression that indicates the relative strength of a signal within its noise environment. A good signal has a high signal-to-noise ratio.

solid-state A term indicating that the radio, TV set, etc., uses semiconductors and not vacuum tubes, but transistors, diodes, etc.

sound bars Thick horizontal lines or bars, usually alternately dark and light, appearing on the TV picture screen due to unwanted sound energy reaching the picture tube. In appearance, the width, number, and position of these bars varies with the nature of the sound. Sound bars are caused by a misadjusted circuit.

subcarrier The color picture information carrier. It is called a subcarrier because it is a secondary carrier in the particular channel. The color subcarrier frequency is 3.58 MHz.

surface-mounted components The surface-mounted parts are soldered into the circuit on the same side as the pc wiring. You may find surface-mounted components mounted under the pc chassis with larger components on top in the latest TV receivers.

sync An abbreviation for a synchronizing signal. It is a timing signal or series of pulses sent by the transmitter and used by the receiver to stay in precise step with the transmitter.

sync clipper *See* clipper.

sync separator A circuit in a TV receiver that separates the sync from the picture information or the vertical sync pulses from the horizontal sync pulses.

sweep marker generator A generator used by the electronic technician for TV receivers alignment.

transistor A solid-state semiconductor used in amplifier, oscillator, and power circuits of the TV chassis. The transistor operates at lower voltage than the vacuum tube. Some chassis may have both npn type and pnp type transistors.

trap An electrical circuit which absorbs or contains a particular electrical signal (also called wave trap). Also a magnet, called an ion trap, used on the neck of some picture tubes for electron beam deflection.

triad The three-color, three-dot group (red, green, and blue) of which the color picture tube phosphor is made. Each group of three dots is a triad, and there are thousands of triads on a modern color tube screen.

triac A solid-state controller device usually located in the low voltage power supply circuits.

tripler A solid-state component made up of capacitors and diodes to triple the applied RF voltage from the flyback or horizontal output transformer. In the latest TV chassis the horizontal output transformer and the high-voltage rectifiers can be molded into one component.

tube shield A metal sleeve that fits snugly over a glass tube and shields it from extraneous electrical impulses. A tube shield is part of the tube's electrical circuit.

tuner The tuner picks up each broadcast TV signal and passes it to the IF circuits for amplification. The tuner may be operated manually or with a remote control.

UHF Ultrahigh frequencies. Radio and TV frequencies from 300 MHz upward. Channels 14 through 83 are all located in the UHF band and are, therefore, called UHF stations.

varactor A semiconductor device with the characteristics of a tuneable device through the application of a voltage. In contrast to the conventional frequency variation through the use of coil and capacitor techniques, the varactor requires only a voltage variation to effect tuning. In some recent TV sets varactor tuners have been used to replace the conventional coil-switching type tuners. Simplicity, greater stability and freedom from deterioration are claimed for this type of tuner.

variac The variac information can raise or lower the power line voltage to locate intermittent, defective flyback transformers and high-voltage shut-down cases.

vertical Pertaining to the circuits and functions associated with the up-down motion or deflection of the electron beam.

vertical amplifier An amplifier following the vertical oscillator used to enlarge the vertical sweep signal.

VHF Very high frequencies. Radio and TV stations located below 300 MHz (down to 50 MHz). TV channels 2 through 13 as well as the FM band are in the VHF spectrum.

video A term applied to picture signals or information (video, circuits, video amplifier, etc.).

VIR Vertical interval reference. Sometimes called broadcast controlled color correction system. It is an automatic, station controlled signal used to initiate color and luminance corrective action in the TV set if so equipped. It is more elaborate and more effective than the automatic color control. *See* ACC. It is found in the more expensive TV receivers.

VOM (volt-ohmmeter) The first pocket-sized VOMs was used for continuity, voltage, resistance, and current tests. The VOM utilizes a meter to display the measured readings.

wave The name given to each recurring variation in alternating electric energy, including radio and TV signals.

width The width of a TV screen may be pulled in at each side, indicating problems within the horizontal deflection system. Poor width may be caused by the HV regulator transistors, SCRs, and zener diodes in the regulator circuit. Poorly soldered pincushion transformer connections may cause width problems.

X demodulator The designation of the red (R-Y) signal demodulator.

yoke Deflection yoke. The electrical assembly, somewhat in the shape of a yoke or collar, mounted on the picture tube neck against the flaring bell of the tube. By electric and magnetic means the yoke imparts to the electron beam the scanning (left to right and top to bottom) or deflection to produce the raster and the image.

Y signal The picture-only (minus color) signal that is fed to the color picture tube. It is sometimes called the brightness signal, meaning the actual brightness and darkness (and all shades in between) of the picture. This brightness signal plus the red signal produce all the red hues in the picture; the same Y signal and blue give all the blue coloration and finally the Y signal plus the green give the green coloration to the scene.

Z demodulator Same as the X demodulator, except it is for the blue (B-Y) signal.

zener diode A special semiconductor diode with a reverse breakdown voltage. The zener diode provides a fixed voltage in power circuits.

Index

D

E

F

G

H

I

R

S

T

V

W

X

Y